Henri Skoda
Jean-Marie Trépreau
(Eds.)

**Contributions to Complex Analysis
and Analytic Geometry**

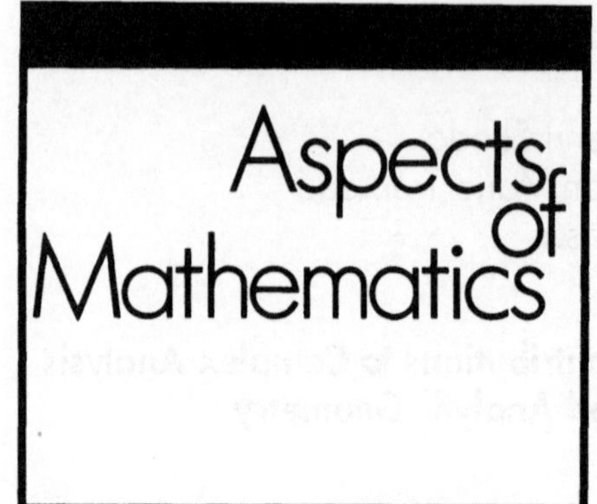

Aspects of Mathematics

Edited by Klas Diederich

*A Publication of the Max-Planck-Institut für Mathematik, Bonn

Henri Skoda
Jean-Marie Trépreau
(Eds.)

Contributions to Complex Analysis and Analytic Geometry

Dedicated to Pierre Dolbeault

Henri Skoda and Jean-Marie Trépreau
Université Paris 6
Département de Mathématiques
Tour 45–46, 5ème étage
4, Place Jussieu
F-75252 Paris Cedex 05, France

Mathematics Subject Classification: 32–06, 32–02, 30Fxx, 32C35, 32H04, 32F15, 32F20, 32F40, 32C30, 32M10, 32M12.

Cover design: Wolfgang Nieger, Wiesbaden

Printed on acid-free paper

ISSN 0179-2156
ISBN 978-3-528-06633-8   ISBN 978-3-663-14196-9 (ebook)
DOI 10.1007/978-3-663-14196-9

Henri Skoda
Jean-Marie Trépreau
(Eds.)

Analyse Complexe et Géométrie Analytique

Mélanges en l'honneur de
Pierre Dolbeault

Pierre Dolbeault

Foreword

Preface of the Editors

Ce volume prend sa source dans le Colloque en l'honneur de Pierre Dolbeault, organisé à l'occasion de son départ à la retraite, à l'initiative des Universités de Paris 6 et de Poitiers. Ce colloque, consacré à l'Analyse Complexe et à la Géométrie Analytique, s'est tenu à Paris, sur le campus de l'Université Pierre et Marie Curie, du 23 au 26 Juin 1992. Il a réuni autour de ces thèmes une centaine de congressistes, dont de nombreux mathématiciens étrangers (Allemagne, Argentine, Canada, Etats-Unis, Islande, Italie, Pologne, Roumanie, Russie, Suède).

Nous avons souhaité prolonger cet hommage par la publication d'un volume dédié à Pierre Dolbeault. Le présent recueil d'articles ne constitue pas strictement les actes du Colloque. Nous avons voulu qu'il rassemble uniquement des articles originaux ou synthétiques, qui illustrent l'œuvre scientifique de Pierre Dolbeault à travers les thèmes abordés ou la personnalité de leurs auteurs. Nous remercions les conférenciers qui ont bien voulu contribuer à cet ouvrage, et Klas Diederich de l'avoir accueilli dans la collection "Aspects of Mathematics" qu'il dirige.

Au nom du Comité d'Organisation du Colloque (C. Laurent-Thiébaut, J. Le Potier, J.B. Poly, J.P. Vigué et nous-mêmes), nous remercions les institutions qui nous ont apporté leur aide financière et matérielle: les Universités Paris 6 et de Poitiers, la Direction de la Recherche et des Études Doctorales, le Centre National de la Recherche Scientifique et le Ministère de la Recherche et de la Technologie.

Enfin nous souhaitons à Pierre Dolbeault une heureuse retraite et nous formons des vœux pour qu'il continue longtemps à contribuer au développement de la science mathématique.

H. Skoda,
J.M. Trépreau.

Acknowledgement

Chers collègues et amis[1],

En cette dernière journée du Colloque d'Analyse Complexe tenu en l'honneur de Pierre Dolbeault, je rappellerai ce qu'il nous a apporté. Cela m'amènera, cher ami, à évoquer votre belle carrière de professeur et de mathématicien. Cela me conduira aussi à rappeler quelques-uns des résultats que vous avez donnés à la science.

Votre jeunesse studieuse se passe à Paris. Né en 1924, vous faites vos études au Collège Lavoisier, puis au Lycée Henri IV, à deux pas d'ici. Vous êtes reçu à l'École Normale de la rue d'Ulm dans la promotion 1944. Elle est fort réduite. En effet, c'est encore la guerre et l'occupation. Vous n'entrerez qu'en Mars de l'année suivante dans cette École, aujourd'hui rajeunie, dont nous fêterons bientôt le bicentenaire. Vous en sortez agrégé de mathématiques. Reste alors à entrer dans une nouvelle carrière, celle de la recherche. Ce n'est pas facile. Certes la paix est revenue, mais les contacts et même les correspondances entre scientifiques ont été presque totalement interrompus pendant cinq ans. Heureusement le C.N.R.S. a été reconstitué, et il vous accueille. Vous y demeurerez six ans, de 1947 à 1953. Mais vous avez l'heureuse idée de passer l'année 1949-1950 à Princeton, à la Graduate School. À l'Université et à l' Institut dirigé par Marston Morse, les mathématiciens de tous les pays se retrouvent. Il y a là Kodaira et Spencer, merveilleuse équipe. Où pourrait-on, mieux que là, se familiariser avec les techniques et les notions qui sont indispensables aux recherches d'une Analyse Complexe en rapide évolution? De retour en France, où des séminaires sont à nouveau actifs, vous fréquentez celui de H.Cartan à l'École Normale et il sera le directeur de votre thèse, que vous soutiendrez en 1955. Entre-temps vous avez été nommé à l'Université de Montpellier, puis à celle de Bordeaux où, là encore, vous demeurerez six ans, de 1954 à 1960. Votre thèse vous a fait connaître, c'est un immense travail d'une centaine de pages qui paraîtra en deux articles publiés en 1956 et 1957 aux Annals of Mathematics. Mais c'est votre épouse, Simone, qui vous amène à l'Université de Poitiers où elle enseigne et où, Professeur, vous exercerez douze ans, au grand avantage des mathématiques qui profiteront de votre double résidence. Vous y avez une maison charmante, vous y formez des élèves et des chercheurs parmi vos étudiants. Vous venez parfois à Paris, le séminaire d'Analyse, que j'ai créé à l'Institut H. Poincaré, vous y voit souvent et vous y donnez des exposés sur les résidus des formes différentielles, publiés dans les volumes du séminaire aux Lecture Notes. Grâce à vous, des colloques d'Analyse Complexe nous amènent à Poitiers. Cependant, en 1972, vous ne résistez pas à l'attraction parisienne. Vous êtes élu ici, à l'Université Paris VI, car la prestigieuse Sorbonne a éclaté en fragments dont deux se trouvent ici même. Vous y donnerez des enseignements d'Analyse dont témoigne l'édition de votre cours de maîtrise d'Analyse Complexe récemment paru, et dont témoignent aussi de nombreuses thèses de troisième cycle ainsi que plusieurs belles thèses de doctorat soutenues à Paris ou à Poitiers.

Il me faut maintenant –mais ce sera trop brièvement– décrire votre itinéraire et votre oeuvre de mathématicien. Peut-être faut-il remonter à la période 1940-1960 pour comprendre vos difficultés et vos mérites. Les historiens des mathématiques noteront, je crois, que, au

---

[1]Allocution de clôture du Colloque en l'honneur de Pierre Dolbeault.

cours de cette double décennie, l'Analyse Complexe a subi une transformation encore plus rapide que celle de l' Analyse Réelle. En fait on passe en quelques années d'une discipline classique, l'étude fine des fonctions analytiques d'une variable complexe, à un monde très riche, englobant sous le nom d'Analyse Complexe l'ensemble des constructions qu' on peut faire sur le corps des nombres complexes. Votre thèse y contribuera, non seulement par des résultats importants, mais aussi, ce qui est essentiel alors, en rassemblant et faisant collaborer des techniques nouvelles. Pendant sa captivité, J. Leray a élaboré la théorie des faisceaux, les formalismes de l'algèbre homologique se répandent, ils se prêtent à l'étude de la d''-cohomologie des formes différentielles, lesquelles donnent un accès à des propriétés géométriques ou topologiques par des moyens analytiques.

Votre thèse: "Formes différentielles et cohomologie sur une variété analytique complexe" établit, à l'aide du lemme de Poincaré-Grothendieck, l'isomorphisme du groupe de cohomologie de dimension q à coefficients dans le faisceau des p-formes différentielles holomorphes et du groupe de d''-cohomologie des formes ou des courants de type (p, q). Ce résultat et d'autres, analogues, ont trouvé aussitôt un très grand champ d'applications en géométrie analytique et algébrique et, d'emblée, ils ont assuré à votre nom une place et, dirais-je, une pérennité dans l'histoire de notre science. On est obligé d'utiliser vos énoncés, on n'y échappe pas, et je puis dire, parlant comme les medias, qu'ils vous rendent incontournable, dût votre modestie en souffrir .

Utilisant la dualité de Poincaré, vous étendez ensuite l'homomorphisme résidu de Leray aux formes différentielles semi-méromorphes fermées et définissez ce que vous appelez un résidu homologique, lequel, peut-être, pose encore des problèmes selon les singularités polaires de la forme. L'étude des résidus des formes semi-méromorphes, leur définition et les propriétés du courant résiduel ne cesseront de vous préoccuper, compte tenu de l'apparition d'autres approches (Hironaka, Coleff et Herrera). D'autre part, vos méthodes sont la source d'applications dans deux directions: aux variétés kählériennes compactes et aux variétés algébriques projectives et à leurs diviseurs.

Attiré comme vous l'êtes par l'étude des singularités, même des plus méchantes, il était fatal que vous portiez attention aux variétés, aux ensembles et aux fonctions analytiques réels. Considérant de telles fonctions, mais à valeurs complexes, vous obtenez des généralisations de vos résultats, puis, avec J. Poly, vous étudiez les formes à singularités sous-analytiques.

Vous donnez alors plusieurs articles sur la difficile recherche des chaînes holomorphes et des chaînes maximalement complexes ayant un bord donné; les premières sont des combinaisons linéaires de diviseurs, représentés par leurs courants d'intégration qui sont des courants positifs fermés; les secondes sont des combinaisons linéaires entières de courants d'intégration sur des sous-variétés réelles de dimension $2p - 1$, à singularités négligeables, de CR-dimension $p - 1$. Vous généralisez pour elles le problème du bord en vous plaçant sur un hyperplan réel, ce qui vous donne une solution et des applications aux espaces projectifs.

Ce trop bref exposé de votre oeuvre de mathématicien montre, cher ami, que vous ne craignez pas les problèmes difficiles, ni ceux où se cumulent les difficultés de l' Analyse et celles d'une situation géométrique qu'il faut aborder dans sa généralité. Cela donne une valeur particulière à vos résultats: la Géométrie y est intimement mêlée à l' Analyse. C'est là un point de vue moderne que vos travaux et ceux de vos élèves ont contribué à répandre en montrant son efficacité. Au terme de ce Colloque qui a réuni des mathématiciens de beaucoup de pays venus vous faire hommage de leurs travaux, vous pouvez, vous et les vôtres ici présents, mesurer combien d'amitié s'ajoute à la reconnaissance qui vous est témoignée et à l' hommage qui vous est rendu aujourd'hui.

Pierre Lelong

Selected Publications by Pierre Dolbeault

Une sélection des publications de Pierre Dolbeault
(préparée par l'auteur)

- *Sur la cohomologie des variétés analytiques complexes.* C.R. Acad. Sci. Paris **236**, 175–177 (1953).

- *Formes différentielles et cohomologie sur une variété analytique complexe, I.* Ann. of Math. **64**, 83–130 (1956); *II.* Ann. of Math. **65**, 282–330 (1957).

- *Formes différentielles méromorphes localement exactes.* Trans. Am. Math. Soc. **82**, 494–518 (1956); *ComplÚments.* Trans. Am. Math. Soc. **91**, 390–398 (1959).

- *Une généralisation de la notion de diviseur.* Atti Convegno Inter. Geometria algebrica, Torino 1961, 125–150.

- *Sur la cohomologie entière de dimension deux d'une variété analytique complexe.* Rendiconti di Matematica 1-2 **21**, 219–239 (1962).

- avec Guy ROBIN. *Sur le faisceau des diviseurs à coefficients complexes.* C.R. Acad. Sci. Paris **262**, 1452–1455 (1966).

- *Theory of residues and homology.* Symposia Mathematica, Istituto di Alta Matematica **III**, 295–304. Academic Press 1970.

- *Résidus et courants.* Questions on algebraic varieties, C.I.M.E. Sept. 1969, 3–28. Ed. Cremonese Roma 1970.

- *Valeurs principales et résidus sur les espaces analytiques complexes, d'après Herrera-Lieberman.* Séminaire P. Lelong 1970-71, Springer Lecture Notes in Math. **275**, 14–26.

- *Valeurs principales et opérateurs différentiels semi-holomorphes.* Fonctions analytiques et analyse complexe, Colloque inter. C.N.R.S. Paris 1972, 35–50. Gauthier-Villars Paris 1974.

- avec J. POLY. *Differential forms with subanalytic singularities; integral cohomology; residues.* Several Complex Variables, Proc. Symp. Pure Math. **30**(1), 255–261. Am. Math. Soc. 1977.

- *On holomorphic chains with given boundary in* $\mathbf{P}^n(\mathbf{C})$. Analytic Functions, Blazejewko 1982, Springer Lecture Notes in Math. **1039**, 118–129.

- *General theory of multidimensional residues.* Itogi nauki i tekhniki **7**, 227–251, VINITI Moscou 1985 (en russe). Encyclopaedia of Math. Sci. **7**, 215–241 (en anglais, avec supplément), Springer Verlag 1990.

- *Sur les chaînes maximalement complexes de bord donné.* Geometric measure theory and the calculus of variations, Proc. Symp. Pure Math. **44**, 171–205. Am. Math. Soc. 1986.

- Analyse Complexe. Coll. Maîtrise de Math. Pures, Masson Paris 1990.

- *On the structure of residual currents.* Several Complex Variables, Math. Notes **38**, 258–273. Princeton Univ. Press 1993.

- *On CR analytic varieties with given boundary.* Complex Analysis and Geometry (Ancona and Silva ed.), 195–207. Plenum New York 1993.

- avec G. HENKIN. *Surfaces de Riemann de bord donné dans* $\mathbf{CP}^2$. C.R. Acad. Sci. Paris **316**, 27–32 (1993).

Authors' addresses

V. Ancona, Università degli Studi, Dipartimento di Mathematica U. Dini, Viale Morgagni 16 / A, 50134 Firenze Italy

B. Berndtsson, Department of Mathematics CTH, S-412 96 Göteborg, Sweden
bob@math.chalmers.se

E.M. Chirka, Steklov Institute of Mathematics, Vavilova 42, 117966 Moscow, Russia

J.-P. Demailly, Université de Grenoble I, Institut Fourier, BP 74 U.R.A. 188 du C.N.R.S., F-38402 Saint-Martin d'Hères

K. Diederich, Bergische Universität-GHS, Mathematik, Gausstr. 20, D-42097 Wuppertal, Germany
klas@wmka1.math.uni-wuppertal.de

P. Dolbeault, Mathématique, Analyse Complexe et Géométrie, Université Paris VI, 4, place Jussieu, F-75252 Paris Cedex 05, France

B. Gaveau, Université Pierre et Marie Curie, Mathmatiques, t. 45-46, $5^{\text{ème}}$ étage, 4, place Jussieu, 75252 Paris Cedex 05, France

G. Henkin, Mathématique, Analyse Complexe et Géométrie, Université Paris VI, 4, place Jussieu, F-75252 Paris Cedex 05 France

G. Herbort, Bergische Universität-GHS, Mathematik, Gausstr. 20, D-42097 Wuppertal, Germany
gregor@math.uni-wuppertal.de

A. Huckleberry, Mathematisches Institut, Ruhr-Universität Bochum, Universitätsstrasse 150, W-44780 Bochum 1, Germany
huckleberry@ruba.rz.ruhr-uni-bochum.dbp.de

M. Passare, Matematiska institutionen, Stockholms universitet, 106 91 Stockholm, Sweden
passare@matematik.su.se

B. Shiffman, Department of Mathematics, Johns Hopkins University, Baltimore, MD 21218, USA

E.L. Stout, Department of Mathematics, Washington University, Seattle, WA 98195, USA
stout@math.washington.edu

A. Tsikh, Department of Mathematics, Krasnoyarsk State University, prospect Svobodnyi 79, Krasnoyarsk 660062, Russia

O. Zhdanov Department of Mathematics, Krasnoyarsk State University, prospect Svobodnyi 79, Krasnoyarsk 660062, Russia

Contents

The de Rham Complex of a Reduced Analytic Space

Vincenzo Ancona and Bernard Gaveau

Abstract

We construct a de Rham complex on a reduced analytic space of fine sheaves of differential forms which are identical to the usual sheaves of C^∞ forms outside the singularities.

Introduction

A natural problem is to extend the definition of differential forms and the de Rham resolution of the constant sheaf to spaces with singularities so that one could describe the usual cohomology by analytic methods. This problem is solved in the present work for a reduced analytic space.

Several authors have generalized the definition of differential forms to analytic spaces, usually by embedding the analytic space in the space $\mathbb{C}^n$ and taking the quotient of the forms of $\mathbb{C}^n$ modulo the equations and the differential of the equations of the analytic set. This definition was used by Grauert [10] for holomorphic differential forms and Ferrari [9] proved that these holomorphic differertial forms (with the d operators) give a resolution of the constant sheaf near an isolated singularity which is "holomorphically contractible" (see also Reiffen [13] and Reiffen, Vetter [14]). In [6], Bloom and Herrera constructed the hypercohomology of the complex of holomorphic forms on a general analytic space and proved that the singular cohomology is a subspace of this hypercohomology. Fary [8] has also described the cohomology of algebraic varieties which have isolated singularities and are complete intersections using the spectral algebra associated to an ample linear system, he does not construct differential forms but he uses the abstract resolution of Leray ([12] and [7]) (called "recouvrement fin, complexe canonique fin"). More recently, we began to construct sheaves of differential forms of type $(0, p)$ to obtain a Dolbeault resolution of the sheaf of holomorphic functions. Our method was restricted to the case of a normal space such that the singular locus was smooth and essentially, the exceptional divisor is irreducible ([1], [2], see also [5] for slight generalizations). These strong restrictions come from technical difficulties related to the infinitesimal formal neighborhood of the exceptional divisor, but these difficulties do not exist for the case of the constant sheaf as we shall see below.

In this work, we shall construct a de Rham resolution

$$0 \longrightarrow \mathbb{C}_S \longrightarrow \Lambda^0_S \xrightarrow{d} \Lambda^1_S \xrightarrow{d} \cdots \xrightarrow{d} \Lambda^p_S \xrightarrow{d} \Lambda^{p+1}_S \xrightarrow{d} \cdots$$

on a general complex analytic space S. We want to construct a sequence of sheaves Λ^p_S on S with the following poperties:

(i) Λ^p_S are fine sheaves on S (and in particular they are acyclic).

(ii) the de Rham complex is a resolution of the constant sheaf $\mathbb{C}_S$.

(iii) the sheaves Λ^p_S are given by geometrical objects which are analogous to forms.

The requirements (i) and (ii) are precise but do not suffice to determine "concrete" sheaves (we can always use flabby sheaves or the "complexe canonique fin" of Leray to resolve $\mathbb{C}_S$). On the other hand, the requirement (iii) is not well defined. In this work, we shall construct the sheaves Λ^p_S, so that at regular points of S, they coincide with the usual C^∞–forms of degree p on the regular part of S (this requirement is still not sufficient to determine the sheaves Λ^p_S in a unique way).

This work is organized in the following manner: in the first two sections, we construct the de Rham complex of a divisor with normal crossings and relate it to the neighborhood of this divisor in the ambient space. In sections 3 and 4, we construct the de Rham complex when the singular locus of S is a smooth manifold. The general case is treated by a kind of recursion on the singular locus, namely, we assume that we can construct the sheaves $\Lambda^p_{\Sigma(S)}$ for the singular locus $\Sigma(S)$ and we deduce Λ^p_S. The way to do this is to construct towers of simultaneous desingularizations of the spaces $S, \Sigma(S), \Sigma(\Sigma(S)), \ldots$ which means that we construct desingularizations of these spaces such that there are natural morphisms of the exceptional space of the desingularization of S into a desingularization of $\Sigma(S)$. The main step of this construction is the construction of the first two floors of this tower. This is done in section 5 and the associated de Rham complex is constructed in section 6. The general case (towers with n floors) is treated in sections 7 and 8.

1   The de Rham complex of a union of transverse hypersurfaces

Let M be a complex analytic manifold of dimension d, Y a connected analytic subset of M which is a union of N smooth transverse hypersurfaces

$$(1.1) \qquad\qquad\qquad Y = \bigcup_{i=1}^{N} Y_i.$$

For any point m in Y, we denote $I(m)$ the subset of indices $i = 1, \ldots, N$ such that $m \in Y_i$. We can find complex analytic coordinates $(z_i)_{i=1,\ldots,d}$ in a neighborhood U of m in M such that

$$(1.2) \qquad Y_i \cap U = \{z \in U : z_i = 0\} \qquad \text{for } i \in I(m).$$

We can now define the sheaves of $\mathcal{C}^\infty$ differential forms on Y in a natural way. We take a point $m \in Y$. A germ of a $\mathcal{C}^\infty$ form of degree p in a neighborhood U of m is the data

$$(1.3) \qquad \pi = (\pi_i)_{i \in I(m)}$$

where π_i is a $\mathcal{C}^\infty$ form of degree p in $U \cap Y_i$ for $i \in I(m)$ where

$$(1.4) \qquad \pi_i|_{Y_j \cap Y_i} = \pi_j|_{Y_j \cap Y_i} \qquad \text{for } i, j \in I(m)$$

where $|_Z$ denotes the restriction (as forms) on Z. We call $\Lambda^p_{Y,m}$ the set of germs of $\mathcal{C}^\infty$ forms of degree p on Y.

It is clear that Λ^p_Y is a fine sheaf and that the sequence of differentials

$$d : \Lambda^p_Y \longrightarrow \Lambda^{p+1}_Y$$

defined by

$$(1.5) \qquad d\pi = (d\pi_i)_{i \in I(m)}$$

is a complex.

Theorem 1.1 *The complex*

$$(1.6) \qquad 0 \longrightarrow \mathbb{C}_Y \longrightarrow \Lambda^0_Y \xrightarrow{\ d\ } \Lambda^1_Y \xrightarrow{\ d\ } \cdots \xrightarrow{\ d\ } \Lambda^p_Y \xrightarrow{\ d\ } \Lambda^{p+1}_Y \xrightarrow{\ d\ } \cdots$$

is a fine resolution of $\mathbb{C}_Y$ (the constant sheaf of Y).

Proof 1. We shall need a special realization of the Poincaré operator P which solves the d complex on a smooth manifold. We start with an open convex neighborhood W of 0 in $\mathbb{R}^d$ with coordinates $x_1, \ldots, x_d$. If α is a p–form on $\mathbb{R}^d$, and $V_1, \ldots, V_p$ p vectors and x a point we denote by

$$< \alpha(x), V_1 \wedge \cdots \wedge V_p >$$

the action of $\alpha(x)$ (the value of α at point x) on the p–vector $V_1 \wedge \cdots \wedge V_p$. Let us now assume that α is a $\mathcal{C}^\infty$ form on W and define a $(p-1)$–form on W denoted $P\alpha$ by

$$(1.7)$$
$$< (P\alpha)(x), V_1 \wedge \cdots \wedge V_{p-1} > = \int_0^1 dt\, t^{p-1} < \alpha(tx), x \wedge V_1 \wedge \cdots \wedge V_{p-1} > .$$

It is easy to see that we have the formula

$$d(P\alpha) + P(d\alpha) = \alpha.$$

In particular, if α is d–closed $dP(\alpha) = \alpha$.

2. Now, let us take a point m in Y and a d–closed p–form π in a neighborhood U of m. We choose the coordinates in U such that (1.2) holds and we assume that $U \cap Y_i$ is convex for $i \in I(m)$. For each i in $I(m)$, we have a form π_i in $U \cap Y_i$ which is d–closed (on the convex subset $U \cap Y_i$) and we can apply the Poincaré operator to obtain $P\pi_i$ such that $dP\pi_i = \pi_i$. Let us now take two different indices $i \neq j$ in $I(m)$. We have to compute $P\pi_i|_{Y_i \cap Y_j}$. By definition of $|_{Y_i \cap Y_j}$ we choose $p-1$ vectors $V_1, \ldots, V_{p-1}$ in the linear space $Y_i \cap Y_j$ and by definition of P a point x in $Y_i \cap Y_j$ and we define $P\pi_i$ as in (1.7) so that

$$< (P\pi_i)\,|_{Y_i \cap Y_j}(x), V_1 \wedge \cdots \wedge V_{p-1} > = \int_0^1 dt\, t^{p-1} < \pi_i|_{Y_i \cap Y_j}(tx), x \wedge V_1 \wedge \cdots \wedge V_{p-1} >$$

so that it is clear that

$$(P\pi_i)\,|_{Y_i \cap Y_j} = P\left(\pi_i|_{Y_i \cap Y_j}\right)$$

and finally

$$P\pi = (P\pi_i)_{i \in I(m)}$$

is in $\Lambda_{Y,m}^{p-1}$ and satisfies $dP\pi = \pi$. $\square$

Definition $\Lambda_Y^{\bullet}$ will denote the preceding resolution and will be called *the de Rham resolution of* Y.

2   The retraction of a neighborhood of Y into Y

We start with the notations of section 1 concerning Y (embedded in the complex manifold M).

Lemma 2.1 *There exist an open neighborhood W of Y in M and a retraction*

$$r : W \longrightarrow Y.$$

We shall denote by $j : Y \longrightarrow W$ the natural injection. This injection induces a natural morphism of the constant sheaves still denoted by j

$$j : \mathbb{C}_W \longrightarrow \mathbb{C}_Y$$

which, in turn, induces an isomorphism between the Čech cohomology groups of W and Y:

$$(2.8) \hspace{3cm} j : H^k(W, \mathbb{C}) \longrightarrow H^k(Y, \mathbb{C})$$

because of the retraction of lemma 2.1.

We shall denote Λ_W^p the sheaf of ordinary $\mathcal{C}^\infty$ forms of degree p on the smooth manifold W. Then j induces by restriction a sequence of morphisms of sheaves

$$j^k : \Lambda_W^k \longrightarrow \Lambda_Y^k$$

which is the restriction of a k–form on W to a k–form on Y defined component by component.

The morphisms $j^\bullet$ induce a morphism of the de Rham resolutions of the topological spaces W and Y, i.e. a commutative diagram

$$
\begin{array}{ccccccccc}
0 & \longrightarrow & \mathbb{C}_W & \longrightarrow & \Lambda_W^0 & \xrightarrow{d} & \Lambda_W^1 & \xrightarrow{d} & \cdots \\
 & & \downarrow{\scriptstyle j} & & \downarrow{\scriptstyle j^0} & & \downarrow{\scriptstyle j^1} & & \\
0 & \longrightarrow & \mathbb{C}_Y & \longrightarrow & \Lambda_Y^0 & \xrightarrow{d} & \Lambda_Y^1 & \xrightarrow{d} & \cdots
\end{array}
$$

which we denote

$$(2.9) \qquad\qquad j^\bullet : \Lambda_W^\bullet \longrightarrow \Lambda_Y^\bullet.$$

Theorem 2.2 *The morphism $j^\bullet$ of the de Rham resolutions $\Lambda_W^\bullet \longrightarrow \Lambda_Y^\bullet$ induces an isomorphism between the de Rham cohomology groups of the resolutions $\Lambda_W^\bullet$ and $\Lambda_Y^\bullet$.*

Proof This theorem has been proved in a more abstract setting in [1] (see theorem 1-1). □

We shall also need the following prolongation lemma:

Lemma 2.3 *If π is a C^∞ form on Y, there exists a C^∞ form $\tilde{\pi}$ on a neighborhood W of Y in M such that $j(\tilde{\pi}) \equiv \tilde{\pi}|_Y$ is π.*

Proof a) We shall prove this first for a C^∞ function f on Y. We decompose $W = \bigcup_j U_j$ such that in each U_j the situation described in (1.2) holds. Using a C^∞ partition of unity, it is then sufficient to see the lemma in U_j.

To simplify the notation, we suppose that we have a point $m \in Y$ and we also assume that

$$I(m) = \{1, \ldots, k\}$$

so that

$$Y_j \cap U = \{z_j = 0\} \qquad \text{for } j = 1, \ldots, k.$$

We also call $\zeta = \{z_{k+1}, \ldots, z_d\}$ the other coordinates. The data of a C^∞ function f on $Y \cap U$ means that we have $f_j = f|_{Y_j \cap U}$ which is a C^∞ function of $\{z_1, \ldots, \hat{z}_j, \ldots, z_k, \zeta\}$ (where $\hat{z}_j$ means that z_j is forgotten in the preceding list) and the f_j satisfy $f_j|_{Y_l} = f_l|_{Y_j}$ for $1 \leq j, l \leq k$. We shall denote

$$(2.10) \qquad\qquad f_{j_1 \ldots j_r} = f|_{Y_{j_1} \cap \cdots \cap Y_{j_r}}$$

for $1 \leq j_1 < \cdots < j_r \leq k$. Then $f_{j_1 \ldots j_r}$ is a C^∞ function of the z_l and of the ζ for $l \neq j_1, \ldots, j_r$. We can consider that the $f_{j_1 \ldots j_r}$ are also C^∞ functions on U. We define

$$(2.11) \quad \widetilde{f} = \sum_{j=1}^{k} f_j - \sum_{1 \leq j_1 < j_2 \leq k} f_{j_1 j_2} + \sum_{1 \leq j_1 < j_2 < j_3 \leq k} f_{j_1 j_2 j_3} + \cdots + (-1)^{k-1} f_{1\ldots k}$$

It is clear that $\widetilde{f}$ is a C^∞ function on U such that $\widetilde{f}|_{Y_j} = f_j$.

b) Let us now take a C^∞ p–form π. We can again restrict ourselves to the case where π is defined on a neighborhood U of a point m and

$$I(m) = \{1, \ldots, k\} \qquad \text{and} \qquad Y_j \cap U = \{z_j = 0\} \qquad \text{for } j = 1, \ldots, k.$$

We again call $\zeta = \{z_{k+1}, \ldots, z_d\}$ and introduce real coordinates x_j, x'_j such that

$$z_j = x_j + ix'_j \qquad \text{for } 1 \leq j \leq d.$$

We consider the decomposition of $\pi = (\pi_j)_{j=1,\ldots,k}$

$$\pi_j = \sum_{|J|+|K|=p} \pi_{J,K,j} dx^J \wedge dx'^K$$

where by definition $\pi_{J,K,j} \equiv 0$ if j belongs to J or K (because π_j is defined on Y_j with equation $x_j = x'_j = 0$) and we shall continue π to U component by component.

First of all, let us consider multiindices J and K such that $|J| + |K| = p$ and if j is in J (resp. l is in K), we have $j \geq k+1$ (resp. $l \geq k+1$). Then the compatibility conditions are

$$\pi_{J,K,j}|_{Y_l} = \pi_{J,K,l}|_{Y_j} \qquad \text{for } 1 \leq j, l \leq k.$$

We can define a C^∞ function $\widetilde{\pi}_{J,K}$ such that

$$\widetilde{\pi}_{J,K}|_{Y_j} = \pi_{J,K,j} \qquad \text{for } 1 \leq j \leq k$$

and we continue $\left(\pi_{J,K,j} dx^J \wedge dx'^K \right)_{1 \leq j \leq k}$ as $\widetilde{\pi}_{J,K} dx^J \wedge dx'^K$ on a neighborhood of m in M. Let us now consider multiindices J, K and define

$$L = (J \cup K) \cap \{1, \ldots, k\}.$$

Then the $\pi_{J,K,j}$ are identically 0 for j in L. We can define now a C^∞ function $\widetilde{\pi}_{J,K}$ such that

$$\widetilde{\pi}_{J,K}|_{Y_j} = \pi_{J,K,j} \qquad \text{for } j \in \{1, \ldots, k\} \setminus L$$

because the compatibility conditions are

$$\pi_{J,K,j}|_{Y_l} = \pi_{J,K,l}|_{Y_j} \qquad \text{for } j \in \{1, \ldots, k\} \setminus L.$$

Then it is clear that

$$(\widetilde{\pi}_{J,K} dx^J \wedge dx'^K)|_{Y_j} = \pi_{J,K,j} dx^J \wedge dx'^K$$

for all $j = 1, \ldots, k$.

Finally we define

$$\widetilde{\pi} = \sum_{|J|+|K|=p} \widetilde{\pi}_{J,K}\, dx^J \wedge dx'^K$$

and we have for $1 \leq j \leq k$

$$\widetilde{\pi}|_{Y_j} = \pi_j.$$

$\square$

3 The de Rham complex for a space with a smooth singular locus

a) Notations and hypothesis

Let S be a complex analytic space and X be its singular locus. We shall assume in this section that

(i) there exists a smooth complex manifold $\widetilde{S}$ and a morphism $\varphi : \widetilde{S} \longrightarrow S$ which is a desingularization of S. We shall denote

$$\widetilde{X} = \varphi^{-1}(X)$$

the exceptional divisor lying above X so that

$$\varphi|_{\widetilde{S}\setminus\widetilde{X}} : \widetilde{S} \setminus \widetilde{X} \longrightarrow S \setminus X$$

is an isomorphism of complex manifolds and $\widetilde{X}$ is a union of smooth transverse hypersurfaces of $\widetilde{S}$.

(ii) X is a smooth complex manifold.

Remark *The hypothesis (i) is a consequence of Hironaka's desingularization theorem. The hypothesis (ii) will be suppressed in the following sections.*

We shall also write the irreducible decomposition of $\widetilde{X}$

$$\widetilde{X} = \bigcup_{j=1}^{N} Y_j$$

where the Y_j are smooth hypersurfaces with transverse intersections as in section 1.

For any complex manifold M, $\Lambda^\bullet_M$ will denote the de Rham resolution of the constant sheaf $\mathbb{C}_M$ of M.

b) Lifting a form from X to $\widetilde{X}$

Lemma 3.1 *We have a natural morphism*

(3.12) $$\varphi^* : \Lambda^\bullet_X \longrightarrow \Lambda^\bullet_{\widetilde{X}}.$$

Proof   Let U be a small open neighborhood of $m \in X$ and let us consider $\pi \in \Gamma(U \cap X, \Lambda_X^\bullet)$. Then we define $\widetilde{\pi}_j = \left(\varphi|_{Y_j \cap \varphi^{-1}(U)}\right)^* (\pi)$ which is an element of $\Gamma\left(\varphi^{-1}(U) \cap Y_j, \Lambda_{Y_j}^\bullet\right)$. It is clear that

$$\widetilde{\pi}_j|_{Y_l} = \widetilde{\pi}_l|_{Y_j}$$

so that $\widetilde{\pi} = (\widetilde{\pi}_j)_j$ is an element of $\Gamma\left(\varphi^{-1}(U) \cap \widetilde{X}, \Lambda_{\widetilde{X}}^\bullet\right)$.   □

Moreover there exists a j such that

$$\varphi|_{Y_j \cap \varphi^{-1}(U)} : Y_j \cap \varphi^{-1}(U) \longrightarrow X \cap U$$

is surjective. This implies that if π is not $0, \widetilde{\pi}_j$ is not 0, and we obtain:

Lemma 3.2 *If U is an open subset of S, then*

$$\varphi^* : \Gamma(U \cap X, \Lambda_X^\bullet) \longrightarrow \Gamma\left(\varphi^{-1}(U) \cap \widetilde{X}, \Lambda_{\widetilde{X}}^\bullet\right)$$

is injective.

Notation 1 *If π is in $\Gamma(U \cap X, \Lambda_X^\bullet)$, we denote simply*

$$\varphi^* \pi \in \Gamma\left(\varphi^{-1}(U) \cap \widetilde{X}, \Lambda_{\widetilde{X}}^\bullet\right)$$

its lifting through φ to $\widetilde{X}$.

c) The de Rham complex of S and the d operator

We can now define the de Rham complex of S denoted $\Lambda_S^\bullet$ as follows.

1. Let m be a point in $S \setminus X$, so that m is not singular, then $\Lambda_{S,m}^p$ will be the germs at m of ordinary C^∞ forms of degree p on S (or what is the same the germs at $\widetilde{m} = \varphi^{-1}(m)$ of ordinary C^∞ forms of degree p on $\widetilde{S}$).

2. Let m be a point in X and U a small neighborhood of m in S. A section π of Λ_S^p on U will be a C^∞ form $\widetilde{\pi}$ of degree p on $\varphi^{-1}(U)$, such that there exists a form α in $\Gamma(U \cap X, \Lambda_X^p)$ with

$$(3.13) \qquad\qquad j(\pi) \equiv \widetilde{\pi}|_{\widetilde{X}} = \varphi^* \alpha.$$

Because of the injectivity of φ^* (lemma 3.2), α is then unique so that we have a natural morphism still denoted by j:

$$j : \Lambda_S^p \longrightarrow \Lambda_X^p$$

which associates to π the element α such that (3.13) holds.

Lemma 3.3 *The d operator of $\widetilde{S}$ maps Λ_S^p into Λ_S^{p+1}*

$$d : \Lambda_S^p \longrightarrow \Lambda_S^{p+1}$$

with the following commutative diagram

(3.14)

$$
\begin{array}{ccc}
\Lambda_S^p & \xrightarrow{\ d\ } & \Lambda_S^{p+1} \\
{\scriptstyle j}\big\downarrow & & \big\downarrow{\scriptstyle j} \\
\Lambda_X^p & \xrightarrow{\ d\ } & \Lambda_X^{p+1}
\end{array}
$$

Proof If π is a section of Λ_S^p on U by definition, it generates a section $\widetilde{\pi}$ of $\Lambda_{\widetilde{S}}^p$ on $\varphi^{-1}(U)$. We define $d\pi$ by

$$\widetilde{d\pi} = d\widetilde{\pi}.$$

Then by (3.13):

$$j(d\pi) \equiv \widetilde{d\pi}\big|_{\widetilde{X}} = \varphi^* d\alpha$$

so that the diagram (3.14) is commutative. $\square$

d) The sheaves $\Lambda_S^\bullet$ are fine sheaves

Theorem 3.4 *The sheaves Λ_S^p are fine sheaves on S.*

Proof   The proof of this theorem is slightly more involved than the proof of the corresponding lemma 3.1 of [1] because the exceptional divisor is reducible. We first notice that it is sufficient to define on S C^∞ partitions of unity subordinated to open covering by sufficiently small open sets of S. Here, a C^∞ partition of unity means a partition of unity by continuous functions on S which are C^∞ in the sense of Λ_S^0.

We cover S by two kinds of open sets

- $(V_j)_{j \in J}$ with V_j disjoint from X and

- $(U_i)_{i \in I}$ with U_i open neighborhoods of points of X such that each $U_i \cap X$ has coordinates.

Then $\widetilde{S}$ is covered by the open sets $\widetilde{V}_j = \varphi^{-1}(V_j)$ and $\widetilde{U} = \bigcup_{i \in I} \varphi^{-1}(U_i)$. We construct a C^∞ partition of unity subordinated to the covering of $\widetilde{S}$ by the $\left(\widetilde{V}_j\right)_{j \in J}$ and $\widetilde{U}$ denoted $\widetilde{\rho}_j$ (support of $\widetilde{\rho}_j$ in $\widetilde{V}_j$) and $\widetilde{\rho}$ (support of $\widetilde{\rho}$ in $\widetilde{U}$). In particular, $\widetilde{\rho}$ is identically 1 on a small neighborhood of $\widetilde{X}$ because the $\widetilde{V}_j$ are disjoint from $\widetilde{X}$.

We construct also on X a C^∞ partition of unity $(\chi_i)_{i \in I}$ subordinated to the covering of X by the open sets $(U_i \cap X)_{i \in I}$.

We also consider $\chi_i' = \chi \circ \varphi|_{\widetilde{X}}$: it is clear that χ_i' is a section of $\Lambda^0_{\widetilde{X}}$ on $\varphi^{-1}(U_i) \cap \widetilde{X}$ and it defines a partition of unity of $\widetilde{X}$ subordinated to the covering $\left(\varphi^{-1}(U_i) \cap \widetilde{X}\right)_{i \in I}$ of $\widetilde{X}$.

Now let us consider an open covering of $\widetilde{U}$ by small open subsets $\left(\widetilde{W}_l\right)_{l \in L}$ such that in each $\widetilde{W}_l$, there exist complex analytic coordinates $(z_1, \ldots, z_n)$ of $\widetilde{S}$, where $Y_i \cap \widetilde{W}_l$ are given by local equations $z_i = 0$. Let us also take a C^∞ partition of unity $(\alpha_l)_{l \in L}$ subordinated to the covering $\left(\widetilde{W}_l\right)_{l \in L}$. Now, we consider $\chi_i'|_{\widetilde{W}_l \cap \varphi^{-1}(U_i) \cap \widetilde{X}}$ which is in $\Lambda^0_{\widetilde{S}}$. We can extend it as a function $(\widetilde{\chi}_i')_l$ on $\widetilde{W}_l$ by formula (2.11) of section 2 and it is clear by this formula and the fact that $\sum\limits_{i \in I} \chi_i' = 1$ that we have for each l

$$(3.15) \qquad \sum_{i \in I} (\widetilde{\chi}_i')_l = 1 \quad \text{on } \widetilde{W}_l.$$

Now we define

$$(3.16) \qquad \widetilde{\chi}_i' = \sum_{l \in L} \alpha_l (\widetilde{\chi}_i')_l .$$

Then because $(\alpha_l)_{l \in L}$ is a partition of unity and because $(\widetilde{\chi}_i')_l|_{\widetilde{X}} = \chi_i'$, we have $\widetilde{\chi}_i'|_{\widetilde{X}} = \chi_i'$ and by (3.15), we have $\sum\limits_{i \in I} \chi_i' = 1$.

Finally, the set of functions $(\widetilde{\chi}_i \widetilde{\rho})_{i \in I}, (\widetilde{\rho}_j)_{j \in J}$ is a C^∞ partition of unity subordinated to the covering of $\widetilde{S}$ by the $\left(\varphi^{-1}(U_i)\right)_{i \in I}, \left(\widetilde{V}_j\right)_{j \in J}$ and $\widetilde{\chi}_i \widetilde{\rho}$ is a section of $\Lambda^0_{\widetilde{S}}$ because $\widetilde{\chi}_i \widetilde{\rho}|_{\widetilde{X}} = \widetilde{\chi}_i|_{\widetilde{X}} = \chi_i' = \chi_i \circ \varphi$.

$\qquad\qquad\qquad\qquad\qquad\qquad\qquad\qquad\qquad\qquad\qquad\qquad\qquad\qquad\qquad\qquad\qquad\square$

4   The de Rham complex is locally exact (X smooth)

We want now to prove:

Theorem 4.1 *Under the hypothesis (i) and (ii) of section 3, the de Rham complex:*

$$(4.17) \qquad 0 \longrightarrow \mathbb{C}_S \longrightarrow \Lambda^0_S \xrightarrow{d} \Lambda^1_S \longrightarrow \cdots \longrightarrow \Lambda^p_S \xrightarrow{d} \Lambda^{p+1}_S \longrightarrow \cdots$$

is a resolution of the constant sheaf $\mathbb{C}_S$ by fine sheaves.

Proof In this proof it is sufficient to consider the case of a point $m \in X$ and to prove that if π is a section in $\Gamma(U, \Lambda^p_S)$ (where U is a small neighborhood of m in S), which is d–closed, there exists α section in $\Gamma\left(U', \Lambda^{p-1}_S\right)$ with $d\alpha = \pi$ for U' a smaller neighborhood of m.

1. Exactness in degree 0

Take f in $\Gamma\left(U, \Lambda_S^0\right)$ with $df = 0$. Then f defines an element $\tilde{f} \in \Gamma\left(\varphi^{-1}(U), \Lambda_{\tilde{S}}^0\right)$ with $d\tilde{f} = 0$, so that $\tilde{f}|_{\varphi^{-1}(U \cap X)}$ is constant on each connected component of $\varphi^{-1}(U)$. On the other hand $\tilde{f}|_{\varphi^{-1}(U \cap X)}$ comes from the function $f|_{U \cap X}$ through φ^* and we can assume that $U \cap X$ is connected, so that $f|_{U \cap X}$ must be a constant so $\tilde{f}|_{\varphi^{-1}(U)}$ is the same constant in each connected component of $\varphi^{-1}(U)$ and f is a constant on U.

2. Exactness in degree 1

Take π in $\Gamma\left(U, \Lambda_S^1\right)$ with $d\pi = 0$. Then π defines a 1–form $\tilde{\pi} \in \Gamma\left(\varphi^{-1}(U), \Lambda_{\tilde{S}}^1\right)$ with $d\tilde{\pi} = 0$ and by definition

$$(4.18) \qquad \tilde{\pi}|_{\tilde{X}} = \varphi^*\alpha$$

where α is in $\Gamma\left(U, \Lambda_X^1\right)$ and $\varphi^* d\alpha = 0$. But because φ^* is injective (lemma 3.2), $d\alpha = 0$ so that

$$\alpha = d\beta$$

and

$$(4.19) \qquad \tilde{\pi}|_{\tilde{X}} = d\varphi^*\beta.$$

Now because of theorem 2.2, the de Rham cohomology class $[\tilde{\pi}]$ of $\tilde{\pi}$ is 0, so that

$$(4.20) \qquad \tilde{\pi} = d\tilde{\gamma}$$

on $\varphi^{-1}(U)$. If we substract (4.19) from (4.20), we obtain

$$(4.21) \qquad d\left(\tilde{\gamma}|_{\tilde{X}} - \varphi^*\beta\right) = 0$$

so that $\tilde{\gamma}|_{\tilde{X}} - \varphi^*\beta$ is a constant on each connected component of $\varphi^{-1}(U \cap \tilde{X})$. By subtracting this constant from $\tilde{\gamma}$ on each component, we can assume that this locally constant function is 0 and we obtain

$$\tilde{\pi} = d\tilde{\gamma} \qquad \tilde{\gamma}|_{\tilde{X}} = \varphi^*\beta$$

so that π is the d of a $\gamma \in \Gamma\left(U, \Lambda_S^0\right)$.

3. Exactness in degree $p > 1$

The proof is the same as in degree $p = 1$ until formula (4.21). At that point, we deduce that $\tilde{\gamma}|_{\tilde{X}} = \varphi^*\beta$ is a de Rham cohomology class of degree $p - 1$ of $\tilde{X}$. By theorem 2.2, this cohomology class comes from a de Rham cohomology class of $\varphi^{-1}(U)$ (maybe restricting U) which means that there exists a $(p - 1)$–form $\tilde{\omega}$ on $\varphi^{-1}(U)$ which is d–closed and a $(p - 2)$–form ψ in $\Gamma\left(\tilde{X}, \Lambda_{\tilde{X}}^{p-2}\right)$ such that

$$(4.22) \qquad \tilde{\omega} = \tilde{\gamma}|_{\tilde{X}} - \varphi^*\beta - d\psi \qquad \text{and} \qquad d\tilde{\omega} = 0.$$

By lemma 2.3, there exists a $(p-2)$–form $\widetilde{\psi}$ on $\varphi^{-1}(U)$ (maybe restricting U still further) extending ψ. Changing $\widetilde{\omega}$ in $\widetilde{\omega}' = \widetilde{\omega} + d\widetilde{\psi}$ and calling $\widetilde{\gamma}' = \widetilde{\gamma} - \widetilde{\omega}'$, we obtain

$$\widetilde{\pi} = d\widetilde{\gamma}' \qquad \text{and} \qquad \widetilde{\gamma}'|_{\widetilde{X}} = \varphi^*\beta$$

which means exactly that π is d of an element γ in $\Gamma\left(U, \Lambda_S^{p-1}\right)$. $\square$

5   Towers of desingularizations: first two floors

For any analytic space S, we shall denote $\Sigma(S)$ its singular locus. In particular, we shall denote $\Sigma^{(k)}(S)$ the singular locus of $\Sigma^{(k-1)}(S)$.

a) <u>Embedding of a resolution of $\Sigma(S)$</u>

Lemma 5.1 *Let S be an analytic space, $X = \Sigma(S)$ its singular locus. One can construct a desingularization $\rho : X' \longrightarrow X$ of X, an analytic space S' and a morphism $\sigma : S' \longrightarrow S$ with a commutative diagram*

(5.23)
$$\begin{array}{ccc} X' & \lhook\joinrel\longrightarrow & S' \\ {\scriptstyle \rho}\downarrow & & \downarrow{\scriptstyle \sigma} \\ \Sigma(S) & \lhook\joinrel\longrightarrow & S \end{array}$$

with the following properties:

(i) One has an isomorphism

$$\sigma : S' \setminus \sigma^{-1}\left(\Sigma\left(X\right)\right) \simeq S \setminus \Sigma\left(X\right).$$

(ii) $\Sigma(S') = X' \cup Y' \subset \sigma^{-1}(\Sigma(S))$ and X' and Y' have distinct components and $\sigma(Y') \subset \Sigma(X)$.

Proof 1. First of all, we know by Hironaka's theorem that we can construct a desingularization of $X_0 \equiv \Sigma(S)$ by successive blowing–up $p_0, \ldots, p_{n-1}$ of center $C_0 \subset X_0, \ldots, C_i \subset X_i, \ldots$ to obtain a desingularization $\rho : X' \longrightarrow X$ where

$$X_n \equiv X' \overset{p_{n-1}}{\longrightarrow} X_{n-1} \longrightarrow \cdots \longrightarrow X_1 \overset{p_0}{\longrightarrow} X_0$$

$$\rho = p_0 \circ \cdots \circ p_{n-1}.$$

Moreover, C_j is in $\Sigma(X_j)$ and at each step:

(5.24) $$p_j : X_{j+1} \setminus p_j^{-1}\left(C_j\right) \simeq X_j \setminus C_j.$$

In particular $p_0 : X_1 \setminus p_0^{-1}\left(C_0\right) \simeq X_0 \setminus C_0$ and $\Sigma(X_1) \subset p_0^{-1}\left(\Sigma(X_0)\right)$, so that $p_0\left(C_1\right) \subset \Sigma\left(X_0\right)$. By recursion $p_j\left(C_{j+1}\right) \subset \Sigma\left(X_j\right)$ and $\Sigma\left(X_j\right) \subset p_{j-1}^{-1}\left(\Sigma\left(X_{j-1}\right)\right)$ and so

$$(5.25) \qquad (p_0 \circ \cdots \circ p_j)(C_{j+1}) \subset \Sigma(X_0).$$

Finally, because $\rho : X' \longrightarrow X$ is a desingularization of X, we have

$$(5.26) \qquad \rho : X' \setminus \rho^{-1}(\Sigma(X)) \simeq X \setminus \Sigma(X).$$

2. Now, we blow–up S along the analytic subspace C_0 to obtain a space S_1 and a morphism $\sigma_0 : S_1 \longrightarrow S$ with a commutative diagram

$$
\begin{array}{ccc}
X_1 & \hookrightarrow & S_1 \\
{\scriptstyle p_0}\downarrow & & \downarrow{\scriptstyle \sigma_0} \\
X & \hookrightarrow & S
\end{array}
$$

and

$$\sigma_0 : S_1 \setminus \sigma_0^{-1}(C_0) \simeq S \setminus C_0.$$

In particular, because $C_0 \subset \Sigma(X)$

$$\sigma_0 : S_1 \setminus \sigma_0^{-1}(\Sigma(X)) \simeq S \setminus \Sigma(X).$$

Then, we blow–up S_1 along $C_1 \subset \Sigma(X_1)$ to obtain a space S_2, a morphism $\sigma_1 : S_2 \longrightarrow S_1$ with

$$\sigma_1 : S_2 \setminus \sigma_1^{-1}(C_1) \simeq S_1 \setminus C_1$$

and so on, until finally we obtain an analytic space S', a morphism $\sigma_{n-1} : S' \longrightarrow S_{n-1}$ and a commutative diagram

$$(5.27) \qquad
\begin{array}{ccc}
X' & \hookrightarrow & S' \\
{\scriptstyle \rho}\downarrow & & \downarrow{\scriptstyle \sigma} \\
X & \hookrightarrow & S
\end{array}
$$

where $\sigma = \sigma_0 \circ \sigma_1 \circ \cdots \circ \sigma_{n-1}$.
Now each blowing–up σ_j modifies only C_j and C_j is in $p_{j-1}^{-1}(\Sigma(X_{j-1}))$ so that

$$(5.28) \qquad \sigma : S' \setminus \sigma^{-1}(\Sigma(X)) \simeq S \setminus \Sigma(X).$$

In particular $S' \setminus \sigma^{-1}(X)$ is isomorphic to $S \setminus X$ so it is smooth and the singularities of S' are in $\sigma^{-1}(X)$

$$(5.29) \qquad \Sigma(S') \subset \sigma^{-1}(X).$$

Moreover, all points of X' in S' which are not in $\rho^{-1}(\Sigma(X))$ are not modified by σ and so they are still singular for S' (because their image by σ is in X and so they are singular for S), so that $\Sigma(S')$ contains $X' \setminus \rho^{-1}(\Sigma(X))$ and so it contains X'. We can thus write

$$(5.30) \qquad \Sigma(S') = X' \cup Y'$$

where Y' is another analytic subspace of S' such that

$$(5.31) \qquad \sigma\left(Y'\right) \subset \Sigma\left(X\right)$$

the reason being that σ is an isomorphism from $X' \setminus \rho^{-1}\left(\Sigma\left(X\right)\right)$ into $X \setminus \Sigma\left(X\right)$ so that all components of $\Sigma\left(S'\right)$ which are not in X' (and which are sent by σ in X by (5.29)) must be sent necessarily in $\Sigma\left(X\right)$. $\qquad\qquad\square$

b) <u>Desingularization of S'</u>

We start now from the diagramm (5.23) of lemma 5.1 and we construct a desingularization of S' $\quad \beta : \widetilde{S} \longrightarrow S'$ such that

$$\widetilde{Z} \equiv \beta^{-1}\left(\Sigma\left(S'\right)\right)$$

is a divisor with normal crossings

$$\widetilde{Z} = \bigcup_i \widetilde{Z}_i.$$

Lemma 5.2 *Either $\beta\left(\widetilde{Z}_i\right)$ is in Y' and not in X' and then $(\sigma \circ \beta)\left(\widetilde{Z}_i\right)$ is in $\Sigma\left(X\right)$ or $\beta\left(\widetilde{Z}_i\right)$ is in X' and $\sigma \circ \beta|_{\widetilde{Z}_i}$ can be decomposed*

$$(5.32)$$

$$\begin{array}{ccc} & \widetilde{Z}_i & \\ h_i \downarrow & \searrow \sigma \circ \beta & \\ X' & \xrightarrow{\rho} & X \end{array}$$

Proof We know that $\beta\left(\widetilde{Z}\right) \subset \Sigma\left(S'\right)$ and $\sigma\left(\Sigma\left(S'\right)\right) \subset X$ by lemma 5.1. This implies that for each i $\quad(\sigma \circ \beta)\left(\widetilde{Z}_i\right)$ is in X. But $\beta\left(\widetilde{Z}_i\right)$ is irreducible and is in $\Sigma\left(S'\right) = X' \cup Y'$. If $\beta\left(\widetilde{Z}_i\right)$ is in X', then we have a diagram (5.32). If $\beta\left(\widetilde{Z}_i\right)$ is not a subset of X', then $\beta\left(\widetilde{Z}_i\right) \cap \left(X' \setminus Y'\right)$ is empty (because if $\beta\left(\widetilde{Z}_i\right) \cap \left(X' \setminus Y'\right)$ is not empty and is not in X', it should have a non empty intersection with Y', and $\beta\left(\widetilde{Z}_i\right) \cap X'$ and $\beta\left(\widetilde{Z}_i\right) \cap Y'$ would contain at least two distinct irreducible components of $\beta\left(\widetilde{Z}_i\right)$). Then $\beta\left(\widetilde{Z}_i\right)$ is a subset of Y' and $\sigma\left(\beta\left(\widetilde{Z}_i\right)\right) \subset \sigma\left(Y'\right) \subset \Sigma\left(X\right)$ (by lemma 5.1). $\qquad\qquad\square$

c) <u>Construction of the first two floors of the tower</u>

We can now obtain the following result:

Theorem 5.3 *Let S be an analytic space, $\Sigma(S) = X$ its singular locus. We can construct a desingularization $\rho : R(X) \longrightarrow X$, an analytic space S' and a morphism $\sigma : S' \longrightarrow S$, a complex manifold $R(S)$ and a morphism $\alpha : R(S) \longrightarrow S'$ such that we have the following properties:*

1. $\sigma \circ \alpha : R(S) \longrightarrow S$ *is a resolution of singularities of S with exceptional divisor $\hat{Z}$ with normal crossings. In particular, we have the isomorphism $\sigma \circ \alpha : R(S) \setminus \hat{Z} \xrightarrow{\sim} S \setminus \Sigma(S)$.*

2. *The exceptional divisor of $\sigma \circ \alpha$ is denoted*

$$\hat{Z} = \left(\bigcup_{j \in J} \hat{Z}_j \right) \cup \left(\bigcup_{k \in K} \hat{Z}_k \right) \cup \left(\bigcup_{l \in L} \hat{Z}_l \right)$$

with the following properties:

a) $\left(\bigcup_{j \in J} \hat{Z}_j \right) \cup \left(\bigcup_{k \in K} \hat{Z}_k \right) = \alpha^{-1} \left(\Sigma \left(S' \right) \right)$

b) $\alpha \left(\hat{Z}_j \right)$ *is in $R(X)$, $\alpha \left(\hat{Z}_k \right)$ is in Y' and $(\sigma \circ \alpha) \left(\hat{Z}_k \right)$ is in $\Sigma(X)$*

c) $\alpha \left(\hat{Z}_l \right)$ *is not in $\Sigma \left(S' \right)$ and $(\sigma \circ \alpha) \left(\hat{Z}_l \right)$ is in $\Sigma(X)$.*

3. *There is an embedding*

$$R \left(\Sigma(S) \right) \hookrightarrow S'.$$

4. *We have the following commutative diagram:*

(5.33)

$$
\begin{array}{ccccccccc}
\hat{Z} & = & \left(\bigcup_{l \in L} \hat{Z}_l \right) & \cup & \left(\bigcup_{k \in K} \hat{Z}_k \right) & \cup & \left(\bigcup_{j \in J} \hat{Z}_j \right) & \hookrightarrow & R(S) \\
& & \downarrow & & \downarrow & & \downarrow & & \downarrow \alpha \\
& & Y'' & \cup & \dfrac{Y' \quad \cup \quad R(\Sigma(S))}{\Sigma \left(S' \right)} & & & \hookrightarrow & S' \\
& & & \searrow & \downarrow & & \downarrow \rho & & \downarrow \sigma \\
& & & & \Sigma^2(S) & \hookrightarrow & \Sigma(S) & \hookrightarrow & S
\end{array}
$$

we denote $\varphi = \sigma \circ \alpha$.
Here Y' and Y'' are divisors (not necessarily with normal crossings) such that

$$\sigma|_{Y' \cup Y''} : Y' \cup Y'' \longrightarrow \Sigma^2(S)$$

and $\Sigma \left(S' \right) = R \left(\Sigma(S) \right) \cup Y'$.

Proof We start from the desingularization

$$\beta : \widetilde{S} \longrightarrow S'.$$

We know by lemma 5.1, that we have an isomorphism

$$\sigma : S' \setminus \sigma^{-1} \left(\Sigma(X) \right) \longrightarrow S \setminus \Sigma(X).$$

Now, some of the components of $\sigma^{-1}\left(\Sigma(X)\right)$ are in $\Sigma\left(S'\right)$ (the union of these components is denoted Y' in lemma 5.1) and other components Y'' are not included in $\Sigma\left(S'\right)$ (although they intersect X' and so $\Sigma\left(S'\right)$). All these components Y'' are obtained by the blowing–up of the construction of S' and so they are hypersurfaces in S'. Then $\beta^{-1}\left(Y'' \setminus \Sigma\left(S'\right)\right)$ can be again blown–up in $\widetilde{S}$, so that we obtain now

$$\alpha : R(S) \longrightarrow S'$$

with a new divisor with normal crossing

$$\hat{Z} = \left(\bigcup_{j \in J} \hat{Z}_j\right) \cup \left(\bigcup_{k \in K} \hat{Z}_k\right) \cup \left(\bigcup_{l \in L} \hat{Z}_l\right)$$

such that

1. $\left(\bigcup_{j \in J} \hat{Z}_j\right) \cup \left(\bigcup_{k \in K} \hat{Z}_k\right) = \alpha^{-1}\left(\Sigma\left(S'\right)\right)$,

2. $\alpha\left(\hat{Z}_j\right)$ is in $X' = R\left(\Sigma\left(S\right)\right)$, $\alpha\left(\hat{Z}_k\right)$ is in Y' and $(\sigma \circ \alpha)\left(\hat{Z}_k\right)$ is in $\Sigma(X)$.

3. $\alpha\left(\hat{Z}_l\right)$ is not in $\Sigma\left(S'\right)$ and $(\sigma \circ \alpha)\left(\hat{Z}_l\right)$ is in $\Sigma(X)$.

Moreover

$$\sigma \circ \alpha : R(S) \longrightarrow S$$

is a resolution of singularities of S, because $\sigma^{-1}(X) = X' \cup Y' \cup Y''$ and $\alpha^{-1}\left(X' \cup Y' \cup Y''\right) = \hat{Z}$ and because

$$\alpha : R(S) \setminus \hat{Z} \simeq S' \setminus \left(X' \cup Y' \cup Y''\right)$$
$$\sigma : S' \setminus \sigma^{-1}(X) \simeq S \setminus X.$$

$\square$

6   The de Rham complex when X has a smooth singular locus

a) <u>Notations and hypothesis</u>

We start again with a complex analytic space S and in this section we assume

(i)   we construct a tower of desingularizations as in theorem 5.3 (diagram (5.33)) and

(ii) $\Sigma(S) \equiv X$ has a smooth singular locus $\Sigma^2(S)$.

b) The de Rham complex of $X \equiv \Sigma(S)$

First of all, we notice that X satisfies the two hypothesis of section 3: namely X is a space with a smooth singular locus and we have a resolution of singlarities $\rho : R(X) \longrightarrow X$. In particular, we can use this resolution $R(X)$ to construct the de Rham complex $\Lambda_X^\bullet$ of X. A section π of Λ_X^p on an open set U of X is a $\mathcal{C}^\infty$ form $\tilde{\pi}$ of degree p on $\rho^{-1}(U)$ such that there exists a form ω in $\Gamma\left(U \cap \Sigma(X), \Lambda_{\Sigma(X)}^p\right)$ with

$$(6.34) \qquad \tilde{\pi}|_{\rho^{-1}(\Sigma(X))} = \rho^*\omega.$$

c) Lifting forms from $\Sigma(S)$ to $\hat{Z}$

Let π be a p–form on an open set U in $\Sigma(S)$, $\tilde{\pi}$ its realization on $\rho^{-1}(U)$ in $R(\Sigma(S))$ and $\rho^*\omega$ its restriction to the exceptional set $\rho^{-1}(\Sigma^2(S))$ where ω is a form on $U \cap \Sigma^2(S)$. We can lift π on the components $\hat{Z}_j$ of $\hat{Z}$ which project on $R(\Sigma(S))$ (5.33) by pulling-back $\tilde{\pi}$, to obtain forms $\hat{\pi}_j$. The other components $\hat{Z}_l$ and $\hat{Z}_k$ project down to Y'' or Y' (diagram (5.33)) and then to $\Sigma^2(S)$. We can lift π on $\hat{Z}_l$ or $\hat{Z}_k$ by pulling-back ω from $\Sigma^2(S)$ to $\hat{Z}_l$ or $\hat{Z}_k$ through the composed map $\varphi : \hat{Z}_l \longrightarrow \Sigma^2(S)$ or $\varphi : \hat{Z}_k \longrightarrow \Sigma^2(S)$, and we obtain forms $\hat{\omega}_l$ or $\hat{\omega}_k$. It is clear that the set of $\hat{\pi}_j$ is in $\Lambda_{\bigcup_j \hat{Z}_j}^p$ and the set of $\hat{\omega}_k$ or $\hat{\omega}_l$ is in

$\Lambda_{\bigcup_l \hat{Z}_l \cup \bigcup_k \hat{Z}_k}^p$. Moreover an intersection $\hat{Z}_k \cap \hat{Z}_j$ is mapped to $\Sigma^2(S)$ in S, and to $\rho^{-1}(\Sigma^2(S))$ in $R(\Sigma(S))$, so that the restriction to $\hat{Z}_k \cap \hat{Z}_j$ of $\hat{\pi}_j$ is $\alpha^*\rho^*\omega$ and the restriction to $\hat{Z}_k \cap \hat{Z}_j$ of $\hat{\omega}_k$ is $\varphi^*\omega$, so they coincide. As a consequence the sets of $\hat{\pi}_j, \hat{\omega}_k, \hat{\omega}_l$ define an element of $\Lambda_{\hat{Z}}^p$.

In conclusion, we have obtained the following lemma:

Lemma 6.1 *There is an injective morphism which we denote φ^**

$$(6.35) \qquad \varphi^* : \Gamma\left(U \cap \Sigma(S), \Lambda_{\Sigma(S)}^\bullet\right) \longrightarrow \Gamma\left(\varphi^{-1}(U) \cap \hat{Z}, \Lambda_{\hat{Z}}^p\right)$$

for any open set U in S.

Remark 1: *The fact that φ^* is injective is due to the fact that φ is surjective on at least one component of $\hat{Z}$.*

Remark 2: *If the open set U of (6.35) does not intersect $\Sigma^2(S)$, $\Lambda_{\Sigma(S)}^p$ is then the usual sheaf of p–forms on the smooth manifold $U \cap \Sigma(S)$ and $\varphi^{-1}(U) \cap \hat{Z}$ does not intersect the extra components $\left(\bigcup_{l \in L} \hat{Z}_l\right) \cup \left(\bigcup_{k \in K} \hat{Z}_k\right)$ and the lifting of a p–form is the usual lifting by $\left(\varphi|_{\hat{Z}_j}\right)^*$ for each component $\hat{Z}_j$ as in section 3.*

d) The de Rham complex of S

We define the de Rham complex of S as follows:

1. Let m be a point in $S \setminus \Sigma(S)$. Then $\Lambda_{S,m}^p$ is the set of germs of ordinary $\mathcal{C}^\infty$ p–forms at m on S (or the set of germs of ordinary $\mathcal{C}^\infty$ p–forms at $\tilde{m} = \varphi^{-1}(m)$ on $R(S)$).

2. Let m be a point in $\Sigma(S)$, U an open neighborhood of m. A section π of Λ_S^p on U is the data of a section $\tilde{\pi}$ of $\Lambda_{R(S)}^p$ on $\varphi^{-1}(U)$ such that there exists a section ω of $\Lambda_{\Sigma(S)}^p$ on $\Sigma(S) \cap U$ with

$$(6.36) \qquad\qquad \tilde{\pi}|_{\hat{Z}} = \varphi^* \omega$$

where φ^* has been defined in lemma 6.1.

In particular, we have a mapping

$$j : \Gamma\left(U, \Lambda_S^p\right) \ni \pi \mapsto \omega \in \Gamma\left(U \cap \Sigma(S), \Lambda_{\Sigma(S)}^p\right)$$

where $j(\pi) = \omega$ with the relation (6.36) between π and ω.

Lemma 6.2 *The d–operator on $R(S)$ defines a complex*

$$d : \Lambda_S^p \longrightarrow \Lambda_S^{p+1}$$

such that the diagram

$$(6.37)$$

$$\begin{array}{ccc} \Lambda_S^p & \overset{d}{\longrightarrow} & \Lambda_S^{p+1} \\ \Big\downarrow{\scriptstyle j} & & \Big\downarrow{\scriptstyle j} \\ \Lambda_{\Sigma(S)}^p & \overset{d}{\longrightarrow} & \Lambda_{\Sigma(S)}^{p+1} \end{array}$$

is commutative.

e) The sheaves Λ_S^p are fine sheaves

Theorem 6.3 *The sheaves Λ_S^p are fine sheaves on the analytic space S.*

Proof   The proof is almost the same as the proof of theorem 3.4. We just give the modifications of this proof. We cover again S by two kinds of open sets:

- $(V_j)_{j \in J}$ with V_j disjoint from $\Sigma(S)$ and

- $(U_i)_{i \in I}$ with U_i open neighborhoods of points in $\Sigma(S)$ such that one can construct a C^∞ partition of unity $(\chi_i)_{i \in I}$ on $\Sigma(S)$ subordinated to the covering $U_i \cap \Sigma(S)$ of $\Sigma(S)$.

This C^∞ partition of unity lives in fact on $R(\Sigma(S))$ but χ_i is an element of $\Gamma\left(U_i \cap \Sigma(S), \Lambda_{\Sigma(S)}^0\right)$ with support in $U_i \cap \Sigma(S)$. It has been constructed in theorem 3.4 applied to $\Sigma(S)$ itself. In particular, we can lift all the χ_i to $\hat{Z}$ by

$$\chi_i' = \varphi^* \chi_i$$

(φ^* given by lemma 6.1) and the $(\chi_i')_{i \in I}$ define a C^∞ partition of unity of $\hat{Z}$ subordinated to the covering $\left(\varphi^{-1}(U) \cap \hat{Z}\right)_{i \in I}$. At that point, the proof continues exactly as in theorem 3.4 (it is a construction only on the complex manifold $R(S)$). $\qquad\square$

f) The de Rham complex is a resolution

Theorem 6.4 *The de Rham complex*

$$(6.38) \qquad 0 \longrightarrow \mathbb{C}_S \longrightarrow \Lambda_S^0 \xrightarrow{\ d\ } \Lambda_S^1 \xrightarrow{\ d\ } \cdots \xrightarrow{\ d\ } \Lambda_S^p \xrightarrow{\ d\ } \Lambda_S^{p+1} \xrightarrow{\ d\ } \cdots$$

is a resolution of the constant sheaf $\mathbb{C}_S$ on S by fine sheaves.

Proof The proof is almost the same as the one of theorem 4.1 of section 4. We have to use the fact that $\Lambda_{\Sigma(S)}^p$ is a resolution of the constant sheaf $\mathbb{C}_{\Sigma(S)}$ so that we can locally solve the d–equation for $\Lambda_{\Sigma(S)}^p$. $\qquad\qquad\square$

7   General construction of towers of desingularizations

a) <u>Notations</u>

In this section, we shall generalize the construction procedure of section 5. In general,

- (i) $\Sigma(S)$ will denote the singular locus of an analytic space, $\Sigma^k(S)$ the singular locus of $\Sigma^{k-1}(S)$ and

- (ii) if S is an analytic space $\rho : R(S) \longrightarrow S$ will denote a resolution of singularities of S obtained by successive blowing–ups and $\mathrm{Ex}(S)$ will denote the exceptional divisor of $R(S)$.

In the construction, we will introduce intermediary analytic spaces which are not necessarily smooth manifolds. These spaces will be indexed by several numbers, the meaning of which will be explained below.

b) The case where $\Sigma^4(S)$ is empty: applying the construction of theorem 5.3

We begin to do our construction when $\Sigma^4(S) \equiv \Sigma^3(\Sigma(S))$ is empty.

First of all, we apply the construction of theorem 5.3 to $\Sigma(S)$ instead of S. This construction leads to the diagram (5.33) of theorem 5.3, but we take more systematic notations and we present the diagram upside down.

(7.39)

$$
\begin{array}{ccccc}
\Sigma^3(S) & \hookrightarrow & \Sigma^2(S) & \hookrightarrow & \Sigma^1(S) \\[2mm]
\text{floor 1} \quad \diagdown & & \uparrow{\scriptstyle p_1^2} & & \uparrow{\scriptstyle \alpha_1^1} \\[2mm]
P\left(\Sigma_1^1(S)\right) \ \cup \ R\left(\Sigma^2(S)\right) & \hookrightarrow & \Sigma_1^1(S) \\[2mm]
\text{floor 2} \quad \uparrow & \uparrow & \uparrow{\scriptstyle \alpha_1^1} \\[2mm]
\mathrm{Ex}_P\left(\Sigma^1(S)\right) \ \cup \ \mathrm{Ex}_R\left(\Sigma^1(S)\right) & \hookrightarrow & \Sigma_1^1(S)
\end{array}
$$

Let us recall that at floor 1, we perform successive blowing–up in $\Sigma^1(S)$ with center projecting in $\Sigma^3(S)$. This introduces:

1. the resolution of singularities of $\Sigma^2(S)$

2. extra divisors denoted $P\left(\Sigma_1^1(S)\right)$ which project in $\Sigma^3(S)$ and the singular locus $\Sigma\left(\Sigma_1^1(S)\right)$ is contained in $P\left(\Sigma_1^1(S)\right) \cup R\left(\Sigma^2(S)\right)$ and contains $R\left(\Sigma^2(S)\right)$.

At floor 2, we perform successive blowing–up in $\Sigma_1^1(S)$ with successive centers projecting in $P\left(\Sigma_1^1(S)\right) \cup R\left(\Sigma^2(S)\right)$ to obtain $R\left(\Sigma^1(S)\right)$ (resolution of $\Sigma^1(S)$) with exceptional divisor a part of which projects on $R\left(\Sigma^2(S)\right)$ (this is the part $\mathrm{Ex}_R\left(\Sigma^1(S)\right)$) and a part of which projects in $P\left(\Sigma_1^1(S)\right)$ (the part $\mathrm{Ex}_P\left(\Sigma_1^1(S)\right)$).

c) <u>Embedding the construction (7.39) in S</u>

We complete the first line of the diagram (7.39) by embedding $\Sigma^1(S)$ in S and we perform the successive blowing–up along the *same* centers. This introduces a morphism $\alpha_1^0 : S_1^0 \longrightarrow S$ and the singular set $\Sigma\left(S_1^0\right)$ contains $\Sigma_1^1(S)$. This procedure introduces new divisors $P\left(S_1^0\right)$ which project in $\Sigma^3(S)$ (some of these divisors are in $\Sigma\left(S_1^0\right)$).

The new floor 1 is

(7.40)

$$\Sigma^3(S) \quad\hookrightarrow\quad \Sigma^2(S) \quad\hookrightarrow\quad \Sigma(S) \quad\hookrightarrow\quad S$$
$$\nwarrow \qquad\qquad\qquad \uparrow p_1^2 \qquad\qquad \uparrow \alpha_1^1 \qquad\qquad \uparrow \alpha_1^0$$
$$P\left(\Sigma_1^1(S)\right) \;\cup\; R\left(\Sigma^2(S)\right) \;\hookrightarrow\; P\left(\Sigma_1^0(S)\right) \cup \Sigma_1^1(S) \;\hookrightarrow\; \Sigma_1^0(S)$$

Then, at floor 2, we perform in the space S_1^0 the successive blowing–up of the former floor 2 of diagram (7.39) with the same centers to construct a morphism $\alpha_2^0 : S_2^0 \longrightarrow S_1^0$. These centers project on $P\left(\Sigma_1^1(S)\right) \cup R\left(\Sigma^2(S)\right)$. Then the singular locus $\Sigma\left(S_2^0\right)$ contains $R\left(\Sigma^1(S)\right)$ and this process introduces new divisors $P\left(S_2^0\right)$ (some of them being in $\Sigma\left(S_2^0\right)$), all projecting to $P\left(\Sigma_1^1(S)\right) \cup R\left(\Sigma^2(S)\right)$. The new floor 2 is

(7.41)

$$P\left(\Sigma_1^1(S)\right) \quad\cup\quad R\left(\Sigma^2(S)\right) \quad\hookrightarrow\quad \Sigma_1^1(S) \cup P\left(S_1^0\right) \quad\hookrightarrow\quad \Sigma_1^0(S)$$
$$\uparrow \qquad\qquad\qquad \uparrow \qquad\qquad\qquad \uparrow \qquad\qquad \uparrow \alpha_2^0$$
$$\mathrm{Ex}_P\left(\Sigma^1(S)\right) \quad\cup\quad \mathrm{Ex}_R\left(\Sigma^2(S)\right) \quad\hookrightarrow\quad R\left(\Sigma_1^1(S)\right) \cup P\left(S_2^0\right) \quad\hookrightarrow\quad \Sigma_2^0(S)$$

Finally, we desingularize S_2^0 and possibly perform other blowing–ups until we obtain a desingularization $R(S) \longrightarrow S_2^0 \longrightarrow S_1^0 \longrightarrow S$ with exceptional divisor $\mathrm{Ex}(S)$ projecting to $R\left(\Sigma^1(S)\right)$ or to $P\left(S_2^0\right)$ and we obtain a third floor

$$R\left(\Sigma^1(S)\right) \quad\cup\quad P\left(\Sigma_2^0(S)\right) \quad\hookrightarrow\quad S_2^0$$
$$\nwarrow \qquad\qquad \uparrow \qquad\qquad\qquad \uparrow$$

(7.42)

$$\mathrm{Ex}(S) \quad\hookrightarrow\quad R(S)$$

Finally, putting the diagrams (7.40), (7.41) and (7.42) together we have obtained a full tower of embedded desingularizations (7.44) where we have denoted S by $\Sigma^0(S)$ and S_k^0 by $\Sigma_k^0(S)$.

Moreover the desingularizations of $\Sigma^k(S)$ are

$$(7.43) \qquad \begin{aligned} p^2 &\equiv p_1^2 : R\left(\Sigma^2(S)\right) \longrightarrow \Sigma^2(S) \\ p^1 &\equiv \alpha_1^1 \circ \alpha_2^1 : R\left(\Sigma^1(S)\right) \longrightarrow \Sigma^1(S) \\ p^0 &\equiv \alpha_1^0 \circ \alpha_2^0 \circ \alpha_3^0 : R\left(\Sigma^0(S)\right) \longrightarrow \Sigma^0(S) \end{aligned}$$

(7.44)

$$\begin{array}{ccccccc}
\Sigma^3(S) & \hookrightarrow & \Sigma^2(S) & \hookrightarrow & \Sigma^1(S) & \hookrightarrow & \Sigma^0(S) \\
 & \nwarrow & \uparrow{p_2^1} & & \uparrow{\alpha_1^1} & & \uparrow{\alpha_1^0} \\
 & P\left(\Sigma_1^1(S)\right) \ \cup\ R\left(\Sigma^2(S)\right) & & \hookrightarrow & P\left(\Sigma_1^0\right)\cup\Sigma_1^1(S) & \hookrightarrow & \Sigma_1^0(S) \\
 & & \nwarrow & & \uparrow{\alpha_2^1} & & \uparrow{\alpha_2^0} \\
 & & & P\left(\Sigma_2^0\right) \ \cup\ R\left(\Sigma^1(S)\right) & & \hookrightarrow & \Sigma_2^0(S) \\
 & & & \nwarrow & \uparrow{\alpha_3^1} & & \uparrow{\alpha_3^0} \\
 & & & & \mathrm{Ex}(S) & \hookrightarrow & R(S)
\end{array}$$

d) <u>General statement</u>

Theorem 7.1 *Let $S \equiv \Sigma^0(S)$ be an analytic space such that $\Sigma^{N+1}(S)$ is empty. One can construct the following spaces and morphisms:*

(i) resolution of singularities

$$p^k : R\left(\Sigma^k(S)\right) \longrightarrow \Sigma^k(S) \qquad \text{for } 0 \le k \le N$$

(ii) analytic spaces $\Sigma_l^k(S)$ for $0 \le k \le N$ and $0 \le k + l \le N$ with

$$\Sigma_0^k(S) \equiv \Sigma^k(S) \qquad \text{and} \qquad \Sigma_{N-k}^k(S) \equiv R\left(\Sigma^k(S)\right)$$

(iii) morphisms $\alpha_l^k : \Sigma_l^k(S) \longrightarrow \Sigma_{l-1}^k(S)$

(iv) divisors $P_l^k \subset \Sigma_l^k$ such that

$$\alpha_l^0|_{P_l^k} : P_l^k \longrightarrow P_{l-1}^{N-l} \cup R\left(\Sigma^{N-l+1}(S)\right)$$

(v) the exceptional divisor $\mathrm{Ex}(S)$ in $R\left(\Sigma^0(S)\right)$ with a projection

$$\mathrm{Ex}(S) \longrightarrow P_{N-1}^0 \cup R\left(\Sigma^1(S)\right)$$

and the following commutative diagram

(7.45)

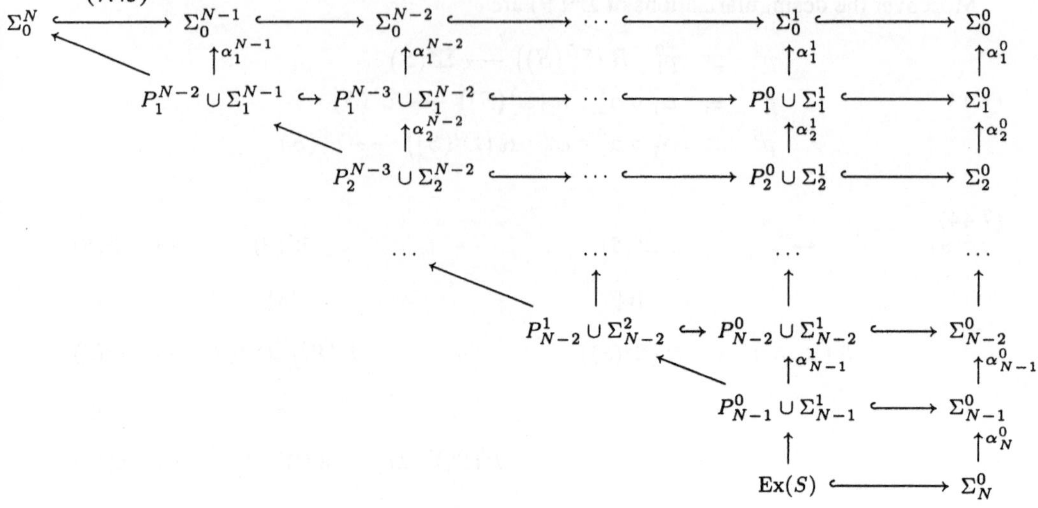

$$
\begin{array}{ccccccccc}
\Sigma_0^N & \hookrightarrow & \Sigma_0^{N-1} & \hookrightarrow & \Sigma_0^{N-2} & \hookrightarrow \cdots \hookrightarrow & \Sigma_0^1 & \hookrightarrow & \Sigma_0^0 \\
& & \big\uparrow \alpha_1^{N-1} & & \big\uparrow \alpha_1^{N-2} & & \big\uparrow \alpha_1^1 & & \big\uparrow \alpha_1^0 \\
& P_1^{N-2} \cup \Sigma_1^{N-1} & \hookrightarrow & P_1^{N-3} \cup \Sigma_1^{N-2} & \hookrightarrow \cdots \hookrightarrow & & P_1^0 \cup \Sigma_1^1 & \hookrightarrow & \Sigma_1^0 \\
& & & \big\uparrow \alpha_2^{N-2} & & & \big\uparrow \alpha_2^1 & & \big\uparrow \alpha_2^0 \\
& & P_2^{N-3} \cup \Sigma_2^{N-2} & \hookrightarrow \cdots \hookrightarrow & & & P_2^0 \cup \Sigma_2^1 & \hookrightarrow & \Sigma_2^0
\end{array}
$$

$$
\begin{array}{ccccc}
\cdots & \cdots & & \cdots & \cdots \\
& \big\uparrow & & \big\uparrow & \big\uparrow \\
P_{N-2}^1 \cup \Sigma_{N-2}^2 & \hookrightarrow & P_{N-2}^0 \cup \Sigma_{N-2}^1 & \hookrightarrow & \Sigma_{N-2}^0 \\
& & \big\uparrow \alpha_{N-1}^1 & & \big\uparrow \alpha_{N-1}^0 \\
& P_{N-1}^0 \cup \Sigma_{N-1}^1 & \hookrightarrow & & \Sigma_{N-1}^0 \\
& \big\uparrow & & & \big\uparrow \alpha_N^0 \\
& \mathrm{Ex}(S) & \hookrightarrow & & \Sigma_N^0
\end{array}
$$

(vi) Moreover the resolutions of singularities are given by:

$$
\begin{aligned}
p^{N-1} &\equiv \alpha_1^{N-1} : \Sigma_1^{N-1} \longrightarrow \Sigma_0^{N-1} \\
p^{N-2} &\equiv \alpha_1^{N-2} \circ \alpha_2^{N-2} : \Sigma_2^{N-2} \longrightarrow \Sigma_0^{N-2} \\
&\;\;\vdots \\
p^1 &\equiv \alpha_1^1 \circ \alpha_2^1 \circ \cdots \circ \alpha_{N-1}^1 : \Sigma_{N-1}^1 \longrightarrow \Sigma^1 \\
p^0 &\equiv \alpha_1^0 \circ \alpha_2^0 \circ \cdots \circ \alpha_N^0 : \Sigma_N^0 \longrightarrow \Sigma^0
\end{aligned}
$$

(vii) The intersection of P_k^{N-k-1} with Σ_k^{N-k} is contained in the exceptional divisor $\left(p^{N-k}\right)^{-1}\left(\Sigma_0^{N-k+1}\right)$ of the resolution $p^{N-k} : \Sigma_k^{N-k} \longrightarrow \Sigma_0^{N-k}$. In general, the exceptional set of p^k is mapped to Σ_{N-k-1}^{k+1} by α_{N-k}^0.

8   The de Rham complex of a general reduced analytic space

a) <u>Hypothesis</u>

We assume the following hypothesis and notations:

  (i) S is a reduced analytic space,

  (ii) we construct a tower of desingularizations as in theorem 7.1 diagram (7.45) and

  (iii) $\Sigma^{N+1}(S)$ is empty.

b) Recursion hypothesis

We shall assume that we have constructed the sheaves $\Lambda^p_{\Sigma^k_0}$ for $1 \leq k \leq N$ using the successive resolutions $\Sigma^k_{N-k} \longrightarrow \Sigma^k_0$. In particular, if U is an open set of Σ^k_0, a section $\pi^{(k)}$ of $\Lambda^p_{\Sigma^k_0}$ on U is the data of a $\mathcal{C}^\infty$ p–form $\widetilde{\pi}^{(k)}$ on $\left(p^k\right)^{-1}(U)$ in $\Lambda^p_{\Sigma^k_{N-k}}$ such that:

$$(8.46) \qquad \widetilde{\pi}^{(k)}\big|_{\mathrm{Ex}\left(\Sigma^k_0\right)} = \left(p^k\right)^* \pi^{(k+1)}$$

with the following conventions: $\pi^{(k+1)}$ is a section on $U \cap \Sigma^{k+1}_0$ of $\Lambda^p_{\Sigma^{k+1}_0}$, $\mathrm{Ex}\left(\Sigma^k_0\right)$ denotes the exceptional set of $p^k : \Sigma^k_{N-k} \longrightarrow \Sigma^k_0$ which projects by α^0_k to Σ^{k+1}_{N-k-1} (see diagram (7.45)) and $\pi^{(k+1)}$ induces $\widetilde{\pi}^{(k+1)}$ on Σ^{k+1}_{N-k-1}. Then $\left(p^k\right)^* \pi^{(k+1)}$ denotes $\left(\alpha^0_{N-k}\right)^* \widetilde{\pi}^{(k+1)}$ so that

$$(8.47) \qquad \left(p^k\right)^* \pi^{(k+1)} \in \Gamma\left(\left(p^k\right)^{-1}\left(U \cap \Sigma^k_0\right) \cap \mathrm{Ex}\left(\Sigma^k_0\right), \Lambda^p_{\mathrm{Ex}\left(\Sigma^k_0\right)} \right)$$

This convention is coherent because in the diagram (7.45) all mappings induce each other by restriction. We also assume that $\Lambda^p_{\Sigma^k_0}$ are fine sheaves on Σ^k_0 and that they give a resolution of $\mathbb{C}_{\Sigma^k_0}$.

c) Lifting forms from $\Sigma^1_0(S)$ to $\mathrm{Ex}(S)$

Let U be an open set of S.

1. We take $\pi^{(1)}$ to be a section of $\Lambda^p_{\Sigma^1_0}$ in $U \cap \Sigma^1_0$, and $\widetilde{\pi}^{(1)}$ the induced form on $\left(p^{(1)}\right)^{-1}\left(U \cap \Sigma^1_0\right) \subset \Sigma^1_{N-1}$. Then $\pi^{(1)}$ induces by (8.46) a form $\pi^{(2)}$ on $U \cap \Sigma^2_0$ which has its realization $\widetilde{\pi}^{(2)}$ on $\left(p^{(2)}\right)^{-1}\left(U \cap \Sigma^2_0\right) \subset \Sigma^2_{N-2}, \ldots$ etc. $\ldots$ and finally, we have a p–form $\pi^{(N-1)}$ on $U \cap \Sigma^{N-1}_0$ which induces a form $\widetilde{\pi}^{(N-1)}$ on $\left(p^{(N-1)}\right)^{-1}\left(U \cap \Sigma^{N-1}_0\right) \subset \Sigma^{N-1}_1$ and by (8.46), $\widetilde{\pi}^{(N-1)}$ induces a form $\pi^{(N)}$ on $U \cap \Sigma^N_0$ (which is a smooth manifold).

2. We want now to lift $\pi^{(1)}$ to an element $\varphi^* \pi^{(1)}$ in $\Lambda^p_{\mathrm{Ex}(S)}$. Let us consider a component Z of $\mathrm{Ex}(S)$ in diagram (7.45). If this component, say Z_1, is mapped to Σ^1_{N-1} by α^0_N, we can pull back $\left(\alpha^0_N\right)^* \widetilde{\pi}^{(1)} = \hat{\pi}^{(1)}$.

If the component, say Z_2, is mapped into P^0_{N-1} by α^0_N, we can map it by α^0_{N-1}, and it goes either in Σ^2_{N-2} or in P^1_{N-2}. If it goes in Σ^2_{N-2}, we pull back $\widetilde{\pi}^{(2)}$ through $\alpha^0_{N-1} \circ \alpha^0_N$ to Z_2, to obtain $\hat{\pi}^{(2)}$. Now $Z_1 \cap Z_2$ is mapped in Σ^1_{N-1} to the exceptional set of the resolution $p^{(1)} : \Sigma^1_{N-1} \longrightarrow \Sigma^1_0$ because Z_2 is mapped to Σ^2_{N-2}, which has its image in Σ^2_0 singular locus of Σ^1_0. But $\hat{\pi}^{(1)}$ restricted to $\left(p^{(1)}\right)^{-1}\left(\Sigma^2_0\right)$ is $\left(p^{(1)}\right)^* \pi^{(2)}$ or by definition in (8.46) $\left(\alpha^0_{N-1}\right)^* \widetilde{\pi}^{(2)}$ so that

$$\begin{aligned}
\hat{\pi}^{(1)}\big|_{Z_1 \cap Z_2} &= \left(\alpha^0_N\right)^* \widetilde{\pi}^{(1)}\big|_{Z_1 \cap Z_2} = \left(\alpha^0_N\right)^* \left(\rho^{(1)}\right)^* \pi^{(2)}\big|_{Z_1 \cap Z_2} \\
&= \left(\alpha^0_N\right)^* \left(\alpha^0_{N-1}\right)^* \widetilde{\pi}^{(2)}\big|_{Z_1 \cap Z_2} = \hat{\pi}^{(2)}\big|_{Z_1 \cap Z_2}.
\end{aligned}$$

3. More generally, a component Z of $\mathrm{Ex}(S)$ will be sent to Σ^k_{N-k} through $\alpha^0_{N-k+1} \circ \cdots \circ \alpha^0_N$ (possibly $k = N$ but we can assume $k > 1$).

Denote Z_k such a component. We can pull back $\widetilde{\pi}^{(k)}$ by $\alpha^0_{N-k+1} \circ \cdots \circ \alpha^0_N$ to Z_k to obtain $\hat{\pi}^{(k)}$. Let us suppose now that such component Z_k has an intersection with a component Z_l, $l < k$. Then

$$(8.48) \qquad\qquad \hat{\pi}^{(k)}|_{Z_k \cap Z_l} = \hat{\pi}^{(l)}|_{Z_k \cap Z_l}.$$

It is sufficient to prove this if $l = k - 1$ and $k > 2$. Now $Z_k \cap Z_{k-1}$ is sent to the exceptional set of the resolution $p^{k-1} : \Sigma^{k-1}_{N-k+1} \longrightarrow \Sigma^{k-1}_0$ and the restriction of $\widetilde{\pi}^{(k-1)}$ to this exceptional set is the pull back of $\widetilde{\pi}^{(k)}$ through α^0_{N-k+1} by (8.46) and the conventions underlying it. But this exactly means that (8.48) holds because both members of this equality are the pull back by

$$\left(\alpha^0_{N-k+1} \circ \cdots \circ \alpha^0_N \right)|_{Z_k \cap Z_{k-1}}$$

of $\widetilde{\pi}^{(k-1)}|_{\mathrm{Ex}\left(\Sigma^{k-1}_0\right)}$.

4. Finally, we have lifted the form $\pi^{(1)}$ on Σ^1_0 to a form $\hat{\pi}^{(1)}$ on $\mathrm{Ex}(S)$, i. e. a form $\hat{\pi}^{(1)}$ which is in the de Rham complex $\Lambda^p_{\mathrm{Ex}(S)}$.

We shall denote this form $\hat{\pi}^{(1)}$ by the notation

$$\left(p^0 \right)^* \pi^{(1)}$$

which is coherent with the notation of (8.46).

Remark *At each intermediary step we do not know if $P^k_{N-k-1} \cup \Sigma^{k+1}_{N-k-1}$ is a divisor with transverse intersections and smooth components as in section 1. Only at the last step for $\mathrm{Ex}(S)$ this is valid. As a consequence, it is not in general possible to define the de Rham complex of $P^k_{N-k-1} \cup \Sigma^{k+1}_{N-k-1}$.*

Finally we have obtained the lemma:

Lemma 8.1 *We can define for any open set U of S an injective mapping*

$$(8.49) \qquad \left(p^0 \right)^* : \Gamma\left(U \cap \Sigma^1_0(S), \Lambda^p_{\Sigma^1_0} \right) \longrightarrow \Gamma\left(\left(p^0 \right)^{-1}(U) \cap \mathrm{Ex}(S), \Lambda^p_{\mathrm{Ex}(S)} \right).$$

d) <u>Definition of the de Rham complex of S</u>

We define the sheaf Λ^p_S of $\mathcal{C}^\infty$ p–forms on S as follows.

1. If m is a regular point of S, $\Lambda^p_{S,m}$ is the set of germs of $\mathcal{C}^\infty$ p–forms on S in a neighborhood of m (or what is the same the set of germs of $\mathcal{C}^\infty$ p–forms on $\Sigma^0_N \equiv R(S)$ in a neighborhood of $\left(p^0 \right)^{-1}(m)$).

2. For a point m in $\Sigma^1(S)$ and an open set U in S containing m, we define a form π in $\Gamma(U, \Lambda^p_S)$ to be a $\mathcal{C}^\infty$ form $\widetilde{\pi}$ on $\left(p^0 \right)^{-1}(U)$ in $R(S)$ such that

$$(8.50) \qquad\qquad \widetilde{\pi}|_{(p^0)^{-1}(U) \cap \mathrm{Ex}(S)} = \left(p^0 \right)^* \omega^{(1)}$$

for $\omega^1 \in \Gamma\left(U \cap \Sigma^1_0(S), \Lambda^p_{\Sigma^1_0(S)} \right)$ where $\left(p^0 \right)^*$ has been defined in the construction leading to lemma 8.1 and we deduce a map

$$(8.51) \qquad\qquad j : \Gamma(U, \Lambda^p_S) \longrightarrow \Gamma\left(U \cap \Sigma^1_0(S), \Lambda^p_{\Sigma^1_0(S)} \right).$$

The Λ^p_S define a complex.

Lemma 8.2 *The d operator on $R(S)$ maps Λ_S^p into Λ_S^{p+1} with a commutative diagram*

(8.52)
$$
\begin{array}{ccc}
\Lambda_S^p & \xrightarrow{\ d\ } & \Lambda_S^{p+1} \\
\downarrow{\scriptstyle j} & & \downarrow{\scriptstyle j} \\
\Lambda_{\Sigma_0^1(S)}^p & \xrightarrow{\ d\ } & \Lambda_{\Sigma_0^1(S)}^{p+1}
\end{array}
$$

where j is defined as in (8.51).

Theorem 8.3 *The sheaves Λ_S^p are fine sheaves on S.*

Proof Same as the proof of theorem 6.3.   □

e) The de Rham resolution of $\mathbb{C}_S$

Theorem 8.4 *The complex*

(8.53) $$0 \longrightarrow \mathbb{C}_S \longrightarrow \Lambda_S^0 \xrightarrow{\ d\ } \Lambda_S^1 \xrightarrow{\ d\ } \cdots \xrightarrow{\ d\ } \Lambda_S^p \xrightarrow{\ d\ } \Lambda_S^{p+1} \xrightarrow{\ d\ } \cdots$$

is a resolution of the constant sheaf $\mathbb{C}_S$ of S by fine sheaves.

Proof The proof is the same as in theorem 4.1 provided we use the recursion hypothesis made above that the sheaves $\Lambda_{\Sigma_0^1(S)}^p$ give a resolution of $\mathbb{C}_{\Sigma_0^1(S)}$ and in particular that the d operator is locally solvable in $\Lambda_{\Sigma_0^1(S)}^p$.   □

Remark *We notice that this construction is also valid for a real reduced analytic space S.*

Bibliography

[1]   V. Ancona and B. Gaveau. Differential forms and resolutions on certain analytic spaces I Irreducible exceptional divisor. *Bull. Sci. Math.*, 116:307–324, 1992.

[2]   V. Ancona and B. Gaveau. Differential forms and resolutions on certain analytic spaces II Flat resolutions. *Canadian J. Math.*, 44:728–749, 1992.

[3]   V. Ancona and B. Gaveau. Formes differentielles et résolutions sur certains espaces analytiques. *C. R. Acad. Sci. Paris*, 314:133–138, 1992.

[4]   V. Ancona and B. Gaveau. Modules plats de differentielles sur certains espaces analytiques. *C. R. Acad. Sci. Paris*, 314:223–225, 1992.

[5]   V. Ancona and B. Gaveau. Differential forms and resolutions on certain analytic spaces III Spectral resolutions. *Annali di Mathematica*, to appear.

[6]   T. Bloom and M. Herrera. De Rham cohomology of an analytic space. *Invent. Math.*, 7:275–296, 1969.

[7]   I. Fary. Valeurs critiques et algèbres spectrales d'une application. *Ann. of Math.*, 63(2):437–490, 1956.

[8]   I. Fary. Cohomologie des variétés algébriques. *Ann. of Math.*, 65(2):21–73, 1957.

[9]   A. Ferrari. Cohomology and holomorphic differential forms on complex analytic spaces. *Ann. Scuola Norm. Sup. Pisa*, 24(3):65–77, 1970.

[10]   H. Grauert. Ein Theorem der analytische Garbentheorie und die Modulräume komplexer Strukturen. *Inst. Hautes Etudes Sci. Publ. Math.*, (5):64 pp., 1960.

[11]   R. Hartshorne. On the de Rham cohomology of algebraic varieties. *Publ. Math. I.H.E.S.*, (45):5–99, 1975.

[12]   J. Leray. L'anneau spectral et l'anneau filtré d'homologie d'un espace localement compact et d'une application continue. *J. Math. Pures Appl.*, 29(9):1–139, 1950.

[13]   H. J. Reiffen. Das Lemma von Poincaré für holomorphe Differentialformen auf komplexen Räumen. *Math. Z.*, 101:269–284, 1967.

[14]   H. J. Reiffen and U. Vetter. Pfaffsche Formen auf komplexen Räumen. *Math. Ann.*, 167:338–350, 1966.

Supported by MURST and GNSAGA of CNR

Some recent results on estimates for the $\bar{\partial}$-equation

Bo Berndtsson

Abstract

We survey some recent work by Sibony, Fornaess-Sibony, and the author concerning L^p and Hölder estimates for solutions to the $\bar{\partial}$-equation. Two new results about Hölder estimates, and L^p-estimates for $\bar{\partial}_b$ are also included.

The L^2-estimates for the $\bar{\partial}$-equation of Hörmander and Kohn (see e g [11], [12] and [13]) have for a long time been a basic tool in several complex variables. In some situations it would be desirable to have a generalization of these estimates to other norms like L^p and Hölder norms, and it is the purpose of this article to survey recent attempts in this direction. In the case of strictly pseudoconvex domains one can obtain a variety of estimates by using explicit formulas for the solution (see the surveys [35] and [16]), but these methods are necessarily less general and precise than the L^2-theory. Here we shall mostly be interested in weakly pseudoconvex domains. Before continuing, let us recall a basic form of Hörmander's theorem.

Theorem 0.1 *Let Ω be a bounded pseudoconvex domain in $\mathbb{C}^n$, and let φ be a plurisubharmonic function in Ω. Let f be a $\bar{\partial}$-closed $(0,1)$-form in Ω. Then there is a solution to the equation*

$$(0.1) \qquad \bar{\partial}u = f$$

such that

$$\|ue^{-\varphi/2}\|_2 \leq C\|fe^{-\varphi/2}\|_2$$

where C is a constant depending only on the diameter of Ω.

The generalization of Theorem 0.1 to other norms is not just a matter of technical difficulties. In [18] Sibony gave an example of a bounded pseudoconvex domain in $\mathbb{C}^2$ where one cannot solve the $\bar{\partial}$-equation with L^∞-estimates. This domain however does not have smooth boundary, but later Sibony gave a smooth example to the same effect in $\mathbb{C}^3$ (cf [19]). After that, in [20], Sibony contructed a smooth domain in $\mathbb{C}^3$ where Hölder estimates for $\bar{\partial}$ don't hold. This is interesting in particular because it shows that the difficulty in the estimates is not just an instance of the familiar phenomenon of a singular integral operator which is bounded on most spaces except the "end-point" cases like L^∞.

In all of the examples mentioned so far the weight function φ is identically zero. In [8] Fornaess and Sibony started to consider the case of weighted estimates, which is a non-trivial problem even in one variable. They proved, in the case of one variable, that the natural analog of Hörmander's theorem does hold when $1 < p \leq 2$, but is in general false for $p > 2$. They also showed that their counterexamples in one dimension could be used to construct examples of smoothly bounded domains in $\mathbb{C}^4$ where L^p-estimates fail for $p > 2$ (even when $\varphi = 0$). The positive results where later generalized in [3] to include the case $p = 1$, and we also showed that in fact $L^1 - L^p$ estimates hold for any $p < 2$. This paper also showed that when $p > 1$ we actually have $L^p - L^2$-estimates, and gave estimates for the solution kernels. In the case when the domain is all of $\mathbb{C}^1$, weighted estimates for any p between 1 and ∞, were proved by Christ, [6], under some regularity assumptions on φ, that are of special relevance in the study of 2-dimensional domains of finite type. A construction of special solution formulas for the weighted problem appears in [1].

On the other hand the case of dimension 2 is rather special, and many results that hold for $n = 2$ have no counterpart in higher dimensions, at least not as yet. It is for instance known that for smooth convex domains in $\mathbb{C}^2$ we have have L^p-estimates for $1 < p < \infty$, [15], and Hölder estimates for $0 < \alpha < 1$, [17], and there is an extensive theory for two-dimensional domains of finite type (see the survey [7]). In [4] we showed however that there is a smoothly bounded domain also in $\mathbb{C}^2$ where L^∞-estimates fail. Shortly after, Fornaess and Sibony [9] showed that the same method gives a smooth counterexample to L^p-estimates in $\mathbb{C}^2$ for any $p > 2$. Here we shall see that the same construction also give counterexamples to Hölder estimates in $\mathbb{C}^2$ (Theorem 2.7).

It is an open question whether L^p-estimates for $\bar{\partial}$ hold in several variables for $p < 2$, but for the induced boundary operator $\bar{\partial}_b$ such estimates are false. Basically the same construction as in the above examples in $\mathbb{C}^2$ give counterexamples to L^p-estimates for $\bar{\partial}_b$ when $p > 2$. But $\bar{\partial}_b$ is a symmetric operator on the boundary of two-dimensional domains, so counterexamples for $p > 2$ also give counterexamples for $q < 2$. Therefore we get an example of a smoothly bounded domain in $\mathbb{C}^2$ where L^p- estimates for $\bar{\partial}_b$ don't hold for any $p \neq 2$ (Theorem 3.1). For $p = 2$ such estimates do hold and were proved by Boas-Shaw [2] and Kohn [14].)We also note that for the interior $\bar{\partial}$- problem , Bonneau and Diederich have proved local L^1-estimates with a logarithmic loss when the right hand side is a (p, q)-form with $q > n - 2$ (see [5] for the precise formulation of this and related results in L^p-norms).

The plan of this paper is as follows. In section 1 we discuss the proof of Theorem 0.1 to see where the difficulty in generalizing it to other norms lies. It turns out that the problem is mainly a question of boundary values, which however disappears in one variable. As a result we get the one-dimensional version of Theorem 0.1 mentioned earlier from [8] and [3]. Here we also give the construction from [8] that disproves these estimates for $p > 2$.

In section 2 we follow the method from [4] and [9] to show how the failure of weighted estimates for very pathological weights in one variable implies the failure of un-weighted estimates in $\mathbb{C}^2$, even for smooth domains. In section 3 we study the implications of these constructions for the tangential operator $\bar{\partial}_b$, and construct in particular a domain in $\mathbb{C}^2$ where L^p-estimates for $\bar{\partial}_b$ fail for all $p \neq 2$. These examples come from a primitive form of microlocal analysis on the boundary. We choose our domain to be a Hartogs domain, i e invariant under rotations in the second variable. Our microlocal analysis then consists in a Fourier series expansion. As it turns out the obstructions to estimates for $p > 2$ come from the positive half of the spectrum, whereas those for $p < 2$ come from the negative half. We explain how this means that the ones for $p > 2$ extend to examples in the interior, but that the examples for $p < 2$ don't extend.

As most of this paper is a survey of known results, we should perhaps point out that Theorems 2.7 and 3.1 appear here for the first time. Finally, we refer to the survey [21] for a general discussion of weakly pseudoconvex domains.

1

Let us first recall the usual proof of Hörmanders theorem. Define the formal adjoint of the $\bar{\partial}$-operator in the weighted L^2-space $L^2(e^{-\varphi})$ by the relation

$$\int \bar{\partial} u . \bar{\alpha} e^{-\varphi} = \int u \overline{\bar{\partial}_\varphi^* \alpha} e^{-\varphi}$$

for all smooth testforms α of bidegree $(0, 1)$ with compact support in Ω. Then the equation

$$(1.2) \qquad\qquad\qquad \bar{\partial} u = f$$

means precisely that

$$\int f . \bar{\alpha} e^{-\varphi} = \int u \overline{\bar{\partial}_\varphi^* \alpha} e^{-\varphi}$$

for all testforms. If Theorem 0.1 holds, then it follows that

$$(1.3) \quad \left| \int f . \bar{\alpha} e^{-\varphi} \right|^2 \leq \int |u|^2 e^{-\varphi} \int |\bar{\partial}_\varphi^* \alpha|^2 e^{-\varphi} \leq C \int |f|^2 e^{-\varphi} \int |\bar{\partial}_\varphi^* \alpha|^2 e^{-\varphi},$$

for all $\bar{\partial}$-closed forms f in $L^2_{(0,1)}(e^{-\varphi})$ and all testforms α. On the other hand, if (1.3) holds, then we can consider the linear subspace of $L^2(e^{-\varphi})$

$$E = \{v = \bar{\partial}_\varphi^* \alpha; \alpha \ \text{testform}\}$$

and define an antilinear functional

$$T : E \to \mathbb{C}$$

by

$$T(\bar{\partial}_\varphi^* \alpha) = \int f . \bar{\alpha} e^{-\varphi}.$$

Then (1.3) implies that T is well defined and bounded with a norm less than

$$(1.4) \qquad C \int |f|^2 e^{-\varphi}.$$

The domain of T can be extended to all of $L^2(e^{-\varphi})$ without increasing the norm, and T must then be given by the scalar product with some element u:

$$\int f.\bar{\alpha} e^{-\varphi} =: T(\bar{\partial}_\varphi^* \alpha) = \int u \overline{\bar{\partial}_\varphi^* \alpha} e^{-\varphi},$$

where the norm of u does not exceed (1.4). This means that u solves (1.2) and satisfies the required estimate. In conclusion, Theorem 0.1 is equivalent to the inequality (1.3). Since, moreover, (1.3) must hold for all $\bar{\partial}$-closed f , what we really need is that

$$(1.5) \qquad \int |P(\alpha)|^2 e^{-\varphi} \leq C \int |\bar{\partial}_\varphi^* \alpha|^2 e^{-\varphi}$$

where $P(\alpha)$ is the orthogonal projection to the subspace of $\bar{\partial}$-closed forms.

To simplify the rest of the discussion, let us assume that $\partial\Omega$ is smooth and is given by

$$\partial\Omega = \{\rho = 0\}$$

where ρ is smooth and satisfies $\partial\rho \neq 0$ on $\partial\Omega$. For α a $(0,1)$-form let us denote by $\alpha_{\vec{n}}$ the normal component of α at the boundary so that on $\partial\Omega$

$$\alpha_{\vec{n}} = \sum \alpha_j \frac{\partial\rho}{\partial z_j} / |\nabla\rho|.$$

Let $\alpha^1 = P(\alpha)$ and let $\alpha^2 = \alpha - \alpha^1$. The fact that α^2 is orthogonal to all forms of the type $\bar{\partial}v$, where v is a function, implies that

$$\bar{\partial}_\varphi^* \alpha^2 = 0$$

and

$$\alpha_{\vec{n}}^2 =: 0$$

on $\partial\Omega$ (in a weak sense). Translated into a condition on α^1 this means that

$$\bar{\partial}\alpha^1 = 0,$$

$$\bar{\partial}_\varphi^* \alpha^1 = \bar{\partial}_\varphi^* \alpha,$$

and

$$\alpha_{\vec{n}}^1 = 0 \text{ on } \partial\Omega$$

(since α has compact support).Therefore the inequality (1.5) is a consequence of

Proposition 1.1 *Let Ω be a smooth pseudoconvex domain in $\mathbb{C}^n$. Let $\varphi \in C^2(\bar{\Omega})$, and let α be a $(0,1)$-form that is smooth on $\bar{\Omega}$ and satisfies $\alpha_{\bar{n}} = 0$ on $\partial\Omega$. Then*

$$\int \Sigma \varphi_{j\bar{k}} \alpha_j \bar{\alpha}_k e^{-\varphi} \leq \int |\bar{\partial}_\varphi^* \alpha|^2 e^{-\varphi} + \int |\bar{\partial}\alpha|^2 e^{-\varphi}.$$

To prove (1.5) from Proposition 1.1 one needs to replace φ by $\varphi + |z|^2$, which just introduces an extra constant since Ω is bounded. One also has to take into account that, a priori, α_1 is not smooth up to the boundary, but we shall not discuss this difficulty here.

The important point for us is that in passing from α to α^1 we loose the property that α vanishes on $\partial\Omega$, the only boundary condition that survives is the weaker $\alpha_{\bar{n}} = 0$. However, when $n = 1$, any $(0,1)$-form is $\bar{\partial}$-closed, so $\alpha = \alpha^1$, and the conditions $\alpha = 0$ and $\alpha_{\bar{n}} = 0$ are of course equivalent.

Proposition 1.1 can be seen as a variant of inequalities that make up the so called Bochner technique (see [22]), and are used to prove vanishing theorems in differential geometry. There are other variants of this technique that give L^p or even pointwise bounds, and they are based on a computation of the Laplacian of $|\alpha|^2 e^{-\varphi}$. However, when we take the norm of α we loose our information at the boundary, except in the case $n = 1$, which we now turn to. To formulate our next technical result we introduce the $\bar{\partial}$-Laplacian

$$\Box_\varphi =: -(\bar{\partial}\bar{\partial}_\varphi^* + \bar{\partial}_\varphi^* \bar{\partial}).$$

Note that when $n = 1$, or more generally when $\bar{\partial}\alpha = 0$

$$\Box_\varphi = -\bar{\partial}\bar{\partial}_\varphi^* \alpha.$$

We also use the convention that $\Delta = \frac{\partial^2}{\partial z \partial \bar{z}}$.

The proof of the following proposition from [3] is a straightforward computation.

Proposition 1.2 *Let α be a $(0,1)$-form in Ω in $\mathbb{C}$. Let $\varphi \in C^2(\Omega)$. Then*

$$(1.6) \quad \Delta(|\alpha|^2 e^{-\varphi}) = 2\operatorname{Re} \Box_\varphi(\alpha)\bar{\alpha} e^{-\varphi} + |\bar{\partial}_\varphi^* \alpha|^2 e^{-\varphi} + |\frac{\partial\alpha}{\partial\bar{z}}|^2 e^{-\varphi} + \Delta\varphi |\alpha|^2 e^{-\varphi}.$$

$(\alpha = \alpha_0 d\bar{z}$, and to interpret the formula one has to make the appropriate identifications of α and α_0).

If we take an α that vanishes on $\partial\Omega$ we can integrate (1.6) over Ω and get

$$\int \Delta\varphi |\alpha|^2 e^{-\varphi} + \int |\frac{\partial\alpha}{\partial\bar{z}}|^2 e^{-\varphi} = \int |\bar{\partial}_\varphi^* \alpha|^2 e^{-\varphi}$$

which implies Proposition 1.1 when $n = 1$. But, we can now do better by integrating against the logarithmic kernel

$$E_a = \frac{1}{2\pi} \log \frac{1}{|z - a|^2},$$

which is positive if we normalize so that $\operatorname{diam}(\Omega) \leq 1$.

If φ is subharmonic it follows that for any $a \in \Omega$

$$|\alpha|^2 e^{-\varphi}(a) \leq -2\mathrm{Re}\int E_a \square_\varphi(\alpha)\bar\alpha e^{-\varphi} = 2\int E_a|\bar\partial_\varphi^*\alpha|^2 e^{-\varphi} - 2\int \frac{\partial E_a}{\partial\bar z}(\bar\partial_\varphi^*\alpha)\bar\alpha e^{-\varphi}.$$

Since $\frac{\partial E_a}{\partial\bar z}$ lies in L^p for any $p < 2$, and E_a lies in any L^p, with a norm that is bounded independently of a, it is not hard to deduce from this that

$$(1.7) \qquad\qquad \sup_\Omega |\alpha| e^{-\varphi/2} \leq C_q \|(\bar\partial_\varphi^*\alpha)e^{-\varphi/2}\|_q$$

for any $q > 2$. On the other hand, repeating the duality argument from the beginning of this section we get:

Proposition 1.3 *Let $1 \leq p, p' \leq \infty$, and let q, q' be the dual exponents. Let Ω be a domain in $\mathbb{C}$. Then we can, for any f in Ω, solve the equation $\bar\partial u = f$ with the estimate*

$$(1.8) \qquad\qquad \|ue^{-\varphi/2}\|_p \leq C\|fe^{-\varphi/2}\|_{p'}$$

if and only if the dual inequality

$$(1.9) \qquad\qquad \|\alpha e^{-\varphi/2}\|_{q'} \leq C\|(\bar\partial_\varphi^*\alpha)e^{-\varphi/2}\|_q$$

holds for any testform. In higher dimensions, (1.9) for all $\bar\partial$-closed forms with $\alpha_{\bar n} = 0$ on $\partial\Omega$, is at least sufficient for having solutions satisfying (1.8).

From here and (1.7) we immediately get

Theorem 1.4 *Let Ω be a domain in $\mathbb{C}$ with $\mathrm{diam}(\Omega) \leq 1$, and let φ be a function that is subharmonic in Ω. Let $1 \leq p < 2$. Then, for any f, there is a solution u to $\bar\partial u = f$ satisfying*

$$\|ue^{-\varphi/2}\|_p \leq C_p\|fe^{-\varphi/2}\|_1$$

where C_p is a constant depending only on p.

In a similar way we can prove that there are solutions satisfying

$$\|ue^{-\varphi/2}\|_2 \leq C_p\|fe^{-\varphi/2}\|_p$$

for any $p > 1$ (see [3] for further results in this direction).

A direct computation of $\Delta(|\alpha|^2 e^{-\varphi})$ does not give Proposition 1.1 in higher dimensions because of the problems with boundary conditions mentioned earlier. There is however an other variant of Proposition 1.2 which is useful in any dimension.

Proposition 1.5 *Let $\alpha = \Sigma\alpha_j d\bar z_j$ be a $\bar\partial$-closed form in Ω in $\mathbb{C}^n$. Let $\varphi \in C^2(\Omega)$. Then*

$$(1.10) \qquad \Sigma\frac{\partial^2}{\partial z_j \partial\bar z_k}(\alpha_j\bar\alpha_k e^{-\varphi}) = -2\,\mathrm{Re}\,\square_\varphi(\alpha).\bar\alpha e^{-\varphi} + |\bar\partial_\varphi^*\alpha|^2 e^{-\varphi}$$

$$+ \Sigma|\frac{\partial\alpha_j}{\partial\bar z_k}|^2 e^{-\varphi} + \Sigma\varphi_{j\bar k}\alpha_j\bar\alpha_k e^{-\varphi}.$$

Again, the proof is a direct computation.

Integration of (1.10) gives Proposition 1.1 (when $\bar{\partial}\alpha = 0$), but when $n > 1$, (1.10) is not sufficiently strong to give L^p-estimates. However, (1.10) does give a little bit more than the usual estimate when we integrate against an arbitrary plurisuperharmonic function.

Proposition 1.6 *Let Ω be a bounded pseudoconvex domain in $\mathbb{C}^n$, let φ be plurisubharmonic in Ω, and let $w \geq 0$ be plurisuperharmonic in Ω. Let f be a $\bar{\partial}$-closed $(0,1)$-form in $L^2(e^{-\varphi})$, and let u be the $L^2(e^{-\varphi})$-minimal solution to $\bar{\partial}u = f$. Then it holds*

$$(1.11) \qquad \int |u|^2 w e^{-\varphi} \leq C \int |f|^2 w e^{-\varphi}$$

where C is a constant depending only on $\mathrm{diam}(\Omega)$.

Any positive plurisuperharmonic function w can be written $w = e^{-\psi}$, where ψ is plurisubharmonic, so we already knew that there is some solution that satisfies (1.11). The point of the statement is that u does not depend on w.

Finally we shall, following Fornaess and Sibony ([8]), show that, even when $n = 1$, there are no weighted estimates for $\bar{\partial}$ in L^p, when $p > 2$.

Proposition 1.7 *Fix $p > 2$ and let C be an arbitrary positive constant. Then there is a subharmonic function, φ, in the disk, which is harmonic near the boundary and a form f such that any solution to $\bar{\partial}u = f$ satisfies*

$$||ue^{-\varphi/2}||_{p'} \geq C||fe^{-\varphi/2}||_{p'}$$

for any $p' \geq p$.

Proof According to Proposition 1.3, we need to find φ such that the estimate

$$(1.12) \qquad ||\alpha e^{-\varphi/2}||_{q'} \leq C||(\bar{\partial}^*_\varphi \alpha)e^{-\varphi/2}||_{q'}$$

does not hold for all testforms with compact support, if $1 \leq q' \leq q$. Since

$$\bar{\partial}^*_\varphi \alpha = -e^\varphi \frac{\partial}{\partial z} e^{-\varphi}$$

we substitute $e^\varphi \alpha$ for α, so (1.12) is equivalent to

$$(1.13) \qquad ||\alpha e^{\varphi/2}||_{q'} \leq C||\frac{\partial \alpha}{\partial z} e^{\varphi/2}||_{q'}.$$

Let

$$\psi(z) = \frac{4}{Nq}\Sigma_1^N \log|z - a_j|$$

where a_j are distinct points with $|a_j| < \delta$, and N is fixed and chosen so large that $q + 4/N < 2$. Take a testform α_ϵ such that

   (i) $\alpha_\epsilon(z) = 1$ if $|z| \leq 1/2$ and $|z - a_j| > 2\epsilon$ for all j,

   (ii) $\alpha_\epsilon(z) = 0$ if $|z - a_j| < \epsilon$ for some j,

and

(iii) $|\partial\alpha_\epsilon| \leq 10/\epsilon$,

where $\epsilon > 0$ will be chosen very small. Moreover we can of course arrange things so that $|\partial\alpha_\epsilon| \leq 10$ if $|z| \geq 1/2$. Then

$$\lim_{\epsilon\to 0}\int_{|z|<1/2}|\frac{\partial\alpha_\epsilon}{\partial z}|^{q'}e^{-\psi q'} = 0,$$

so

$$||\frac{\partial\alpha_\epsilon}{\partial z}e^{-\psi}||_{q'}$$

is bounded by a fixed constant, independent of δ, if ϵ is small enough. On the other hand

$$\int |\alpha_\epsilon|^{q'}e^{-\psi q'} \geq \int_{2\delta<|z|<1/2}\frac{1}{|z|^2} \sim \log 1/\delta \geq C$$

if δ is small enough. This is almost what we want, except that $-\psi$ is not subharmonic. To remedy this, all we need to do is to replace ψ by

$$\psi - \Sigma_1^N \log|z - a_j|$$

and change α_ϵ to $\alpha_\epsilon/\overline{\Pi(z - a_j)}$. $\square$

Proposition 1.7 shows that the constant for the L^p-estimates for $p > 2$ can be arbitrarily large. As in [8] one can combine weights so as to get a weight for which there is no finite constant, but for the applications in the next section, what we have is sufficient.

2

In this section we construct some smoothly bounded pseudoconvex domains in $\mathbb{C}^2$ with bad properties for the $\bar\partial$-equation. We consider Hartogs domains of the form

$$(2.14) \qquad \Omega = \{(z, w) \in \mathbb{C}^2; |z| < 1, |w|^2 < e^{-\varphi(z)}\}.$$

Here φ is a smooth subharmonic function in the disk that behaves like

$$\log\frac{1}{1 - |z|^2}$$

near the boundary. Then Ω is a smooth pseudoconvex domain.

Lemma 2.1 *Let $1 \leq p \leq \infty$, and let C be a constant. Suppose that for any $\bar\partial$-closed $(0, 1)$-form, f, in Ω, there is a solution to $\bar\partial u = f$ satisfying*

$$(2.15) \qquad\qquad ||u||_p \leq C||f||_p.$$

Then for any $F \in L^p(\Delta)$, and any $n \in N$, there is a solution to $\frac{\partial U}{\partial z} = F$ such that

$$||ue^{-(n+2/p)\varphi/2}||_p \leq C||Fe^{-(n+2/p)\varphi/2}||_p.$$

Proof Take $F \in L^p(\Delta)$ and let

$$f = f_n = F(z)w^n d\bar{z}.$$

Then $\bar{\partial} f = 0$ and

$$\|f\|_p^p = \frac{2\pi}{np+2} \int |F|^p e^{-(np+2)\varphi/2}$$

if $p < \infty$, while

$$\|f\|_\infty = \|Fe^{-n\varphi/2}\|_\infty.$$

Let u be a solution to $\bar{\partial} u = f$ satisfying (2.15). Then u is holomorphic in w, so u can be expanded in a power series

$$u(z, w) = \Sigma_0^\infty U_k(z)w^k.$$

Clearly

$$\frac{1}{2\pi} \int_{-\pi}^{\pi} u(z, we^{i\theta}) e^{-in\theta} d\theta = U_n(z)w^n$$

solves the same equation and satisfies the same estimate. Hence

$$\frac{\partial U_n}{\partial \bar{z}} = F$$

and

$$\int |U_n|^p e^{-(np+2)\varphi/2} \leq C^p \int |F|^p e^{-(np+2)\varphi/2},$$

if $p < \infty$, while

$$\|U_n e^{-n\varphi/2}\|_\infty \leq C\|Fe^{-n\varphi/2}\|_\infty.$$

This completes the proof. $\qquad\qquad\qquad\qquad\qquad\qquad\qquad\qquad\qquad\qquad\qquad$ $\square$

For any subharmonic function in the disk we now let $C_{n,p}(\varphi)$ denote the best constant in the inequality

$$\inf_{\bar{\partial} u = f} \|ue^{-n\varphi/2}\|_p \leq C_{n,p}\|fe^{-n\varphi/2}\|_p.$$

Lemma 2.2 *There is a subharmonic function φ in $C^\infty(\bar{\Delta})$, which is harmonic near the boundary, and is such that*

$$\limsup_{n \to \infty} C_{n,p}(\varphi) = \infty,$$

for any $p > 2$.

Proof By the same reasoning as in the proof of Proposition 1.7, $C_{n,p}$ is also the best constant in the inequality

$$\|\alpha e^{\varphi/2}\|_q \leq C\|\frac{\partial \alpha}{\partial z}e^{\varphi/2}\|_q$$

where α ranges over all testforms, and q is the dual exponent. Observe that $C_{n,p}(\varphi)$ only depends on $\Delta\varphi$, since adding a harmonic function to φ just corresponds to multiplying α by the exponential of an antiholomorphic function. Now, Proposition 1.7 says that for any $p > 2$ we can find a (non-smooth) φ such that $C_{1,p}(\varphi)$ is arbitrarily large, and moreover φ is harmonic near the boundary. Regularizing we get a sequence of smooth subharmonic functions φ_k such that $\Delta\varphi_k \in C_c^\infty(\Delta)$, and

$$(2.16) \qquad \lim_{k\to\infty} C_{1,p}(\varphi_k)2^{-k} = \infty,$$

for all $p > 2$. If χ is any function in Δ and if Δ' is a disk that is compactly included in Δ, we denote by

$$\tau(\Delta', \chi)$$

the transportation of χ to Δ' by scaling and translation (so that $\tau(\Delta', \chi)$ becomes a function in Δ'.

Note that, if φ and ψ are two functions in the disk such that on Δ'

$$(2.17) \qquad \varphi = \tau(\Delta', \psi) + \text{harmonic}$$

then

$$(2.18) \qquad C_{n,p}(\varphi) \geq rC_{n,p}(\psi),$$

where r is the radius of the smaller disk. This can be seen by considering testforms with support in Δ'. The condition (2.17) of course means that

$$(2.19) \qquad \Delta\varphi = 1/r^2\tau(\Delta', \Delta\psi).$$

Take a sequence of disjoint disks Δ_k of radius 2^{-k} that are compactly included in Δ and converge to an interior point. Chose a sequence of $n_k \in N$ such that

$$\|\Delta\varphi_k/n_k\|_{C^k} \leq 2^{-k^2}.$$

Define

$$v = \Sigma 2^{2k}\tau(\Delta_k, \Delta\varphi_k/n_k),$$

and let φ be the Greens potential of v. We claim that φ satisfies the conclusion of the lemma. It is clear that φ is smooth up to the boundary and is harmonic near the boundary. Moreover, by (2.18), we have that for each k and n

$$C_{n,p}(\varphi) \geq 2^{-k}C_{n,p}(\varphi_k/n_k).$$

Taking $n = n_k$ we find that

$$C_{n_k,p}(\varphi) \geq 2^{-k} C_{n_k,p}(\varphi_k/n_k) = 2^{-k} C_{1,p}(\varphi_k),$$

so the conclusion

$$\lim_{k \to \infty} C_{n_k,p}(\varphi) = \infty$$

follows from (2.16). $\qquad\qquad\square$

From lemmas 2.1 and 2.2 we deduce

Theorem 2.3 *There is a smoothly bounded pseudoconvex Hartogs domain in $\mathbb{C}^2$ where L^p-estimates for the $\bar{\partial}$-equation don't hold for $p > 2$.*

Proof Take the function constructed in Lemma 2.2 and scale it down to the disk with radius $1/2$. Then continue it to a subharmonic function in Δ that behaves like $\log 1/(1 - |z|^2)$ near the boundary. Then Ω defined by (2.14) satisfies the conclusion of the theorem by Lemma 2.1. $\qquad\qquad\square$

Theorem 2.3 was first proved in [4] in the case $p = \infty$, but counterexamples in higher dimensions were found earlier in [18], and [8]. Then Fornaess and Sibony, [9] observed that the same reasoning using the function from Proposition 1.7 worked for any $p > 2$. We shall now indicate how the same kind of construction gives counterexamples to $C^{k+\alpha}$-estimates. Again, such examples were previously found by Sibony (see [19]) in higher dimensions, using a different method. In the sequel $\Delta_{1/2}$ denotes the disk with radius $1/2$.

Lemma 2.4 *Let Ω be a domain of the form (2.14). Assume that for any $\bar{\partial}$-closed $(0,1)$-form f in $C^\infty(\bar{\Omega})$ there is a solution to $\bar{\partial}u = f$, satisfying*

$$(2.20) \qquad \|u\|_{C^{k+\alpha}} \leq C\|f\|_{C^{k+\alpha}}.$$

Then, for any F in $C_c^\infty(\Delta_{1/2})$, and any $n \in N$ there is a solution to $\partial U/\partial \bar{z} = F$ satisfying

$$\|Ue^{-n\varphi/2}\|_{L^\infty(\Delta_{1/2})} \leq C_{\varphi,k,\alpha} C\|Fe^{-n\varphi/2}\|_{C^{k+\alpha}}.$$

Proof Given F and n let

$$f = F(z)w^n d\bar{z}.$$

Then

$$\|f\|_{C^{k+\alpha}} \leq C_{\varphi,k,\alpha} n^{k+\alpha}\|Fe^{-n\varphi/2}\|_{C^{k+\alpha}}.$$

Let u be any solution to $\bar{\partial}u = f$, satisfying (2.14). As in the proof of Lemma 2.1 we can assume that

$$u = U(z)w^n.$$

Then

$$\|u\|_{C^{k+\alpha}} \geq C_{\varphi,k,\alpha} n^{k+\alpha}\|Ue^{-n\varphi/2}\|_{L^\infty(\Delta_{1/2})}$$

from which the lemma follows. $\qquad\qquad\square$

We next give the statement corresponding to Proposition 1.7 in this context.

Lemma 2.5 *There is a subharmonic function in the disk, harmonic near the boundary, and a form f such that*

$$fe^{-\varphi} \in C_c^\infty(\Delta_{1/2}),$$

but the equation $\bar{\partial}u = f$ has no solution such that

$$\|ue^{-\varphi}\|_\infty < \infty$$

Proof   Take any function φ which is subharmonic in the disk, is smooth on $\Delta_{1/2}$, and equals $-\infty$ on an infinite compact set. Take any f which is smooth and has compact support in $\Delta_{1/2}$. Clearly $fe^{-\varphi} \in C_c^\infty(\Delta_{1/2})$. If u solves $\bar{\partial}u = f$ and $|u| \le Ce^\varphi$, then u is holomorphic in the annulus $\Delta \setminus \Delta_{1/2}$, and vanishes on an infinite compact. Hence u has compact support. This is impossible if, e g , the mean of f is different from zero. $\square$

Given a smooth subharmonic function φ in the disk, we now define $C_{n,k+\alpha}(\varphi)$ as the best constant in the inequality

$$\inf_{\bar{\partial}u=f} \|ue^{-n\varphi/2}\|_\infty \le C_{n,k+\alpha}\|fe^{-n\varphi/2}\|_{C^{k+\alpha}},$$

where f ranges over smooth forms with compact support.

Lemma 2.6 *There is a smooth subharmonic function in the disk, harmonic near the boundary, such that*

$$\limsup_{n\to\infty} C_{n,k+\alpha}(\varphi) = \infty$$

for any k, α.

Proof   This is very similar to the proof of Lemma 2.2 so we will be brief. The previous lemma implies that there is a (non-smooth) φ, such that $C_{1,k+\alpha}(\varphi) = \infty$ for any k, α. Regularizing we find a sequence of smooth $\varphi_m : s$ such that

$$C_{1,k+\alpha}(\varphi_m)$$

tends to infinity as fast as we want when m tends to infinity. Patching them together as in the proof of Lemma 2.2 we get φ. $\square$

From Lemma 2.4 we get as before

Theorem 2.7 *There is a smoothly bounded pseudoconvex Hartogs domain in $\mathbb{C}^2$ such that $C^{k+\alpha}$-estimates for the $\bar{\partial}$-equation don't hold for any $k \in N, 0 \le \alpha \le 1$.*

3

In this section we shall analyze the implications of the previous constructions for the tangential operator $\bar{\partial}_b$. If Ω is given as

$$(3.21) \qquad \Omega = \{(z, w); |z| < 1, \ |w|^2 < e^{-\varphi(z)}\}$$

where $\varphi \sim \log 1/(1 - |z|^2)$ near $\partial\Delta$, we can define $\bar{\partial}_b$ by

$$\bar{\partial}_b u = w(\frac{\partial}{\partial\bar{z}} - \bar{w}\varphi_{\bar{z}}\frac{\partial}{\partial\bar{w}})u.$$

On $\partial\Omega$ we choose as coordinates (z, θ) where $\theta = \arg(w)$. Expressed in these coordinates $\bar{\partial}_b$ becomes

$$\bar{\partial}_b u = e^{-\varphi/2 + i\theta}(\frac{\partial}{\partial\bar{z}} - \frac{i}{2}\varphi_{\bar{z}}\frac{\partial}{\partial\theta})u.$$

Note also that surface measure on $\partial\Omega$ is equivalent to

$$d\lambda(z)d\theta$$

(where $d\lambda$ is Lebesgue measure), so that in particular, for functions depending only on z integration over $\partial\Omega$ and over Δ are equivalent. Any function in $L^1(\partial\Omega)$ can be expanded in a Fourier series with respect to θ

$$u(z, \theta) \sim \Sigma \hat{u}(n, z)e^{in\theta}.$$

The equation $\bar{\partial}_b u = g$ means

$$(\frac{\partial}{\partial\bar{z}} + \frac{n}{2}\varphi_{\bar{z}})\hat{u}(n, z) = e^{\varphi/2}\hat{g}(n + 1, z),$$

or

$$\frac{\partial}{\partial\bar{z}}(e^{n\varphi/2}\hat{u}(n, z)) = e^{(n+1)\varphi/2}\hat{g}(n + 1, z).$$

The operator $\bar{\partial}_b$ is a densely defined closed operator from $L^p(\partial\Omega)$ to itself for any $1 \le p < \infty$, and also from $C(\partial\Omega)$ to itself. The statement that there is a constant C such that one can solve $\bar{\partial}_b u = g$ with

$$\|u\|_p \le C\|g\|_p$$

for any g in $R(\bar{\partial}_b)$ - the image space of $\bar{\partial}_b$- means precisely that $\bar{\partial}_b$ has closed range in L^p. A theorem of Boas-Shaw [2] and Kohn [14] says that this is always the case for $p = 2$.

Theorem 3.1 *There is a smoothly bounded pseudoconvex Hartogs domain in $\mathbb{C}^2$ such that $\bar{\partial}_b$ does not have closed range in L^p for any $p \neq 2$.*

Proof Take the function from Lemma 2.2, scale it to $\Delta_{1/2}$, and then continue it to a smooth subharmonic function in Δ that behaves like $\log 1/(1 - |z|^2)$ near the boundary. Call the resulting function φ and define Ω by (3.21). We claim that Ω has the stated property. To see this, we consider right hand sides g whose Fourier expansion consists of one single term

$$g(z, \theta) = \hat{g}(n + 1, z)e^{i(n+1)\theta},$$

where $\hat{g}$ lies in $L^p(\Delta)$ and has support in $\Delta_{1/2}$. As in the proof of Lemma 2.1 we see that if $g = \bar{\partial}_b u$ where $u \in L^p$, we can in fact choose u of the form

$$u(z, \theta) = \hat{u}(n, z)e^{in\theta}$$

without increasing the L^p-norm. Then the equation $\bar{\partial}_b u = g$ means that

$$(3.22) \qquad \frac{\partial}{\partial \bar{z}} e^{n\varphi/2}\hat{u}(n, z) = e^{(n+1)\varphi/2}\hat{g}(n + 1, z).$$

We can always solve this with $e^{n\varphi/2}\hat{u}$ in in L^p, so if $n \geq 0$ we also get $\hat{u}$ in L^p. Thus, when $n \geq 0$, any g of this form lies in $R(\bar{\partial}_b)$. However, when $n < -1$, $\hat{u}$ can lie in L^p only if $e^{n\varphi/2}\hat{u}$ vanishes on the boundary. Since $e^{n\varphi/2}\hat{u}$ is holomorphic in $\Delta \setminus \Delta_{1/2}$, this means that $\hat{u}$ has support in $\Delta_{1/2}$. Thus, when $n < 0$, g lies in $R(\bar{\partial})$ if and only if

$$e^{(n+1)\varphi/2}\hat{g}$$

is $\frac{\partial}{\partial \bar{z}}$ of a function with compact support. In this case we also see that the solution is unique.

Now, take $p > 2$ and $n \geq 0$. If $\bar{\partial}_b$ has closed range in L^p, the preceeding discussion shows that there is a constant C such that we can solve (3.22) with

$$\|\hat{u}\|_{L^p(\Delta_{1/2})} \leq C\|\hat{g}\|_{L^p(\Delta_{1/2})}.$$

Substituting

$$U = e^{n\varphi/2}\hat{u}, \quad G = e^{(n+1)\varphi/2}\hat{g}$$

this means that $\frac{\partial U}{\partial \bar{z}} = G$ and

$$\|Ue^{-n\varphi/2}\|_{L^p(\Delta_{1/2})} \leq C\|Ge^{-(n+1)\varphi/2}\|_{L^p(\Delta_{1/2})}.$$

But φ was chosen so that this is not possible. Hence $R(\bar{\partial})$ is not closed in L^p for $p > 2$.

Next take $q < 2$ and and $n < 0$, and take $\hat{g}$ to be in $C_c^\infty(\Delta_{1/2})$. Then g lies in $R(\bar{\partial})$ if and only if there is an α in $C_c^\infty(\Delta_{1/2})$ such that

$$\frac{\partial \alpha}{\partial \bar{z}} = e^{(n+1)\varphi/2}\hat{g}.$$

The only possible solution to (3.22) in L^p is then

$$\hat{u}(n, z) = e^{-n\varphi/2}\alpha.$$

Hence, if $\bar{\partial}_b$ has closed range in L^q there is a constant such that

$$\|\alpha e^{-n\varphi/2}\|_{L^q(\Delta_{1/2})} \leq C\|\frac{\partial \alpha}{\partial \bar{z}}e^{-(n+1)\varphi/2}\|_{L^q(\Delta_{1/2})}.$$

But, this is precisely the dual formulation of L^p-estimates for $\frac{\partial}{\partial \bar{z}}$ (see Proposition 1.3 and the proof of Proposition 1.7). Hence this does not hold either, and the proof is complete. $\square$

Note that the counterexamples to L^p-estimates for $p > 2$ come from the positive half of the spectrum, whereas for $p < 2$ they come from the negative half. This means that the examples for $p > 2$ extend to counterexamples for $\bar{\partial}$ in Ω in the following way.

Let us say that a function , g , on $\partial\Omega$ extends to Ω as a $\bar{\partial}$-closed form, f , if

$$f = f_1 d\bar{z} + f_2 d\bar{w}$$

and

$$g = w(f_1 - \bar{w}\varphi_{\bar{z}}f_2)$$

on $\partial\Omega$. This definition is of course made so that if $\bar{\partial}u = f$ on $\bar{\Omega}$ then $\bar{\partial}_b u = g$ extends to f. Thus, if

$$g = \hat{g}(n + 1, z)e^{i(n+1)\theta}$$

with $n > 0$ then g extends to

$$f = \hat{g}(n + 1, z)e^{(n+1)\varphi/2}w^n d\bar{z}.$$

This extension operator satisfies good L^p-estimates. Our final remark is that such extension is not possible when $n < 0$.

For a function u on $\partial\Omega$ we let

$$u_+(z, \theta) = \Sigma_0^\infty \hat{u}(n, z)e^{in\theta}$$

and

$$u_-(z, \theta) = \Sigma_{-\infty}^{-1} \hat{u}(n, z)e^{in\theta}.$$

Proposition 3.2 *Let $g \in L^p(\partial\Omega)$, $1 \leq p \leq \infty$, and assume that g extends to Ω as a $\bar{\partial}$-closed form $f \in L^p(\Omega)$. Then any solution to $\bar{\partial}_b u = g$ satisfies $u_- \in L^p(\partial\Omega)$.*

Proof $u_+(z, w)$ is the limit of

$$\frac{1}{2\pi}\int_{|\zeta|=e^{-\varphi(z)/2}} u(z, \zeta)\frac{d\zeta}{\zeta - w},$$

as w tends to the boundary. By Cauchy's formula

$$u_-(z, w) = \frac{1}{i\pi}\int_{|\zeta|<e^{-\varphi(z)/2}} \frac{\partial u}{\partial \bar{\zeta}}(z, \zeta)\frac{d\lambda(\zeta)}{\zeta - w} = \frac{1}{i\pi}\int_{|\zeta|<e^{-\varphi(z)/2}} f_2(z, \zeta)\frac{d\lambda(\zeta)}{\zeta - w}.$$

By a simple estimate

$$\|u_-\|_{L^p(\partial\Omega)} \leq C\|f_2\|_{L^p(\Omega)}.$$

$\square$

Comparing this to Theorem 3.1, we see that elements in $R(\bar{\partial})$ don't in general extend to closed forms in $L^p(\Omega)$, when $p < 2$.

Bo Berndtsson

Bibliography

[1]   Amar E, L^p-estimates for $\bar\partial$ in $\mathbb{C}$. In Complex Analysis, ed K Diederich. Vieweg Verlag 1991

[2]   Boas H and Shaw M, Sobolev estimates for the Lewy operator on weakly pseudoconvex boundaries. Math Ann 274 (1986)

[3]   Berndtsson B, Weighted estimates for the $\bar\partial$-equation in domains in $\mathbb{C}$. Duke Math J 1992.

[4]   Berndtsson B, A smoothly bounded pseudoconvex domain in $\mathbb{C}^2$ where L^∞ don't hold. To appear in Arkiv för Matematik

[5]   Diederich K and Bonneau P, Integral solution operators for the Cauchy Riemann equations on pseudoconvex domains. Math Ann 286 (1990)

[6]   Christ M, On the $\bar\partial$-equation in weighted L^2-norms in $\mathbb{C}^1$. J Geom Anal 1 (1991)

[7]   Christ M, Precise analysis of $\bar\partial$ and $\bar\partial_b$ on domains of finite type in $\mathbb{C}^2$. Proceedings of the ICM Kyoto 1990. Springer Verlag 1991

[8]   Fornaess J and Sibony N, L^p-estimates for $\bar\partial$. Proc Symp Pure Math 52.3, A M S 1990

[9]   Fornaess J and Sibony N, Pseudoconvex domains in $\mathbb{C}^2$, where the Corona Theorem and L^p-estimates for $\bar\partial$ don' hold. Preprint

[10] Henkin G M and Leiterer J, Theory of Functions on Complex Manifolds. Akademie-Verlag Berlin 1984.

[11] Hörmander L , An Introduction to complex analysis in several variables,

[12] Hörmander L, L^2-estimates and existence theorems for the $\bar\partial$- operator, Acta Math 113 (1965)

[13] Kohn J and Folland G, The Neumann Problem for the Cauchy-Riemann complex. Annals of Mathematics Studies , Princeton 1972

[14] Kohn J, The range of the tangential Cauchy-Riemann operator. Duke Math J 53 (1986)

[15] Polking J, The Cauchy-Riemann equations on convex sets. Proc Symp Pure Math 52.3, A M S 1990

[16] Range R M, Holomorphic Functions and integral representations in several complex variables. Springer Verlag 1986

[17] Range R M, On Hölder and BMO estimates for $\bar\partial$ on convex domains in $\mathbb{C}^2$, to appear

[18] Sibony N, Prolongement analytique des fonctions holomorph bornées et metrique de Caratheodory. Inv Math 29 (1975).

[19] Sibony N, Un example de domaine pseudoconvexe regulier ou l'equation $\bar\partial u = f$ n'admet pas de solution bornée pour f bornée. Inv Math 62 (1980)

[20] Sibony N, On Hölder estimates for $\bar\partial$, Annals of Math Studies, to appear

[21] Sibony N, Some aspects of weakly pseudoconvex domains. Preprint, abbreviated version in proc ICM Kyoto 1990.

[22] Wu H, The Bochner technique, Proc 1980 Beijing conference on Differential geometry and Differential equations, Science Press 1982

Supported by the Natural Science Research Council

Removable singularities in the boundary

Evgeni M. Chirka and Edgar Lee Stout

This paper is devoted to the theory of removable singularities in the boundary of a domain in $\mathbb{C}^n$, $n \geq 2$, or in a complex manifold of dimension at least two.

The notion of removability with which we are concerned may be formulated as follows. Let D be a relatively compact domain in a complex manifold $\mathcal{M}$, let K be a compact subset of bD, and put $\Gamma = bD \setminus K$. Under the assumption that Γ is locally the graph of a Lipschitz function, it is meaningful to speak of CR-functions on Γ. These are the locally integable functions on Γ with the property that $\int_\Gamma f\bar{\partial}\beta = 0$ for every smooth form β on $\mathcal{M}$ the support of which misses K. Given such a form, the integral in question does exist. If the real hypersurface Γ is smooth, then we can speak of CR-distributions on Γ. The condition that the set K be removable is the condition that for every CR-function f on Γ there is a function $F \in \mathcal{O}(D)$ that assumes the boundary values f along Γ. This formulation is not complete as it stands, because we have not stated explicitly the sense in which the function F is to assume the boundary values f. Various situations occur. To begin with, if f is continuous, we require that F assume the boundary values f continuously. If f is bounded, the condition should be that F is to be bounded and assume the boundary values f nontangentially almost everywhere. If f belongs to an L^p class, $0 < p < \infty$, then F should take the boundary values f nontangentially almost everywhere and in the appropriate L^p-sense. Similarly, we can deal with functions in the classes L^p_{loc} on Γ. If bD is smooth, we can discuss f's that are distributions. In principle, all of these conditions must be spelled out in the statement of theorems, and, presumably, the class of sets that are removable for, say, continuous boundary values differs from the class of sets removable for L^∞ boundary values, *etc.* In the results we give below, we are careful to be explicit about this point.

It turns out, though, that in the case of domains with strictly pseudoconvex boundaries of class $\mathcal{C}^2$, there is essentially no difficulty in this connection: In this case, it is sufficient to discuss the question of removability in the context of functions holomorphic on a neighborhood in the manifold $\mathcal{M} \setminus K$ of the hypersurface Γ. This is a consequence of the simple local structure of strictly pseudoconvex boundaries: If Γ is of class $\mathcal{C}^2$ and strictly pseudoconvex, then given a point $p \in \Gamma$, there are local holomorphic coordinates in a neighborhood Ω of p with respect to which Γ is strictly convex. Thus, locally the picture is this: We have a strictly convex domain Δ in C^n that contains the origin. Denote by Σ^+ the part of $b\Delta$ contained in the real halfspace Π given by $\{y_n > 0\}$ and by Δ^+ the part of Δ

contained in this halfspace. Then every CR-function f on Σ^+ continues holomorphically into Δ^+ as, say, the function F. It is known that F assumes the boundary values f in the appropriate sense. In terms of the initial strictly pseudoconvex domain D, this means that there is a fixed open set Ω in D such that $\Gamma \cup \Omega$ is a neighborhood of Γ in $D \cup \Gamma$ into which every CR-function on Γ extends holomorphically, and the extended holomorphic function assumes the boundary values f in the appropriate sense along Γ. If now D' is the strictly pseudoconvex domain obtained from D by pushing Γ slightly into Ω and leaving the rest of bD fixed, then we see that the set K is removable for the domain D if it is removable for D'. This means that the question of removability of K with respect to the domain D is equivalent to the question of the removability of the set K for continuous CR-functions on Γ' with respect to the domain D'. Indeed, it is sufficient to consider functions holomorphic on a neighborhood of Γ'. The conclusion of this is that, in the case of strictly pseudoconvex domains, to discuss questions of removability, it is entirely sufficient to consider continuous CR-functions on Γ or even functions holomorphic on a neighborhood of Γ.

The preceding analysis depends explicitly on the rather simple local structure of the strictly pseudoconvex domains, which yields local continuation of CR-functions into D. In the case of general domains, this local analysis is not possible, and, consequently, the question of the way in which the extended function F assumes the boundary values f must be addressed with care.

One aspect of our work should be noted at the beginning. We work systematically on domains for which the set Γ introduced above is locally a Lipschitz graph. (Often we will abbreviate this and say simply that Γ is locally Lipschitz.) This implies, for example, that much of our work applies to polyhedral domains and to arbitrary bounded convex domains in C^N, for instance to polydiscs. We should also note that our setting is often a bounded domain D in a complex manifold for which $bD = K \cup \Gamma$ with K compact, Γ open in bD and with Γ locally a Lipschitz hypersurface. As such, Γ has *locally* finite area; we do not suppose Γ to have finite area.

A related notion of removability with which we shall deal is that of *weak removability*. With $\mathcal{M}, D, K$, and Γ as above, the set K is said to be weakly removable if for every $f \in L^1_{loc}(\Gamma)$ orthogonal to every $\overline{\partial}$-closed form β of bidegree $(n, n-1)$ on a neighborhood of $\overline{D}$ with the support disjoint from K, i.e., $\int_\Gamma f\beta = 0$, there is a function F holomorphic on D that assumes the boundary values f along Γ. Since among the admitted forms β the $\overline{\partial}$-exact forms occur, it follows that the functions considered in this definition are automatically CR-functions. In general not every CR-function on Γ satisfies the condition, *e.g.*, when Γ is not connected. The theory of weakly removable sets is generally parallel to that of removable sets.

The subject of removable sets in the boundary has seen considerable development in the last few years. See the work of Alexander [2], Anderson and Cima [3], Battelli [7], Dini and Parrini [22], [23], Duval [25], Forstnerič and Stout [28], Hatziafratis [35], Jöricke [41], [42], Kytmanov [43], [43], [44], Kytmanov and Nikitina [45], Laurent-Thiebaut [46], Laurent-Thiebaut and Leiterer [47], Lawrence [48], Lupacciolu [49], [50], [51], [52], [53], [54], [55], [56], [57], Lupacciolu and Stout [58], Lupacciolu and Tomassini [59], Rosay and Stout [63], and Stout [80] [81].

In the present paper our purpose is two-fold. We give new, simpler proofs of some known results, and we present some new results.

The contents of the paper can be outlined as follows. In Section 1 we recall certain results from the theory of Dolbeault cohomology that will be used in the subsequent parts of the paper. This material is mostly known in one form or another, but we have found it convenient to summarize the results at the beginning of our work, because the sources are scattered. We give proofs for some of these results, for others we give explicit references to the literature. In Section 2, we treat removability questions for strictly pseudoconvex domains in Stein manifolds, and, in particular, we rederive characterizations due to Lupacciolu of removable and weakly removable sets in strictly pseudoconvex boundaries. It is to be emphasized that this section is nearly independent of the cohomological machinery assembled in Section 1; one of the principal goals of this section is to demonstrate how simply the results can be derived. This section also contains a new characterization of weakly removable sets in terms of a convexity property. It implies, as a special case, that compacta in strictly pseudoconvex boundaries that are convex with respect to one-dimensional varieties are weakly removable. Section 3 is devoted to new derivations of some sufficient conditions for removability and weak removability in domains in general complex manifolds. Section 4 contains several subsections that treat sufficient conditions for removability and weak removability. In part, it extends to more general settings results previously known in the context of strictly pseudoconvex domains or domains in C^n. It also contains some essentially new results. For example, in Section 4.7, the removable manifolds of bD of real codimension three are characterized as the submanifolds that are not maximally complex cycles, and Section 4.8 discusses removable manifolds of codimension two. This overlaps with some results recently announced by Jöricke [42]. Kytmanov and Nikitina [45] have also considered related questions. In the final section of the paper we discuss some problems that remain open.

We conclude this introduction with a general remark. Our subject is in essence the theory of the boundary behavior of holomorphic functions of several complex variables. This theory is remarkably rich and draws on many topics in modern function theory as is illustrated by our having found it necessary to devote an appendix to a point concerning the embedding of open Riemann surfaces in Stein manifolds. On the face of it, this topic seems far removed from boundary behavior. Nevertheless, the point in question is used in an essential way in the development of a result on weak removability.

We must express our debt to the work of Guido Lupacciolu. His ideas have been decisive in bringing the theory we study to its present state.

We want also to express our thanks to Lutz Bungart for some helpful discussions.

During the preparation of this paper, the first-named author was a visitor at the University of Washington in Seattle. He regards it as his pleasant duty to thank his colleagues and the staff in the Mathematics Department for their support and hospitality.

1 Preliminaries on Dolbeault cohomology

In this section we present a coherent resumé of those parts of the theory of Dolbeault cohomology that we need in the rest of the paper. Our context will generally be that of a noncompact, connected complex manifold countable at infinity, though occasionally we shall restrict our attention to Stein manifolds.

It is to be emphasized that for the work in Section 2, which concerns strictly pseudoconvex domains, the reader will have little need for this section, aside from the definition of the cohomology $H_\Phi^{p,q}$ given below. Most of the material of the present section is essential, though, for Section 3, which treats questions of removability in general manifolds.

Fix a complex manifold $\mathcal{M}$ of complex dimension n. On this manifold we have the spaces $\mathcal{E}^p(\mathcal{M})$ and $\mathcal{E}^{p,q}(\mathcal{M})$ of smooth p-forms and smooth (p, q)-forms, respectively. By $\mathcal{E}(\mathcal{M})$ we shall understand $\mathcal{E}^0(\mathcal{M})$, i.e., $\mathcal{C}^\infty(\mathcal{M})$, the space of smooth functions on $\mathcal{M}$. These spaces have natural topologies; they are Fréchet spaces. The topologies can be characterized by saying that a sequence $\{\alpha_j\}_{j=1}^\infty$ in $\mathcal{E}^p(\mathcal{M})$ converges to zero if and only if for every open set $U \subset \mathcal{M}$ that admits global smooth real coordinates $(x_1, \cdots, x_{2n})$ in terms of which α_j has the expression $\alpha_j = \sum_{|J|=p} \alpha_J^j dx^J$, each of the sequences $\{\alpha_J^j\}_{j=1}^\infty$ converges, together with all sequences of derivatives, uniformly on compacta in U to zero. The topology on $\mathcal{E}^{p,q}(\mathcal{M})$ may be defined similarly; $\mathcal{E}^{p,q}(\mathcal{M})$ is a closed subspace of $\mathcal{E}^{p+q}(\mathcal{M})$. For each choice of p and q, we have the differential operator $\bar{\partial} : \mathcal{E}^{p,q}(\mathcal{M}) \to \mathcal{E}^{p,q+1}(\mathcal{M})$, which is continuous. Its kernel, which, by continuity, is closed, is denoted by $\mathcal{Z}^{p,q}(\mathcal{M})$. The Dolbeault cohomology group $H^{p,q}(\mathcal{M})$ is the quotient space $\frac{\mathcal{Z}^{p,q}(\mathcal{M})}{\bar{\partial}\mathcal{E}^{p,q-1}(\mathcal{M})}$ endowed with the natural factor topology. In general, $\bar{\partial}$ does not have closed range, so $H^{p,q}(\mathcal{M})$ will not, in general, be a Hausdorff space. In those cases in which $\bar{\partial}\mathcal{E}^{p,q-1}(\mathcal{M})$ is closed, $H^{p,q}(\mathcal{M})$ is itself a Fréchet space. Topological vector spaces with Hausdorff topology are often said to be *separated*.

We shall have to deal with cohomology with supports. A family, Θ, of closed subsets of a manifold $\mathcal{M}$ is called a *paracompactifying family* [12], [72] if

i)   Each closed subset of a member of Θ is in Θ.

ii) Θ is closed under the formation of finite unions.

iii) Each element of Θ has a closed neighborhood that is a member of Θ.

Given a paracompactifying family, Θ, the *dual family* is the collection, Δ, of all closed subsets F of $\mathcal{M}$ such that $F \cap E$ is compact for every $E \in \Theta$. The dual family may not be a compactifying family. For an example, see [5].

The two examples most important for us are these: $\Theta = \mathcal{F}$, the collection of all closed subsets of $\mathcal{M}$ and $\Delta = \mathcal{K}$, the collection of all compact subsets of $\mathcal{M}$. These are dual paracompactifying families. The second example is this. Fix a closed set X in $\mathcal{M}$. Let Φ be the family of all closed subsets of $\mathcal{M}\backslash X$ that have compact closure in $\mathcal{M}$, and let Ψ be the family of all closed subsets of $\mathcal{M}\backslash X$ that are closed in $\mathcal{M}$. Then Φ and Ψ are dual paracompactifying families in the complex manifold $\mathcal{M}\backslash X$. We shall use this notation Φ and Ψ consistently.

These two examples have a special property not shared by all paracompactifying families: An arbitrary paracompactifying family is directed by inclusion. In the case of each of the families Φ, Ψ, and $\mathcal{K}$ and $\mathcal{F}$, there is a *cofinal sequence*. For example, in the case of $\mathcal{K}$, $\mathcal{M}$ is the union of a sequence $\{K_j\}_{j=1}^\infty$ of compact sets such that every $K \in \mathcal{K}$ is contained in all but finitely many of the K_j's. For the family $\mathcal{F}$, there is a final element, *viz.*, the manifold $\mathcal{M}$ itself.

In the sequel, we restrict our attention to paracompactifying families with cofinal sequences. As we shall see, this has an important effect on the structure of the topological vector spaces we consider. We shall also consider only paracompactifying families Θ with $\cup\{A : A \in \Theta\} = \mathcal{M}$. Every such paracompactifying family contains the family $\mathcal{K}$ of compact sets.

Fix a paracompactifying family Θ in the complex manifold $\mathcal{M}$, Θ with a cofinal sequence, say $\{F_j\}_{j=1}^{\infty}$. Denote by $\mathcal{E}_{\Theta}^{p}$ the family of all smooth p-forms on $\mathcal{M}$ with support in Θ. When $p = 0$, we shall write simply $\mathcal{E}_{\Theta}$ or, exceptionally, C_{Θ}^{∞}. In addition to $\mathcal{E}_{\Theta}^{p}$, we have the spaces $\mathcal{E}_{\Theta}^{p,q}$ of (p, q)-forms with support in Θ. If $F \in \Theta$, then by $\mathcal{E}_{F}^{p}$ and $\mathcal{E}_{F}^{p,q}$ we mean the spaces of smooth forms with support in F. Thus $\mathcal{E}_{\Theta}^{p,q} = \bigcup_{F \in \Theta} \mathcal{E}_{F}^{p,q}$. Each of the spaces $\mathcal{E}_{F}^{p,q}$ is a closed subspace of the space $\mathcal{E}^{p,q}$ and so is a Fréchet space, and, as Θ has a cofinal sequence say, $\{F_j\}_{j=1}^{\infty}$, $\mathcal{E}_{\Theta}^{p,q}$ is the union of a sequence of Fréchet spaces. The space $\mathcal{E}_{\Theta}^{p,q}$ is given the finest topology with respect to which each of the inclusions $\mathcal{E}_{F_j}^{p,q} \hookrightarrow \mathcal{E}_{\Theta}^{p,q}$ is continuous. In this way, $\mathcal{E}_{\Theta}^{p,q}$ is exhibited as the *strict inductive limit* of the sequence $\{\mathcal{E}_{F_j}^{p,q}\}_{j=1}^{\infty}$ of Fréchet spaces. As such, it is a locally convex topological vector space of type LF. For the theory of these spaces, one may consult [11], [39], and [68]. The topology in $\mathcal{E}_{\Theta}^{p,q}$ is characterized by the condition that a set $A \subset \mathcal{E}_{\Theta}^{p,q}$ is open if and only if $A \cap \mathcal{E}_{F_j}^{p,q}$ is open for each j.

A linear map φ from $\mathcal{E}_{\Theta}^{p,q}$ to a locally convex topological vector space S is continuous if and only if $\varphi|\mathcal{E}_{F_j}^{p,q}$ is continuous for each j. See [11], p. II.27 or [39], p. 159. The *bounded sets* in $\mathcal{E}_{\Theta}^{p,q}$ are the subsets X that are contained in one of the subspaces $\mathcal{E}_{F_i}^{p,q}$ as bounded sets. This is a theorem of Dieudonné and Schwartz [21], [39], p. 161.

The space $\mathcal{E}_{\Theta}^{p,q}$ is a locally convex, Hausdorff, complete topological vector space. In general, it is not metrizable.

The operator $\overline{\partial}$ carries $\mathcal{E}_{\Theta}^{p,q}$ continuously into $\mathcal{E}_{\Theta}^{p,q+1}$. We define $\mathcal{Z}_{\Theta}^{p,q}$ to be $\mathcal{Z}^{p,q} \cap \mathcal{E}_{\Theta}^{p,q}$, and we then have the Dolbeault cohomology groups with support in Θ given by

$$H_{\Theta}^{p,q}(\mathcal{M}) = \frac{\mathcal{Z}_{\Theta}^{p,q}(\mathcal{M})}{\overline{\partial}\mathcal{E}_{\Theta}^{p,q-1}(\mathcal{M})} \, .$$

This group inherits in a natural way the structure of a locally convex topological vector space, which generally is not a Hausdorff space.

A special case of this construction is that in which we deal with compact supports. When Θ is $\mathcal{K}$, the family of all compact subsets of $\mathcal{M}$, we write $\mathcal{D}$, $\mathcal{D}^{p}$, $\mathcal{D}^{p,q}$ and $H_{c}^{p,q}$ for $\mathcal{E}_{\Theta}$, $\mathcal{E}_{\Theta}^{p}$, $\mathcal{E}_{\Theta}^{p,q}$, $H_{\Theta}^{p,q}$, respectively.

We will use certain spaces of currents on $\mathcal{M}$. By $\mathcal{D}'^{p,q}$ we denote the space of continuous linear functionals on the space $\mathcal{D}^{n-p,n-q}$. There are two natural topologies on $\mathcal{D}'^{p,q}$. The first is the weak* topology in which a net $\{T_{\iota}\}_{\iota \in I}$ converges to zero if and only if for each $\alpha \in \mathcal{D}^{n-p,n-q}(\mathcal{M})$, the net $\{T_{\iota}(\alpha)\}_{\iota \in I}$ in $\mathbb{C}$ converges to zero. The other natural topology is that of uniform convergence on bounded sets, which is finer than the weak*-topology. This topology is called the *strong* topology. In the sequel, *we shall always assume $\mathcal{D}'^{p,q}$ to be endowed with the strong topology.* The dual space of $\mathcal{D}'^{p,q}$, endowed with this strong topology, is $\mathcal{D}^{n-p,n-q}$ (See [19].) The topology on $\mathcal{D}^{p,q}$ has the property that a $\mathbb{C}$-linear map $T : \mathcal{D}^{p,q} \to \mathbb{C}$ is continuous if and only if $T(\alpha_j) \to 0$ for every sequence $\{\alpha_j\}_{j=1}^{\infty}$

in $\mathcal{D}^{p,q}$ that converges to zero in $\mathcal{E}^{p,q}(\mathcal{M})$ and for which supp $\alpha_j \subset K$ *for some fixed K* independent of the index j.

By duality, $\overline{\partial}$ extends to a map $\overline{\partial} : \mathcal{D}'^{p,q} \to \mathcal{D}'^{p,q-1}$: If $T \in \mathcal{D}'^{p,q}$ and $\alpha \in \mathcal{D}^{n-p,n-q-1}$, then $\overline{\partial}T(\alpha) = (-1)^{p+q-1}T(\overline{\partial}\alpha)$.

Cohomology groups can be defined in terms of currents:

$$H'^{p,q} = \frac{\{T \in \mathcal{D}'^{p,q} : \overline{\partial}T = 0\}}{\{\overline{\partial}S : S \in \mathcal{D}'^{p,q-1}\}} .$$

This space is again a topological vector space, generally non-Hausdorff.

If $\beta \in \mathcal{E}^{n-p,n-q}$, then β defines a current $[\beta] \in \mathcal{D}'^{p,q}$ by way of the pairing $\langle [\beta], \alpha \rangle = \int_{\mathcal{M}} \beta \wedge \alpha$ for all $\alpha \in \mathcal{D}^{p,q}$. Stokes's theorem implies that $\beta \mapsto [\beta]$ gives rise to a well-defined map of cohomology groups $j : H^{p,q}(\mathcal{M}) \to H'^{p,q}(\mathcal{M})$. This mapping is an isomorphism [24].

In addition to the Dolbeault cohomologies we have described, we have also certain sheaf cohomologies. By $\mathcal{O} = \mathcal{O}_{\mathcal{M}}$ we denote the sheaf of germs of holomorphic functions on $\mathcal{M}$, and by $\Omega^q = \Omega^q_{\mathcal{M}}$ the sheaf of germs of holomorphic q-forms. Associated with these sheaves are the Čech cohomology groups $H^p(\mathcal{M}, \mathcal{O})$ and $H^p(\mathcal{M}, \Omega^q)$. The sheaves $\mathcal{O}$ and Ω^p are coherent analytic sheaves on $\mathcal{M}$, and these cohomology groups are endowed in a natural way with topologies and are topological vector spaces. The details of these topologies are treated in [30] and [31]. A fundamental theorem is the Dolbeault isomorphism: $H^{p,q}(\mathcal{M})$ and $H^q(\mathcal{M}, \Omega^p)$ are isomorphic. See [24], [38] or [31].

For an n-dimensional Stein manifold $\mathcal{M}$, we have the following general facts:

1) $H^{p,q}(\mathcal{M}) = H^q(\mathcal{M}, \Omega^p) = 0$ *for* $p \geq 0, q > 0$.

2) $H_c^{p,q}(\mathcal{M}) = 0$ *for* $p \geq 0, 0 \leq q < n$.

3) $H_c^{p,n}(\mathcal{M}) \simeq (\Omega^{n-p}(\mathcal{M}))^*$ *and for $p \geq 0$.*

The assertion 1) follows from Cartan's Theorem B or, alternatively, can be derived directly as in [38]. The vanishing theorem 2) is due to Cartan and Schwartz. See [71]. The star in 3) indicates topological dual space; the isomorphic $\cong$ is algebraic. The pairing that effects the isomorphism is given by $\langle \varphi, h \rangle = \int_{\mathcal{M}} \varphi \wedge h$. This is a simple corollary of the Serre duality theorem [72].

We shall use repeatedly the following result [26].

4) *Let $\mathcal{M}$ be a noncompact complex manifold with $H_c^{p,1}(\mathcal{M}) = 0$. If $X \subset \mathcal{M}$ is compact and $\mathcal{M}\backslash X$ is connected, then given $f \in \Omega^p(\mathcal{M}\backslash X)$, there is $F \in \Omega^p(\mathcal{M})$ with $f = F|\mathcal{M}\backslash X$.*

Proof Let $\lambda \in \mathcal{D}(\mathcal{M})$ be identically one on a neighborhood of X. The $(p,1)$-form $\overline{\partial}\lambda \wedge f$, extended as zero through X, is $\overline{\partial}$-closed and so $\overline{\partial}$-exact, say $\overline{\partial}\lambda \wedge f = \overline{\partial}g$, $g \in \mathcal{D}^{p,0}(\mathcal{M})$. Set $F = (1 - \lambda)f + g$, where $(1 - \lambda)f$ is extended through X as zero. Then $F \in \Omega^p(\mathcal{M})$, and $F = f$ outside a compact set. Thus $F = f$ in all of $\mathcal{M}\backslash K$, and the assertion is proved. $\square$

A compact set X in a complex manifold $\mathcal{M}$ is said to be a *Stein compactum* if $X = \overset{\infty}{\underset{j=1}{\cap}} U_j$ where each U_j is a Stein open set in $\mathcal{M}$.

5) *Let X be a Stein compactum in a Stein manifold $\mathcal{M}$ of dimension n, $n \geq 2$. Then*

    (a) $H^{p,q}(\mathcal{M}\backslash X) = 0$ *for all* $p \geq 0$, $0 < q < n - 1$.

    (b) $H_c^{p,q}(\mathcal{M}\backslash X) = 0$ *for all* $p \geq 0$, $1 < q < n$.

    (c) $H^{p,n-1}(\mathcal{M}\backslash X)$ *and* $H_c^{p,n}(\mathcal{M}\backslash X)$ *are Hausdorff (separated) topological vector spaces.*

All of this is well-known, but we wish, for the sake of completeness, to give proofs that are based directly on $\bar{\partial}$-methods.

For a), let $X = \bigcap\limits_{j=1}^{\infty} U_j$ where each U_j is a smoothly bounded, strictly pseudoconvex domain, and $U_{j+1} \Subset U_j$. Let α be an arbitrary smooth, $\bar{\partial}$-closed (p,q)-form on $\mathcal{M}\backslash X$, and let $\lambda_j \in \mathcal{D}(\mathcal{M})$, $\lambda_j = 1$ on a neighborhood of $\overline{U}_{j+1}$, $\lambda_j = 0$ on a neighborhood of $\mathcal{M}\backslash U_j$. Then $\bar{\partial}\lambda_j \wedge \alpha$, extended through X as zero, is a $\bar{\partial}$-closed, $(p, q+1)$-form with compact support in U_j. As U_j is a Stein domain and $q + 1 < n$, there is $\beta_j \in \mathcal{D}^{p,q}(U_j)$ such that $\bar{\partial}\lambda_j \wedge \alpha = \bar{\partial}\beta_j$. We extend $(1 - \lambda_j)\alpha$ by zero in $\overline{U}_{j+1}$. Then $\bar{\partial}(\beta_j + (1 - \lambda_j)\alpha) = 0$ on $\mathcal{M}$, so $\beta_j + (1 - \lambda_j)\alpha = \bar{\partial}\gamma_j$ for some $\gamma_j \in \mathcal{E}^{p,q-1}(\mathcal{M})$.

If $q = 1$, the $(p,0)$-form $\gamma_{j+1} - \gamma_j$ is holomorphic in $\mathcal{M}\backslash\overline{U}_j$, so by 4), there is $h_j \in \Omega^p(\mathcal{M})$ with $\gamma_{j+1} - \gamma_j = h_j$ on $\mathcal{M}\backslash\overline{U}_j$. (Note that this complement is connected, for U_j is a Stein domain.) We set

$$\gamma = \gamma_1 + \sum_{j=1}^{\infty}(\gamma_{j+1} - \gamma_j - h_j).$$

This sum is locally finite on $\mathcal{M}\backslash X$, so γ is a smooth $(p,0)$-form defined there. We have

$$\bar{\partial}\gamma = (1 - \lambda)\alpha + \beta_1 + \sum_{j=1}^{\infty}((\lambda_j - \lambda_{j+1})\alpha + \beta_{j+1} - \beta_j) = \alpha + \lim \beta_j = \alpha.$$

Thus, we have a) for $q = 1$.

For $n \geq 3$ and $q > 1$, use induction: As $\bar{\partial}(\gamma_{j+1} - \gamma_j) = 0$ in $\mathcal{M}\backslash\overline{U}_j$ and $\overline{U}_j$ is a Stein compactum, there is $g_j \in \mathcal{E}^{p,q-2}(\mathcal{M}\backslash\overline{U}_j)$ such that $\gamma_{j+1} - \gamma_j = \bar{\partial}g_j$ in $\mathcal{M}\backslash\overline{U}_j$. We extend $(1 - \lambda_{j-1})g_j$ by zero to U_j and obtain $\gamma_{j+1} - \gamma_j = \bar{\partial}((1 - \lambda_{j-1})g_j)$ on $\mathcal{M}\backslash\overline{U}_{j-1}$. We set $\gamma = \gamma_1 + \sum\limits_{j=1}^{\infty}(\gamma_{j+1} - \gamma_j - \bar{\partial}((1 - \gamma_{j-1})g_j))$, and have then that $\gamma \in \mathcal{E}^{p,q-1}(\mathcal{M}\backslash X)$ and $\bar{\partial}\gamma = \alpha$ as above.

To prove b), let $\alpha \in \mathcal{Z}_c^{p,q}(\mathcal{M}\backslash X)$, $1 < q < n$. If we extend α by zero through X, then there is $\beta \in \mathcal{D}^{p,q-1}(\mathcal{M})$ such that $\bar{\partial}\beta = \alpha$. Thus, $\bar{\partial}\beta = 0$ in U_j for j big enough, and there is $\gamma_j \in \mathcal{E}^{p,q-2}(U_j)$ with $\beta = \bar{\partial}\gamma_j$. Set $\beta_0 = \beta - \bar{\partial}(\lambda_j\gamma_j)$–taking $\lambda_j\gamma_j$ to be zero outside U_j. Then $\bar{\partial}\beta_0 = \bar{\partial}\beta = \alpha$, and $\beta_0 = (1 - \lambda_j)\beta - \bar{\partial}\lambda_j \wedge \gamma_j \in \mathcal{D}^{p,q-1}(\mathcal{M}\backslash X)$.

For c), we are to prove that the range of $\bar{\partial} : \mathcal{D}^{p,n-1}(\mathcal{M}\backslash X) \to \mathcal{D}^{p,n}(\mathcal{M}\backslash X)$ is closed, and to do that, it suffices to prove that $\bar{\partial}(\mathcal{D}^{p,n-1}) \cap \mathcal{D}_K^{p,n}$ is closed for each compact set K in $\mathcal{M}\backslash X$. Thus, let $\alpha \in \mathcal{D}_K^{p,n}$ be the limit $\alpha = \lim \bar{\partial}\beta_j$ with $\beta_j \in \mathcal{D}^{p,n-1}(\mathcal{M}\backslash X)$, $\bar{\partial}\beta_j \in \mathcal{D}_K^{p,n}$. We are to see that $\alpha = \bar{\partial}\beta$ for some $\beta \in \mathcal{D}^{p,n-1}(\mathcal{M}\backslash X)$. To do this, extend α through X by zero. Then if $\eta \in \Omega^{n-p}(\mathcal{M})$, we have $\int \alpha \wedge \eta = \lim \int \bar{\partial}\beta_j \wedge \eta = 0$ so that α is orthogonal to $\Omega^{n-p}(\mathcal{M})$. Thus, by 3), there is $\beta \in \mathcal{D}^{p,n-1}(\mathcal{M})$ with $\overline{\partial}\beta = \alpha$. Arguing as in b), we get a $\beta_0 \in \mathcal{D}^{p,n-1}(\mathcal{M}\backslash X)$ with $\overline{\partial}\beta_0 = \alpha$. Thus, as we wished to show, $\overline{\partial}\mathcal{D}^{p,n-1}(\mathcal{M}\backslash X)$ is closed.

The statement about $H^{p,n-1}(\mathcal{M}\backslash X)$ follows by Serre duality. See [72] or 13) below.

Concerning cohomology with supports in the family Φ introduced above, we have the following result.

6) *If $\mathcal{M}$ is a Stein manifold and X is a Stein compactum in $\mathcal{M}$, then $H^{p,q}_{\Phi}(\mathcal{M}\backslash X) = 0$ for $p \geq 0, 0 \leq q < n - 1$, and $H^{p,n-1}_{\Phi}(\mathcal{M}\backslash X) = 0$ if X is $\mathcal{O}(\mathcal{M})$-convex.*

To prove this for $q < n - 1$, we can simply repeat the proof of 5a): The form constructed there belongs to $\mathcal{E}^{p,q-1}_{\Phi}$ when $\alpha \in \mathcal{Z}^{p,q}_{\Phi}$.

If $q = n - 1$, the forms $\overline{\partial}\lambda_j \wedge \alpha = \overline{\partial}((\lambda_j - 1)\alpha_j)$ from 5a) are orthogonal to holomorphic $(n - p)$-forms on $\mathcal{M}$. When X is $\mathcal{O}(\mathcal{M})$-convex, we can take the U_j's to be Runge domains in $\mathcal{M}$. Then $\Omega^p(\mathcal{M})|U_j$ is dense in $\Omega^p(U_j)$, and thus the form $\overline{\partial}\lambda_j \wedge \alpha$, which is compactly supported in U_j, is orthogonal to $\Omega^p(U_j)$. It follows that $\overline{\partial}\lambda_j \wedge \alpha = \overline{\partial}\beta_j$ for a $\beta_j \in \mathcal{D}^{p,n-1}(U_j)$. The rest of the argument follows as in the proof of 5a).

7) *Let $\mathcal{M}$ be a complex manifold with $H^{p,q}_c(\mathcal{M}) = 0$ for some $p \geq 0$, $q \geq 1$. If X is a compact subset of $\mathcal{M}$, then the natural map*

$$\iota : H^{p,q}_{\Phi}(\mathcal{M}\backslash X) \to H^{p,q}(\mathcal{M}\backslash X)$$

is injective. In particular, if $H^{p,q}(\mathcal{M}\backslash X) = 0$, then $H^{p,q}_{\Phi}(\mathcal{M}\backslash X) = 0$.

The proof of this essentially repeats the proof of 5b). Let $\alpha \in \mathcal{Z}^{p,q}_{\Phi}$, and let $\iota[\alpha] = 0$. (Here $[\alpha] \in H^{p,q}_{\Phi}(\mathcal{M}\backslash X)$ is the cohomology class determined by α.) This means that $\alpha = \overline{\partial}\beta$ for some $\beta \in \mathcal{E}^{p,q-1}(\mathcal{M}\backslash X)$. Let $\lambda \in \mathcal{D}(\mathcal{M})$ be identically one on a neighborhood of $X \cup \text{supp }\alpha$. Then $\overline{\partial}\lambda \wedge \beta$, extended by zero through X, lies in $\mathcal{Z}^{p,q}_c(\mathcal{M})$, and thus there is $\gamma \in \mathcal{D}^{p,q-1}(\mathcal{M})$ such that $\overline{\partial}\lambda \wedge \beta = \overline{\partial}\gamma$. If we extend $(1 - \lambda)\beta$ by zero through X, we obtain a form $h = (1 - \lambda)\beta + \gamma$, which is $\overline{\partial}$-closed on $\mathcal{M}$. ($\overline{\partial}h = \overline{\partial}\gamma - \overline{\partial}\lambda \wedge \beta = 0$, for $\lambda = 1$ where β is not $\overline{\partial}$-closed.) As λ and γ are compactly supported, this h coincides with β outside a compact set. If we put $\beta_0 = \beta - h$, then $\overline{\partial}\beta_0 = \overline{\partial}\beta = \alpha$, and supp β_0 has compact closure in $\mathcal{M}$, i.e., $\beta_0 \in \mathcal{E}^{p,q-1}_{\Phi}$. Thus, $[\alpha] = 0$ in $H^{p,q}_{\Phi}(\mathcal{M}\backslash X)$ as we wish to prove.

The following is a result of Malgrange [60].

8) *If $\mathcal{M}$ is a noncompact complex manifold of dimension $n \geq 1$, then $H^{p,n}(\mathcal{M}) = 0$ for all $p \geq 0$.*

We use only the case $p = n$ of this result, in which the $\overline{\partial}$-problem reduces to the Poisson equation for the Laplace operator as in $\mathbb{C}^n$.

Given a closed set X in the complex manifold $\mathcal{M}$, we have $\mathcal{E}^{p,q}(X)$, the space of germs of (p,q)-forms on X. By definition, this is the direct limit of the family $\mathcal{E}^{p,q}(U)$ where U runs through the collection of open subsets of $\mathcal{M}$ that contain X. The mappings $\iota_{UV} : \mathcal{E}^{p,q}(U) \to \mathcal{E}^{p,q}(V)$ for $V \subset U$ are simply the restriction maps. The space of germs $\mathcal{E}^{p,q}(X)$ is given the inductive limit topology so that if $\iota_U : \mathcal{E}^{p,q}(U) \to \mathcal{E}^{p,q}(X)$ is the natural mapping, then a subset B of $\mathcal{E}^{p,q}(X)$ is open if and only if $\iota_U^{-1}(B)$ is open in $\mathcal{E}^{p,q}(U)$ for every choice of the open set U that contains X.

We have that $\overline{\partial} : \mathcal{E}^{p,q}(X) \to \mathcal{E}^{p,q+1}(X)$ is a well-defined, continuous map. There is then the cohomology group

$$H^{p,q}(X) = \frac{\mathcal{Z}^{p,q}(X)}{\overline{\partial}\mathcal{E}^{p,q-1}(X)}$$

if we denote by $\mathcal{Z}^{p,q}(X)$ the closed subspace $\{u \in \mathcal{E}^{p,q}(X) : \overline{\partial}u = 0\}$ of $\mathcal{E}^{p,q}(X)$. Granted the topology on $\mathcal{E}^{p,q}(X)$ and on its closed subspace $\mathcal{Z}^{p,q}(X)$, $H^{p,q}(X)$ has a natural topology.

Compact cohomology spaces $H_c^{p,q}(X)$ are defined similarly: For $U \supset X$ we denote by $\mathcal{E}_{rc}^{p,q}(U)$ the subspace of $\mathcal{E}^{p,q}(U)$ of forms with support relatively compact in $\mathcal{M}$, and we set

$$Z_{rc}^{p,q}(U) = Z^{p,q}(U) \cap \mathcal{E}_{rc}^{p,q}(U) \, .$$

Then

$$H_c^{p,q}(X) = \frac{\mathcal{Z}_c^{p,q}(X)}{\overline{\partial}\mathcal{E}_c^{p,q-1}(X)}$$

where $\mathcal{E}_c^{p,q}(X)$ and $\mathcal{Z}_c^{p,q}(X)$ are the inductive limits of $\mathcal{E}_{rc}^{p,q}(U)$ and $Z_{rc}^{p,q}(U)$, respectively.

Some of our arguments below will proceed by duality. They will be based on the relation of the cohomology of a closed set to that of its complement, which is given by certain exact cohomology sequences.

9) *Let M be a complex manifold, and let X be a closed subset. For each $p \geq 0$, there is a long exact sequence of continuous linear maps*

$$\cdots \to H_c^{p,q}(\mathcal{M}) \xrightarrow{r} H_c^{p,q}(X) \xrightarrow{\delta} H_c^{p,q+1}(\mathcal{M}\backslash X) \xrightarrow{\iota} H_c^{p,q+1}(\mathcal{M}) \to \cdots$$

Here, the mapping r is induced by restriction. If $\varphi \in \mathcal{Z}_c^{p,q}(\mathcal{M})$, set $\widetilde{r}(\varphi) = \varphi_X$, the germ of φ on X . We have $\widetilde{r}(\overline{\partial}\varphi) = \overline{\partial}(\widetilde{r}(\varphi))$, so $\widetilde{r}$ induces a map, r, evidently continuous, from $H_c^{p,q}(\mathcal{M})$ to $H_c^{p,q}(X)$.

To define the map δ, let $\mathcal{F}$ be a fundamental family of neighborhoods of X directed by inclusion. Given U, $V \in \mathcal{F}$ with $\overline{V} \subset U$, let $\lambda_{UV} \in \mathcal{C}^\infty(U)$, $\lambda_{UV} = 1$ on a neighborhood of $\overline{V}$ in U, $\lambda_{UV} = 0$ near bU. Define $\delta_{UV} : Z_{rc}^{p,q}(U) \to H_c^{p,q+1}(\mathcal{M}\backslash X)$ by $\delta_{UV}(\varphi) = [\overline{\partial}\lambda_{UV} \wedge \varphi]$ where we take $\overline{\partial}\lambda_{UV} \wedge \varphi$ as extended by zero outside U. If $W \in \mathcal{F}$ and $\overline{W} \subset V$, then $\delta_{UV}(\varphi) - \delta_{UW}(\varphi) = [\overline{\partial}((\lambda_{UV} - \lambda_{UW}) \wedge \varphi)] = 0$, for $(\lambda_{UV} - \lambda_{UW})\varphi$ has compact support in $\mathcal{M}\backslash X$. Thus, the cohomology class in $H_c^{p,q}(\mathcal{M}\backslash X)$ determined by $\delta_{UV}(\varphi)$ does not depend on the choice of V. We define $\widetilde{\delta}(\varphi_X)$ to be the class determined by $\delta_{UV}(\varphi)$ if φ_X has representative $\varphi \in Z_{rc}^{p,q}(U)$. It does not depend on U, because for $U_1, U_2 \in \mathcal{F}$, there is $V \in \mathcal{F}$ with $\overline{V} \subset U_1 \cap U_2$. Then $\widetilde{\delta}$ is a continuous linear mapping. As $\widetilde{\delta}(\overline{\partial}\psi_X) = [(\delta_{UV}(\overline{\partial}\psi))_X] = -[\overline{\partial}(\overline{\partial}\lambda_{UV} \wedge \psi)] = 0$, this map induces a contnuous map of cohomology classes.

The mapping ι is induced by the embedding. For $\varphi \in \mathcal{Z}_c^{p,q}(\mathcal{M}\backslash X)$, denote by $\widetilde{\iota}(\varphi)$ the form on $\mathcal{M}$ which is zero on X and is φ on $\mathcal{M}\backslash X$. Then $\widetilde{\iota}$ carries $\overline{\partial}$-exact forms to $\overline{\partial}$-exact forms and induces a continuous homomorphism of cohomology.

We verify exactness as follows. If $\varphi_X \in ker\delta$, then there is $\varphi \in Z^{p,q}(U)$ for some $U \in \mathcal{F}$ with germ φ_X on X such that $[\overline{\partial}\lambda_{UV} \wedge \varphi] = 0$ for some $V \in \mathcal{F}$. That is, $\overline{\partial}\lambda_{UV} \wedge \varphi = \overline{\partial}\psi$ for a $\psi \in \mathcal{D}^{p,q}(\mathcal{M}\setminus X)$. Thus, we have $\overline{\partial}(\lambda_{UV}\varphi - \psi) = 0$, i.e., $\lambda_{UV}\varphi - \psi \in Z_c^{p,q}(\mathcal{M})$. This means $(\lambda_{UV}\varphi - \psi)_X$ is in the range of $\widetilde{r}$. As $\lambda_{UV} = 1$ and $\psi = 0$ in a neighborhood of X, it follows that φ_X is in the range of $\widetilde{r}$ and thus $ker\delta \subset im\, r$. By construction, $\delta \circ r = 0$ because $\overline{\partial}\lambda_{UV} \wedge \varphi = \overline{\partial}(\lambda_{UV}\varphi)$. Thus, $ker\delta = im\, r$.

Let now $\varphi \in ker\widetilde{\iota}$, that is $\varphi = \overline{\partial}\psi$ for some $\psi \in \mathcal{D}^{p,q}(\mathcal{M})$. Then $[\varphi] = [\overline{\partial}(\lambda_{UV}\psi) + \overline{\partial}((1 - \lambda_{UV})\psi)] = [\overline{\partial}(\lambda_{UV}\psi)]$, because $(1 - \lambda_{UV})\psi \in \mathcal{D}^{p,q}(\mathcal{M}\setminus X)$. But $[\overline{\partial}(\lambda_{UV}\psi)] = [\overline{\partial}\lambda_{UV} \wedge \psi] \in im\,\delta$, and thus $ker\iota \subset im\,\delta$. Also, $\iota \circ \delta = 0$, for $[\overline{\partial}\lambda_{UV} \wedge \varphi] = [\overline{\partial}(\lambda_{UV}\varphi)] = 0$ in $H_c^{p,q+1}(\mathcal{M})$ for $\varphi \in Z^{p,q}(U_j)$. Hence, $ker\iota = im\,\delta$.

It remains to verify that $im\,\iota = kerr$. To do this, let $\alpha \in \mathcal{Z}_c^{p,q+1}(\mathcal{M}\setminus X)$ be extended by zero through X, and let $[\alpha] \in H_c^{p,q+1}(\mathcal{M})$ be its cohomology class. By definition, $r \circ \iota[\alpha] \in H_c^{p,q+1}(X)$. As α has compact support in $\mathcal{M}\setminus X$ and r arises from the restriction map, it follows that $r \circ \iota[\alpha] = 0$. Thus, $im\,\iota \subset kerr$. If $\alpha \in \mathcal{Z}_c^{p,q+1}(\mathcal{M})$ and $r[\alpha] = 0$, then, as r arises from restriction, it follows that $\alpha = \overline{\partial}\beta$ on a neighborhood U of X for some $\beta \in \mathcal{E}_{rc}^{p,q}(U)$. (Recall that $H_c^{p,q+1}(X)$ is computed in terms of germs.) Set $\alpha_0 = \alpha - \overline{\partial}(\lambda\beta)$ where $\lambda = 1$ in a neighborhood of X and $\lambda = 0$ on a neighborhood of $\mathcal{M}\setminus U$ (and $\lambda\beta$ is extended by zero in $\mathcal{M}\setminus U$). Then $[\alpha] = [\alpha_0]$ in $H_c^{p,q+1}(\mathcal{M})$, but $\alpha_0 \in \mathcal{Z}_c^{p,q+1}(\mathcal{M}\setminus X)$, so $[\alpha_0] \in im\,\iota$.

This concludes the proof of 9). $\qquad\qquad\square$

Now, let X be a closed set in $\mathcal{M}$, and let, as above, $\Phi = \Phi(X)$ be the family of closed subsets of $\mathcal{M}\setminus X$ with compact closure in $\mathcal{M}$. Let $\Psi = \Psi(X)$ be the dual family of closed sets in $\mathcal{M}\setminus X$ that are, as well, closed in $\mathcal{M}$. The following is an analogue of 9).

10) *Let X be a closed subset of a complex manifold $\mathcal{M}$. For an arbitrary $p \geq 0$, there is a long exact sequence of continuous linear mappings*

$$\cdots \to H^{p,q}(\mathcal{M}) \xrightarrow{r} H^{p,q}(X) \xrightarrow{\delta} H_{\Psi}^{p,q+1}(\mathcal{M}\setminus X) \xrightarrow{\iota} H^{p,q+1}(\mathcal{M}) \to \cdots .$$

The mappings r, δ, and ι are defined precisely as in 9), and the proof follows, *mutatis mutandis*, as in the proof of 9).

We shall need a version of the Serre duality theorem for cohomology with supports. We have introduced above the spaces $\mathcal{D}'^{p,q}$ of currents: $\mathcal{D}'^{p,q}$ is the topological dual of the space $\mathcal{D}^{n-p,n-q}$. If we are given a paracompactifying family Θ with dual Δ, assumed to be paracompactifying, the topological dual space of $\mathcal{E}_{\Theta}^{n-p,n-q}$ can be identified with the space $\mathcal{D}_{\Delta}'^{p,q} = \{\varphi \in \mathcal{D}'^{p,q} : \text{ supp}\,\varphi \in \Delta\}$.

By the Dolbeault isomorphism [24] or directly by regularization [16], there is an isomorphism

$$H_{\Theta}^{p,q-1}(\mathcal{M}) \cong \mathcal{Z}_{\Theta}'^{p,q}/\overline{\partial}\mathcal{D}_{\Theta}'^{p,q-1} = H_{\Theta}'^{p,q}(\mathcal{M}) .$$

If E is a topological vector space, then $^{\sigma}E$, *the associated separated space*, is the quotient space $E/\overline{0}$ where $\overline{0}$ denotes the closure of the origin in E. If E is a Hausdorff space, then $^{\sigma}E = E$; in general, there is a continuous linear surjection $E \to {}^{\sigma}E$.

The following variant of the Serre Duality Theorem is a basic for our later work. We shall give a proof that uses $\overline{\partial}$ methods.

11) *Let $\mathcal{M}$ be an arbitrary connected complex manifold of dimension n. Let Θ and Δ be dual paracompactifying families in $\mathcal{M}$. For all integers $p, q \geq 0$, there is an algebraic isomorphism*

$$^{\sigma}H^{p,q}_{\Theta}(\mathcal{M}) \simeq (^{\sigma}H^{n-p,n-q}_{\Delta}(\mathcal{M}))^{*}.$$

As seen from the following proof, we need not assume in this proposition that Θ or Δ has a cofinal sequence, but it is assumed in 12) for Δ and in 13) for Θ and Δ.

The star denotes topological dual space.

Proof   Set $(r, s) = (n - p, n - q)$. A current $T \in \mathcal{Z}^{\prime p,q}_{\Theta}$ defines a continuous linear functional on $\mathcal{E}^{r,s}_{\Delta}$, which annihilates $\overline{\partial}\mathcal{E}^{r,s-1}_{\Delta}$ and hence its closure. Thus, T defines a continuous linear functional on the space $\mathcal{Z}^{r,s}_{\Delta}/cl\overline{\partial}\mathcal{E}^{r,s-1}_{\Delta} = {}^{\sigma}H^{r,s}_{\Delta}$. In this way, we have a natural continuous linear mapping $\tau : \mathcal{Z}^{\prime p,q}_{\Theta} \to (^{\sigma}H^{r,s}_{\Delta})^{*}$. The map τ is surjective.

Let L denote the closure in $\mathcal{Z}^{\prime p,q}_{\Theta}$ of the space $\overline{\partial}\mathcal{D}^{\prime p,q-1}$, closure with respect to the strong topology.[1] Then $ker\tau \supset L$. Indeed, $ker\tau = L$: If $T \notin L$, there is a form $\varphi \in \mathcal{E}^{r,s}_{\Delta}$ such that $\langle T, \varphi \rangle \neq 0$ but $\langle \Lambda, \varphi \rangle = 0$ for all $\Lambda \in L$. In particular, $\langle \overline{\partial}S, \varphi \rangle = 0$ for all $S \in \mathcal{D}^{\prime p,q-1}_{\Theta}$. Thus, $\varphi \in \mathcal{Z}^{r,s}_{\Delta}$ and $T \notin ker\tau$. We thus have a continuous algebraic isomorphism[2]

$$\mathcal{Z}^{\prime p,q}_{\Theta}/ker\tau = {}^{\sigma}H^{p,q}_{\Theta} \to (^{\sigma}H^{r,s}_{\Delta})^{*}.$$

$\square$

As a corollary, there is the following:

12) *With assumptions on $\mathcal{M}$, Θ and Δ as in 11), if $H^{n-p,n-q+1}_{\Delta}(\mathcal{M})$ is Hausdorff, then*
$$H^{p,q}_{\Theta}(\mathcal{M}) \simeq (^{\sigma}H^{n-p,n-q}_{\Delta}(\mathcal{M}))^{*}.$$

In this proof, it is not necessary for Θ to have a cofinal sequence.

Proof   We show that $H^{p,q}_{\Theta}(\mathcal{M})$ is separated. To this end, let the current T lie in the closure of $\overline{\partial}\mathcal{D}^{\prime p,q-1}_{\Theta}$. It is then orthogonal to $\mathcal{Z}^{r,s}_{\Delta}$ and so defines a continuous linear functional $\widetilde{T}$ on $\mathcal{E}^{r,s}_{\Delta}/\mathcal{Z}^{r,s}_{\Delta}$ by $\langle \widetilde{T}, [\varphi] \rangle = \langle T, \varphi \rangle$. The mapping

$$(1.1) \qquad\qquad \overline{\partial} : \mathcal{E}^{r,s}_{\Delta}/\mathcal{Z}^{r,s}_{\Delta} \to \overline{\partial}\mathcal{E}^{r,s}_{\Delta}$$

is continuous. As $H^{r,s+1}_{\Delta}(\mathcal{M})$ is separated, $\overline{\partial}\mathcal{E}^{r,s}_{\Delta}$ is a closed subspace of $\mathcal{E}^{r,s+1}_{\Delta}$ (or $\mathcal{Z}^{r,s+1}_{\Delta}$). Thus, by the open mapping theorem[3], the map (1.1) is open and so is a topological isomorphism. Consequently, $\widetilde{T}$ induces a continuous linear functional S_0 on $\overline{\partial}\mathcal{E}^{r,s}_{\Delta}$ by

$$\langle S_0, \overline{\partial}\varphi \rangle = (-1)^{q}\langle \widetilde{T}, [\varphi] \rangle$$
$$= (-1)^{q}\langle T, \varphi \rangle.$$

---

[1] This is the same as the closure in the weak* topology because of reflexivity. See [19].

[2] It is natural to suppose that this isomorphism is topological. However, the necessary open mapping theorem does not seem to be available.

[3] A version of the open mapping theorem sufficient in this context is that given by Dieudonné and Schwartz [21]: A continuous linear surjective map from one LF space onto another is open. The space $\mathcal{E}^{r,s}_{\Delta}$ is an LF space, for we have assumed Δ to have a cofinal sequence, say $\{F_j\}^{\infty}_{j=1}$. The space $\overline{\partial}\mathcal{E}^{r,s}_{\Delta}$ is a closed subspace of the LF-space $\mathcal{E}^{r,s+1}$; it is the strict inductive limit of the spaces $F_j \cap \overline{\partial}\mathcal{E}^{r,s}_{\Delta}$ and so is an LF-space.

Again, as $\overline{\partial}\mathcal{E}_\Delta^{r,s}$ is closed, the Hahn-Banach theorem provides an extension, S, of S_0 to all of $\mathcal{E}_\Delta^{r,s+1}$. Thus, we obtain a current $S \in \mathcal{D}_\Theta'^{p,q-1}$ such that $\langle S, \overline{\partial}\varphi \rangle = (-1)^q \langle T, \varphi \rangle$, i.e., such that $T = \overline{\partial}S$. It follows that $\overline{\partial}\mathcal{D}_\Theta'^{p,q-1}$ is closed as claimed. It follows that $H_\Theta^{p,q}(\mathcal{M})$ is separated, and, by 11), we get 12). $\qquad\square$

The results 11) and 12) jointly may be considered as a version of the Serre duality theorem for Dolbeault cohomology (with scalar coefficients).

Another corollary is this.

13) *With $\mathcal{M}$, Θ, Δ as in 11) $H_\Theta^{p,q}(\mathcal{M})$ is Hausdorff if and only if $H_\Delta^{n-p,n-q+1}(\mathcal{M})$ is Hausdorff.*

The proof in one direction is contained in 12). But $(r, s) = (n - p, n - q + 1)$, and we have $(n - r, n - s + 1) = (p, q)$. Thus, if in 12), we interchange Θ and Δ and interchange (p, q) and (r, s) and apply 12), we get the other implication. $\qquad\square$

There are two further consequences that bear directly on removability questions.

14) *Let X be a closed set in a noncompact complex manifold $\mathcal{M}$ of dimension n. If $H^{n,n-1}(X) = 0$, then $H_\Phi^{0,1}(\mathcal{M}\backslash X)$ is Hausdorff.*

By the exact sequence 10) and Malgrange's theorem, point 8), we have the exact sequence

$$0 \to H_\Psi^{n,n}(\mathcal{M}\backslash X) \to H^{n,n}(\mathcal{M}) = 0.$$

By 13) with $(p, q) = (0, 1)$, $\Theta = \Phi$ and $\Delta = \Psi$, it follows that $H_\Phi^{0,1}(\mathcal{M}\backslash X)$ is separated.

15) *Let X be a compact set in a noncompact complex manifold $\mathcal{M}$ of dimension n with $H^{n,n-1}(\mathcal{M}) = 0$. If $H^{n,n-1}(X) = {}^\sigma H^{n,n-2}(X) = 0$ for $n \geq 3$, or if $H^{n,n-1}(X) = 0$ and $\mathcal{Z}^{n,n-2}(\mathcal{M})$ is dense in $\mathcal{Z}^{n,n-2}(X)$ in the sense of germs for $n \geq 2$, then $H_\Phi^{0,1}(\mathcal{M}\backslash X) = 0$.*

Proof If $H^{n,n-1}(X) = 0$, then $H_\Phi^{0,1}(\mathcal{M}\backslash X)$ is separated by 14), and it is enough to prove, by 12), that ${}^\sigma H_\Psi^{n,n-1}(\mathcal{M}\backslash X) = 0$.

As $H^{n,n-1}(\mathcal{M}) = 0$, we have from the exact sequence 10) the exact sequence

$$H^{n,n-2}(\mathcal{M}) \xrightarrow{r} H^{n,n-2}(X) \xrightarrow{\delta} H_\Psi^{n,n-1}(\mathcal{M}\backslash X) \to 0.$$

Thus, δ is surjective. If $\mathcal{Z}^{n,n-2}(\mathcal{M})|X$ is dense in $\mathcal{Z}^{n,n-2}(X)$, then $im\,r$ is dense in $H^{n,n-2}(X)$, so $0 = im(\delta \circ r)$ is dense in $H_\Psi^{n,n-1}(\mathcal{M}\backslash X)$, for δ and r are continuous. In this case, ${}^\sigma H_\Psi^{n,n-1}(\mathcal{M}\backslash X) = H_\Psi^{n,n-1}(\mathcal{M}\backslash X)/\overline{0} = 0$.

The first case is even simpler: If ${}^\sigma H^{n,n-2}(X) = 0$, then 0 is dense in $H^{n,n-2}(X)$, so $0 = \delta(0)$ is dense in $H_\Psi^{n,n-1}(\mathcal{M}\backslash X)$ by the continuity of δ. $\qquad\square$

The following three lemmas are useful in studying removability in local and general, noncompact situations.

16) *Let X be a locally closed[4] set with compact closure $\overline{X}$ and boundary $bX = \overline{X} \setminus X$ in a complex manifold $\mathcal{M}$. Then*

    (a) *If $H^{p,q}(\overline{X}) = H^{p,q-1}(bX) = 0$ for some $p \geq 0, q > 1$, then $H_c^{p,q}(X) = 0$.*

    (b) *If $H^{p,q}(\overline{X}) = 0$ and the restriction mapping $\mathcal{Z}^{p,q-1}(\mathcal{M}) \rightarrow \mathcal{Z}^{p,q-1}(bX)$ has dense range for some $p \geq 0, q \geq 1$, then $^{\sigma}H_c^{p,q}(X) = 0$.*

Proof Let φ be a $\overline{\partial}$-closed smooth (p,q)-form in a neighborhood of X with $X \cap supp\ \varphi$ compact. We can extend φ to a $\overline{\partial}$-closed form in a neighborhood of $\overline{X}$ by setting $\varphi = 0$ in a neighborhood of bX. As $H^{p,q}(\overline{X}) = 0$, there is a smooth $(p, q-1)$-form ψ in a neighborhood of $\overline{X}$ such that $\overline{\partial}\psi = \varphi$ there. In particular, $\overline{\partial}\psi = 0$ in a neighborhood of bX.

If $H^{p,q-1}(bX) = 0$, there is a $(p, q-2)$-form g in a neighborhood U of bX, U disjoint from $supp\ \varphi$ such that $\overline{\partial}g = \psi$ there. Let $\lambda \in \mathcal{D}(U)$, $\lambda = 1$ in a neighborhood of bX, and then define $\psi_o = \psi - \overline{\partial}(\lambda g) = (1 - \lambda)\psi + \overline{\partial}\lambda \wedge g$. Then $\overline{\partial}\psi_o = \varphi$, but now $X \cap supp\ \psi_o$ is compact.

If the germs of $\mathcal{Z}^{p,q-1}(\mathcal{M})$ are dense in $\mathcal{Z}^{p,q-1}(bX)$, then there are a neighborhood U of bX and a sequence of forms $\psi_j \in \mathcal{Z}^{p,q-1}(\mathcal{M})$ such that $\psi = lim\ \psi_j$ in $\mathcal{E}^{p,q-1}(U)$. Then in a neighborhood of $\overline{X}$

$$\varphi = \overline{\partial}((1 - \lambda)\psi) + \overline{\partial}\lambda \wedge \psi = \overline{\partial}((1 - \lambda)\psi) + lim\ \overline{\partial}\lambda \wedge \psi_j$$
$$= \overline{\partial}((1 - \lambda)\psi) - lim\ \overline{\partial}((1 - \lambda)\psi_j) = lim\ \overline{\partial}((1 - \lambda)(\psi - \psi_j)).$$

As $V \cap supp\ (1 - \lambda)$ has compact closure for some neighborhood $V \supset X$ and $X \cap supp\ (1 - \lambda)$ is compact, it follows that $^{\sigma}H_c^{p,q}(X) = 0$.   □

17) *Let $\mathcal{M}$ be a complex manifold of dimension $n \geq 2$ such that $H_c^{n,n}(\mathcal{M})$ is Hausdorff. If X is a closed subset of $\mathcal{M}$ with $H_c^{n,n-1}(X) = 0$, then $H^{0,1}(\mathcal{M} \setminus X)$ is Hausdorff.*

Proof From the exact sequence – see 9) above

$$0 = H_c^{n,n-1}(X) \xrightarrow{\delta} H_c^{n,n}(\mathcal{M} \setminus X) \xrightarrow{\iota} H_c^{n,n}(\mathcal{M})$$

we find that ι is one-to-one. Thus, $0 = \iota^{-1}(0)$ is closed in $H_c^{n,n}(\mathcal{M} \setminus X)$, because 0 is closed in $H_c^{n,n}(\mathcal{M})$) and ι is continuous, whence the result by 13).   □

18) *Let $\mathcal{M}$ be a complex manifold of dimension $n \geq 3$ with $H_c^{n,n-1}(\mathcal{M}) = 0$. If $H_c^{n,n}(\mathcal{M})$ is Hausdorff and if X is a closed subset of $\mathcal{M}$ such that $H_c^{n,n-1}(X) = {}^{\sigma}H_c^{n,n-2}(X) = 0$, then $H^{0,1}(\mathcal{M} \setminus X) = 0$.*

---

[4]Recall that a *locally closed* set in $\mathcal{M}$ is a closed subset of an open subset of $\mathcal{M}$.

Proof From the exact sequence – see 9)

$$H_c^{n,n-2}(X) \xrightarrow{\delta} H_c^{n,n-1}(\mathcal{M} \setminus X) \xrightarrow{\iota} H_c^{n,n-1}(\mathcal{M}) = 0,$$

we obtain that δ is surjective. If $^\sigma H_c^{n,n-2}(X) = 0$, then 0 is dense in $H_c^{n,n-2}(X)$. As δ is continuous and surjective, it follows that $0 = \delta(0)$ is dense in $H_c^{n,n-1}(\mathcal{M} \setminus X)$, that is $^\sigma H_c^{n,n-1}(\mathcal{M} \setminus X) = 0$. Since $H_c^{n,n-1}(X) = 0$, the space $H^{0,1}(\mathcal{M} \setminus X)$ is Hausdorff by 17). Thus, by Serre duality, we obtain

$$H^{0,1}(\mathcal{M} \setminus X) \simeq (^\sigma H_c^{n,n-1}(\mathcal{M} \setminus X))^* = 0.$$

This completes the proof. $\square$

We shall need a regularity result for solutions of $\overline{\partial}$ for which we have no reference, though the result is probably well known.

19) *Let $\mathcal{M}$ be a complex manifold. If α is a $\overline{\partial}$-exact $(0,q)$-form, $q \geq 1$, on $\mathcal{M}$ with real-analytic coefficients, say $\overline{\partial}\beta = \alpha$ for a smooth form β, then there is a $(0, q-1)$-form $\widetilde{\beta}$ on $\mathcal{M}$ with real-analytic coefficients such that $\overline{\partial}\widetilde{\beta} = \alpha$.*

The following proof was suggested to us by Lutz Bungart.

Denote by $\mathcal{E}_\omega^{0,p}$ the sheaf of germs of real-analytic forms of bidegree $(0,q)$ on $\mathcal{M}$. Thus $\mathcal{E}_\omega^{0,0}$ is the sheaf of germs of real-analytic functions on $\mathcal{M}$. We have an acyclic resolution of the sheaf $\mathcal{O}$ of germs of holomorphic functions on $\mathcal{M}$

$$0 \rightarrow \mathcal{O} \rightarrow \mathcal{E}_\omega^{0,0} \xrightarrow{\overline{\partial}} \mathcal{E}_\omega^{0,1} \xrightarrow{\overline{\partial}} \mathcal{E}_\omega^{0,2} \xrightarrow{\overline{\partial}} \cdots .$$

This *is* a resolution, *i.e.*, the sequence is exact, because the lemma is known to be true locally as follows from explicit formulas that are available for solving $\overline{\partial}$. That it is acyclic, *i.e.*, that the cohomology groups $H^p(\mathcal{M}, \mathcal{E}_\omega^{p,q}) = 0$ for $p \geq 0, q \geq 1$ is true because Cartan's Theorem B is true in the real-analytic setting. Cf. [13]. [5]

Consequently the formal deRham theorem implies that

$$H^q(\mathcal{M}, \mathcal{O}) \simeq \frac{\{\alpha \in \mathcal{E}_\omega^{0,q}(\mathcal{M}) : \overline{\partial}\alpha = 0\}}{\{\overline{\partial}\beta : \beta \in \mathcal{E}_\omega^{0,q-1}(\mathcal{M})\}} .$$

We also have the fine resolution of the sheaf $\mathcal{O}$ by *smooth* forms

$$0 \rightarrow \mathcal{O} \rightarrow \mathcal{E}^{0,0} \xrightarrow{\overline{\partial}} \mathcal{E}^{0,1} \xrightarrow{\overline{\partial}} \mathcal{E}^{0,2} \xrightarrow{\overline{\partial}} \cdots .$$

---

[5] The case we are using is, in fact, very simple. The sheaves $\mathcal{E}_\omega^{p,q}$ are locally free over the sheaf $\mathcal{A} = \mathcal{E}_\omega^{(0,0)}$, so what is to be seen is that $H^q(\mathcal{M}, \mathcal{A}) = 0$. Let us ignore the complex structure on $\mathcal{M}$ and consider $\mathcal{M}$ as a real-analytic submanifold of a complexification $\mathcal{M}^*$ of $\mathcal{M}$. The sheaf $\mathcal{A}$ may be identified with the sheaf $\mathcal{O}_{\mathcal{M}^*}|\mathcal{M}$-every real analytic function f on an open subset W of $\mathcal{M}$ extends in a unique way to a holomorphic function on neighborhood of W in $\mathcal{M}^*$. Thus, we have to see that $H^q(\mathcal{M}, \mathcal{O}_{\mathcal{M}^*}) = 0$. This is clear, however, for $\mathcal{M}$ has a fundamental neighborhood basis of Stein domains Ω_α [33]. For each α we have $H^q(\Omega_\alpha, \mathcal{O}_{\mathcal{M}^*}) = 0$, and $H^q(\mathcal{M}, \mathcal{O}_{\mathcal{M}^*})$ is the limit of the groups $H^q(\Omega_\alpha, \mathcal{O}_{\mathcal{M}^*})$.

which yields the isomorphism

$$H^q(\mathcal{M}, \mathcal{O}) \simeq \frac{\{\alpha \in \mathcal{E}^{0,q}(\mathcal{M}) : \overline{\partial}\alpha = 0\}}{\{\overline{\partial}\beta : \beta \in \mathcal{E}^{0,q-1}(\mathcal{M})\}}.$$

By hypothesis we have a form α of bidegree $(0, q)$ with real-analytic coefficients that is $\overline{\partial}$-exact so that there is $\beta \in \mathcal{E}^{0,q-1}(\mathcal{M})$ with $\overline{\partial}\beta = 0$. The two isomorphisms above then imply that there is $\widetilde{\beta} \in \mathcal{E}_\omega^{(0,q-1)}(\mathcal{M})$ with $\overline{\partial}\beta = \alpha$. This completes the proof. $\square$

2 Characterizations of removable sets in strictly pseudoconvex boundaries

In the case of strictly pseudoconvex domains, it is possible to give precise characterizations of removable and weakly removable sets. We emphasize that the sufficient cohomological conditions given below for the strictly pseudoconvex case are sufficient in considerably more general settings. This will be developed in Section 3 below.

2.1 Extrinsic Characterizations

There is a very simple characterization of removable sets in terms of the complementary domain.

Theorem 2.1.1 *Let D be a relatively compact strictly pseudoconvex domain with C^2 boundary in the n-dimensional Stein manifold $\mathcal{M}$. For a compact set $K \subset bD$, the following conditions are equivalent:*

1) K is removable.

2) K is $\mathcal{O}(\overline{D})$-convex if $n = 2$ or $H^{0,1}(\mathcal{M} \backslash K) = 0$ if $n \geq 3$.

3) $H_\Phi^{0,1}(\mathcal{M} \backslash K) = 0$, if $n \geq 3$.

A few explanations are required in connection with this statement.

If X is a complex manifold or variety, then $\mathcal{O}(X)$ denotes the algebra of functions holomorphic on X. A compact set $E \subset X$ is $\mathcal{O}(X)$-convex if

$$E = \hat{E}_{\mathcal{O}(X)} = \{z \in X : |f(z)| \leq \sup_{x \in E} |f(x)| \text{ for every } f \in \mathcal{O}(X)\}.$$

Similary, a subset E of $\overline{D}$ is $\mathcal{O}(\overline{D})$-convex if

$$E = \hat{E}_{\mathcal{O}(\overline{D})} = \{z \in \overline{D} : |f(z)| \leq \sup_{x \in E} |f(x)| \text{ for every } f \in \mathcal{O}(\overline{D})\}.$$

Granted that D is strictly pseudoconvex, the condition that $E \subset \overline{D}$ be $\mathcal{O}(\overline{D})$-convex is equivalent to the condition that E be $\mathcal{O}(U)$-convex for some neighborhood U of $\overline{D}$. The condition that $H^{0,1}(\mathcal{M}\backslash K)$ vanish is simply the condition that each smooth $\overline{\partial}$-closed $(0,1)$-form on $\mathcal{M}\backslash K$ be $\overline{\partial}$-exact. By the Dolbeault isomorphism, the space $H^{0,1}(\mathcal{M}\backslash K)$ is isomorphic to the Čech cohomology space $H^1(\mathcal{M}\backslash K)$. The cohomology group $H^{0,1}_\Phi(\mathcal{M}\backslash K)$ is the quotient of the space of smooth $\overline{\partial}$-exact $(0,1)$-forms on $\mathcal{M}\backslash K$ with support a relatively compact subset of $\mathcal{M}$ modulo the subspace $\{\overline{\partial}g : g \in \mathcal{C}^\infty(\mathcal{M}\backslash K),$ supp g relatively compact in $\mathcal{M}\}$. Recall Section 1. While there is an evident linear map $H^{0,1}_\Phi(\mathcal{M}\backslash K) \to H^{0,1}(\mathcal{M}\backslash K)$, it is not obviously injective or surjective in general.

The removability result stated above is the best possible for strictly pseodoconvex domains: The conditions 2) and 3) are necessary in order that every function holomorphic on a neighborhood of $\Gamma = bD\backslash E$ should extend holomorphically through D. Conversely, they are sufficient to guarantee the continuation through D of CR-functions of arbitrary classes on Γ, including CR-distributions. In essence, Theorem 2.1.1 characterizes a situation in which D is the envelope of holomorphy of Γ. (See [54].)

The equivalence of 1) and 2), in case $n = 2$ is in [81] and [63]. The result for $n \geq 3$ is due to Lupacciolu [53]. Lupacciolu [53] also proved the following result on weakly removable sets.

Theorem 2.1.2 *Let D be a relatively compact strictly pseudoconvex domain with C^2 boundary in the Stein manifold $\mathcal{M}$ of dimension $n \geq 2$. For a compact subset $K \subset bD$, the following conditions are equivalent:*

1)$_w$ K is weakly removable.

2)$_w$ $H^{0,1}(\mathcal{M}\backslash K)$ is separated.

3)$_w$ $H^{0,1}_\Phi(\mathcal{M}\backslash K)$ is separated.

In 2)$_w$ and 3)$_w$, we are dealing with topological vector spaces, and the condition that they be separated is synonymous with the condition that they be Hausdorff spaces.

The condition 2)$_w$ means that if $\{\alpha_j\}_{j=1}^\infty$ is a sequence of $\overline{\partial}$-exact smooth $(0,1)$-forms that converges in the space $\mathcal{E}^{0,1}(\mathcal{M}\backslash K)$ to the form β, then β is exact. Alternatively put, $\overline{\partial} : \mathcal{E}^{0,0} \to \mathcal{E}^{0,1}$ has closed range.

The condition 3)$_w$ means that if $\{g_j\}_{j=1}^\infty$ is a sequence in $\mathcal{E}^{0,0}(\mathcal{M}\backslash K)$ *with the support of all the g_j's contained in some fixed compact subset of $\mathcal{M}$* and if $\{\overline{\partial}g_j\}_{j=1}^\infty$ converges in $\mathcal{E}^{0,1}(\mathcal{M}\backslash K)$ to β, then $\beta = \overline{\partial}g$ for some $g \in \mathcal{E}^{0,0}(\mathcal{M}\backslash K)$ with support a relatively compact subset of $\mathcal{M}$.

The removability result of Theorem 2.1.2 is the best possible in the strictly pseudoconvex case, in the same sense that Theorem 2.1.1 was explained above to be the best possible result of its kind.

The cohomological characterizations of removable and weakly removable compact sets were established by Lupaccilou [53], who worked with continuous CR-functions and sheaf cohomology, drawing in essential ways on duality theory as treated in [6]. The proofs we give for Theorems 2.1.1 and 2.1.2 seem somewhat simpler than those of Lupacciolu. An element of duality theory does remain, as we do draw on a version of the Serre duality theorem.

The characterizations given in Theorems 2.1.1 and 2.1.2 may seem, at first glance, excessively abstract and complicated, but the language in which they are expressed is very natural in treating continuation problems. One of the principal contents of these two theorems is that the condition that K be removable for $n \geq 3$ or weakly removable, in a strictly pseudoconvex boundary, does not depend on the domain in question but rather on the set K itself and its situation as a subset of the ambient manifold $\mathcal{M}$. In Section 2.3, we present certain intrinsic characterizations of removable sets. This cohomological approach to the question of removability yields in a unified way a series of applications, which we give in Section 4 below.

2.2 Proofs

We begin with the proof of Theorem 2.1.1.

First, 3) implies 1). As D is strictly pseudoconvex, there is a neighborhood U of $\Gamma = bD \backslash K$ such that every CR-function or distribution on Γ extends holomorphically to a function $\widetilde{f} \in \mathcal{O}(D \cap U)$. Recall the discussion in the introduction. Let λ be a $\mathcal{C}^\infty$ function on $\mathcal{M} \backslash K$ with $\lambda = 1$ on a neighborhood of $\mathcal{M} \backslash D$, $\lambda = 0$ in a neighborhood of $D \backslash U$. The $(0,1)$-form ψ, which is $\widetilde{f}\,\overline{\partial}\lambda$ extended by zero outside $D \cup U$ is $\overline{\partial}$-closed in $\mathcal{M} \backslash K$, and its support is relatively compact in $\mathcal{M}$. As $H_\Phi^{0,1}(\mathcal{M} \backslash K) = 0$, there is a smooth function g on $\mathcal{M} \backslash K$ with support relatively compact in $\mathcal{M}$ such that $\overline{\partial}g = \psi$. This function is holomorphic on $\mathcal{M} \backslash \overline{D}$. As it has relatively compact support in $\mathcal{M}$ and as $\mathcal{M} \backslash D$ is connected, g vanishes in a neighborhood of $\mathcal{M} \backslash D$ in $\mathcal{M} \backslash K$ by the uniqueness theorem for holomorphic functions. In D we have $\overline{\partial}(\lambda \widetilde{f} - g) = 0$, so the function $F = \lambda \widetilde{f} - g$ is holomorphic in D. As $\lambda = 1$ and $g = 0$ near Γ, the function F is the desired analytic continuation of f into D.

Next, 1) implies 2) in the case $n = 2$. We repeat the argument from [81]. The strict pseudoconvexity of D implies that $\hat{K}_{\mathcal{O}(\overline{D})} \cap bD = K$. There is a strictly pseudoconvex domain D' in $\mathcal{M}$ with $D \subset D'$, with $bD' \cup \overline{D} = K$, and with $\hat{K}_{\mathcal{O}(\overline{D'})} = \hat{K}_{\mathcal{O}(\overline{D})}$. The open set $D' \backslash \hat{K}_{\mathcal{O}(\overline{D'})}$ is pseudoconvex by a result of Słodkowski [77] and so is the domain of holomorphy of some function, say f. On the other hand, if K is removable, then f, being holomorphic on a neighborhood of Γ in $\mathcal{M} \backslash K$, extends holomorphically into D. This implies that $D \subset D' \backslash \hat{K}_{\mathcal{O}(\overline{D})}$, whence $\hat{K}_{\mathcal{O}(\overline{D})} = K$ as desired.

For the case $n \geq 3$, the argument depends on the standard result that if X is a Stein compactum in the Stein manifold $\mathcal{M}$ so that $X = \bigcap_{j=1}^{\infty} V_j$ where each V_j is a Stein open set in $\mathcal{M}$, then $H^{0,1}(\mathcal{M} \backslash X) = 0$ provided $\mathcal{M}$ has dimension at least three [71]. This is contained in 1.5), for the closure of a strictly pseudoconvex domain in a Stein manifold is a Stein compactum.

Since D is strictly pseudoconvex, there are strictly pseudoconvex domains D' and D'' with $D'' \subset D \subset D'$ such that $\overline{D''} \cap bD = K = \overline{D} \cap bD'$.

We are to show given a smooth, $\overline{\partial}$-closed $(0,1)$-form on $\mathcal{M} \backslash K$, there is smooth function β on $\mathcal{M} \backslash K$ with $\overline{\partial}\beta = \alpha$. As D' is a Stein domain, and as $\overline{D''}$ is a Stein compactum, there are functions $g_1 \in \mathcal{C}^\infty(D')$ and $g_2 \in \mathcal{C}^\infty(\mathcal{M} \backslash \overline{D''})$ with $\overline{\partial}g_1 = \alpha$ on D_1 and $\overline{\partial}g_2 = \alpha$ on $\mathcal{M} \backslash \overline{D''}$. The function $f = g_1 - g_2$ on $D' \backslash \overline{D''}$ is holomorphic and is defined on a neighborhood of Γ. As K is removable, there is a function $F \in \mathcal{O}(D)$ that extends $g_1 - g_2$ into D. Thus, if β agrees with $g_1 - F$ on D' and with g_2 on $\mathcal{M}\backslash\overline{D''}$, then β is a well-defined smooth function on $\mathcal{M}\backslash K$ that satisfies $\overline{\partial}\beta = \alpha$ there.

The implication 2) implies 3) in dimension $n \geq 3$ follows from 1.7).

The proof of Theorem 2.1.1 is complete. $\square$

We now take up the proof of Theorem 2.1.2. First, $3)_w$ implies $1)_w$. Let f be a CR-function on Γ that is orthogonal to all $\overline{\partial}$-closed $(n, n-1)$-forms on $\mathcal{M}$ with support disjoint from K. Denote by $\widetilde{f}$ a holomorphic continuation of f into a neighborhood U of Γ in $\overline{D}$. As in the proof that 3) implies 1) above, let λ be a C^∞ function on $\mathcal{M}\backslash K$ that is identically one on a neighborhood of $\mathcal{M}\backslash(D \cup K)$ and identically zero in $D\backslash U$, supp λ a relatively compact subset of $\overline{D}$. Denote by L the closure of the space of $\overline{\partial}$-exact $(0, 1)$-forms in $\mathcal{E}_\Phi^{0,1}(\mathcal{M}\backslash K)$. We show that $\widetilde{f}\overline{\partial}\lambda \in L$. If not, there is a continuous linear functional T on $\mathcal{E}_\Phi^{0,1}(\mathcal{M}\backslash K)$ such that $\langle T, \widetilde{f}\overline{\partial}\lambda \rangle \neq 0$ but $T|L = 0$. Each continuous linear functional on $\mathcal{E}_\Phi^{0,1}(\mathcal{M}\backslash K)$ is a current of bidegree $(n, n-1)$ on $\mathcal{M}\backslash K$ with support closed in $\mathcal{M}$. Smooth $(n, n-1)$-forms are dense in this space of currents in the weak* topology, so there is a smooth $(n, n-1)$-form φ in $\mathcal{M}\backslash K$, e.g., a regularization of T, see [16], with supp φ closed in $\mathcal{M}$ and with $\int \widetilde{f}\overline{\partial}\lambda \wedge \varphi \neq 0$ but $\int \ell \wedge \varphi = 0$ for all $\ell \in L$. In particular, $\int \overline{\partial}g \wedge \varphi = 0$ for every $g \in C^\infty(\mathcal{M}\backslash K)$ with supp g relatively compact in $\mathcal{M}$. This means that φ is $\overline{\partial}$-closed on $\mathcal{M}\backslash K$. Then, by Stokes's formula,

$$\int_D \widetilde{f}\,\overline{\partial}\lambda \wedge \varphi = \int_\Gamma f\varphi = 0,$$

by the hypothesis on f. This contradiction shows that $\widetilde{f}\overline{\partial}\lambda$ belongs to L. By hypothesis, $H_\Phi^{0,1}(\mathcal{M}\backslash K)$ is separated (condition $3)_w$), so $\widetilde{f}\overline{\partial}\lambda$ is itself of the form $\widetilde{f}\overline{\partial}\lambda = \overline{\partial}g$ for some $g \in C^\infty(\mathcal{M}\backslash K)$ with support relatively compact in $\mathcal{M}$. As g is holomorphic in a neighborhood of $\mathcal{M}\backslash(D \cup K)$ and $\mathcal{M}$ is Stein, there is $\widetilde{g} \in \mathcal{O}(\mathcal{M})$ with $\widetilde{g} = g$ in a neighborhood of $\mathcal{M}\backslash(D \cup K)$. By setting $F = \lambda\widetilde{f} - g + \widetilde{g}$, we obtain a function holomorphic in D that agrees with $\widetilde{f}$ near Γ and that therefore has boundary values f on Γ.

Next, we show that $1)_w$ implies $2)_w$. Let $D'' \subset D \subset D'$ be as above in the proof that 1) implies 2). Let α be a closed $(0, 1)$-form on $\mathcal{M}\backslash K$ with $\alpha = \lim \overline{\partial}\beta_j$ for some $\beta_j \in C^\infty(\mathcal{M}\backslash K)$. As D' is Stein, there is a function $g_1 \in C^\infty(D')$ such that $\alpha = \overline{\partial}g_1$ there. Since $\mathcal{M}$ is a Stein manifold and $\overline{D''}$ is a Stein compactum, we have that $H^{0,1}(\mathcal{M}\backslash\overline{D''})$ is zero if $n \geq 3$ and is separated if $n = 2$. In either case, we can conclude that $\alpha = \overline{\partial}g_2$ in $\mathcal{M}\backslash\overline{D''}$ for some $g_2 \in C^\infty(\mathcal{M}\backslash\overline{D''})$. The function $f = g_1 - g_2$ is holomorphic on the neighborhood $D'\backslash\overline{D''}$ of Γ, and if φ is a $\overline{\partial}$-closed $(n, n-1)$-form with compact support in $\mathcal{M}\backslash K$, then

$$\int_\Gamma f\varphi = \int_\Gamma g_1\varphi - \int_\Gamma g_2\varphi$$

$$= \int_D \alpha \wedge \varphi + \int_{\mathcal{M}\backslash D} \alpha \wedge \varphi = \int_\mathcal{M} \alpha \wedge \varphi$$

$$= \lim \int_\mathcal{M} \overline{\partial}\beta_j \wedge \varphi = 0.$$

If now φ is an arbitrary $\overline{\partial}$-closed, smooth $(n, n-1)$-form on a neighborhood of $\overline{D}$ with support disjoint from K, then, as D is strictly pseudoconvex, there is a smooth $(n, n-2)$-form ψ on a relatively compact neighborhood V of $\overline{D}$ such that $\varphi = \overline{\partial}\psi$ in V. Let $\lambda_0 \in \mathcal{C}^\infty(\mathcal{M})$ vanish on $\overline{D}$ and be identically one on a neighborhood of $\mathcal{M}\backslash V$. Put $\varphi_0 = \varphi - \overline{\partial}(\lambda_0\psi) = (1 - \lambda_0)\varphi - \overline{\partial}\lambda_1 \wedge \psi$. Then φ_0 extended by zero to all of $\mathcal{M}\backslash V$ is $\overline{\partial}$-closed and compactly supported in $\mathcal{M}\backslash K$. It follows that

$$\int_\Gamma f\varphi = \int_\Gamma f\varphi_0 + \int_\Gamma f\overline{\partial}(\lambda_0\psi) = 0,$$

for $\lambda_0 = 0$ in a neighborhood of $\overline{D}$.

Consequently, f satisfies the condition for weak removability, and hence, there is a function F holomorphic on D' with $F = f$ on $D'\backslash\overline{D''}$. If we set

$$\beta = \begin{cases} g_1 - F & \text{on } D' \\ g_2 & \text{on } \mathcal{M}\backslash\overline{D''} \end{cases},$$

then we obtain a smooth function in $\mathcal{M}\backslash K$ such that $\overline{\partial}\beta = \alpha$. We have proved that $H^{0,1}(\mathcal{M}\backslash K)$ is separated.

To prove that $2)_w$ implies $3)_w$, let α be a smooth $\overline{\partial}$-closed $(0,1)$-form on $\mathcal{M}\backslash K$ that can be represented as $\lim \overline{\partial}\beta_j$ where $\beta_j \in \mathcal{C}^\infty(\mathcal{M}\backslash K)$, and there is a fixed compact subset, E, of $\mathcal{M}$ that contains the support of all the β_j. Since $H^{0,1}(\mathcal{M}\backslash K)$ is assumed to be separated, (condition $2)_w$), we have $\alpha = \overline{\partial}\beta$ for a $\beta \in \mathcal{C}^\infty(\mathcal{M}\backslash K)$. The function β is holomorphic in $\mathcal{M}\backslash(\overline{E} \cup K)$, and $\mathcal{M}$ is a Stein manifold, so there is a function $\widetilde{\beta} \in \mathcal{O}(\mathcal{M})$ such that $\widetilde{\beta} = \beta$ off a compact subset of $\mathcal{M}$. If $\beta_0 = \beta - \widetilde{\beta}$, then $\alpha = \overline{\partial}\beta_0$, and supp β_0 is relatively compact in $\mathcal{M}$. Thus, $H_\Phi^{0,1}(\mathcal{M}\backslash K)$ is separated, as we wished to show. $\qquad\square$

2.3 Intrinsic Characterizations

It is convenient to have characterizations of removable sets and weakly removable sets in terms of K itself rather than in terms of its complement. Such characterizations were provided by Lupacciolu [53]. We give here alterative proofs.

The criteria are formulated in terms of cohomological properties of the compact set K. The Dolbeault cohomology $H^{p,q}(K) = \mathcal{Z}^{p,q}(K)/\overline{\partial}\mathcal{E}^{p,q}(K)$ (see 1) is isomorphic to the Čech cohomology $H^q(K, \Omega^p)$.

Theorem 2.3.1 *Let D be a relatively compact strictly pseudoconvex domain with C^2 boundary in the Stein manifold $\mathcal{M}$ of dimension $n \geq 3$. For a compact set $K \subset bD$, the following are equivalent.*

1) K is removable.

4) $H^{n,n-1}(K) = {}^\sigma H^{n,n-2}(K) = 0$.

Recall that ${}^\sigma E$ denotes the separated space associated with the topological vector space E.

We should note that it is impossible to give an intrinsic characterization of removability in case $n = 2$, because the condition of $\mathcal{O}(\overline{D})$-convexity in Theorem 2.1.1 depends essentially on D. For instance, the torus $\mathbb{T}^2 = \{|z_1| = |z_2| = 1\}$ in $\mathbb{C}^2$ is not removable for the ball $\mathbb{B} = \{|z_1|^2 + |z_2|^2 < 2\}$, but it is $\mathcal{O}(\overline{D}_\theta)$-convex for the strictly pseudoconvex domain

$$D_\theta = \{(|z_1| - \theta)^2 + (|z_2| - \theta)^2 < 2(1 - \theta)^2\}$$

with $\frac{1}{2} < \theta < 1$, and thus $\mathbb{T}^2$ is removable with respect to this domain. It is not difficult to construct a strictly pseudoconvex Reinhardt domain D with $\mathcal{C}^\infty$ boundary in $\mathbb{B}$ such that $bD = bD_\theta$ in a neighborhood of $\mathbb{T}^2$. Therefore, an intrinsic characterization of removability is in principle not possible, even in terms of bD as an ambient manifold given near the set in question. In this connection, notice that for $n = 2$, the second condition in 4) is never satisfied: $H^{2,0}(K) = \Omega^2_K$ is separated and is never zero.

For weak removability, we have the following result, in which $n = 2$ is not an exceptional case.

Theorem 2.3.2 *Let D be a relatively compact strictly pseudoconvex domain with $\mathcal{C}^2$ boundary in a Stein manifold $\mathcal{M}$ of dimension $n \geq 2$. For a compact set $K \subset bD$, the following conditions are equivalent:*

1)$_w$ K *is weakly removable.*

4)$_w$ $H^{n,n-1}(K) = 0$.

Again, in the sense explained after Theorem 2.1.1, these results are the best possible.

Theorems 2.3.1 and 2.3.2 were proved by Lupacciolu [53]. We give here direct proofs based on integral formulas of Bochner-Martinelli type. Other proofs of the implications 4) implies 1) and 4)$_w$ implies 1)$_w$ are given in Section 3 in a more general situation.

2.4   Proofs

We first prove Theorem 2.3.2. To prove that $H^{n,n-1}(K) = 0$ suffices for weak removability, we consider $\mathcal{M}$ as a complex submanifold of a suitable C^N. Denote by Δ the Laplace operator on $\mathcal{M}$ with respect to the real-analytic metric induced on $\mathcal{M}$ by the Euclidean metric on $\mathbb{C}^N$. For an arbitrary point $a \in \mathcal{M}$, we have the Green function $G_a(\zeta)$ with pole at a, which is harmonic on $\mathcal{M} \setminus \{a\}$ and satisfies $-\Delta G_a = \delta_a$, δ_a the unit point mass at a. Set

$$B_a = c_n \partial G_a \wedge \omega^{n-1}$$

where $\omega = \frac{i}{2} \sum_{j=1}^{N} d\zeta_j \wedge d\overline{\zeta}_j | \mathcal{M}$. Then the $(n, n-1)$-form on $\mathcal{M}$ satisfies the current equation

$\overline{\partial} B_a = \delta_a$ provided the constant c_n is chosen properly. (In case $\mathcal{M} = \mathbb{C}^N$, this process yields the Bochner-Martinelli kernel $c_n \partial |\zeta - a|^{2-2n} \wedge (\partial \overline{\partial} |\zeta|^2)^{n-1}$.)

Let K be a compact set in bD with $H^{n,n-1}(K) = 0$. If $a \in D$, then B_a is $\overline{\partial}$-closed in a neighborhood of K, so there is a smooth $(n, n-2)$-form θ_a in a neighborhood U_a of K such that $B_a = \overline{\partial}\theta_a$ in U_a. Let $\lambda_a \in \mathcal{C}^\infty(\mathcal{M})$, $\lambda_a = 1$ on a neighborhood of K, $\lambda_a = 0$ in a neighborhood of $\mathcal{M} \setminus U_a$. Then the form

$$B_a^0 = B_a - \overline{\partial}(\lambda_a \theta_a) = (1 - \lambda_a)B_a - \overline{\partial}\lambda_a \wedge \theta_a$$

is a $\overline{\partial}$-closed, smooth form on $\mathcal{M}\backslash\{a\}$ with $\overline{\partial}B_a^0 = \delta_a$ and supp $B_a^0 \cap K = \emptyset$.

Let now f be a CR-function on $\Gamma = bD\backslash K$ such that $\int_\Gamma f\varphi = 0$ for every $\overline{\partial}$-closed smooth $(n, n-1)$-form φ on a neighborhood of $\overline{D}$ with $K \cap$ supp $\varphi = \emptyset$. (In the case that f is a CR-distribution, the integral is understood to be $\langle -f[\Gamma]^{0,1}, \varphi\rangle$.) For some neighborhood V of Γ there is a holomorphic function $\tilde{f}$ on $V \cap D$ that takes the boundary values of f along Γ.

As the support of B_a^0 is disjoint from K for $a \notin K$, we may define F on $\mathcal{M}\backslash K$ by

$$F(z) = \int_\Gamma f B_z^0.$$

This function vanishes outside $\overline{D}$ by the orthogonality relations satisfied by f. The function F is well-defined, i.e., the value $F(z)$ does not depend on the choice of θ_a or λ_a: If θ_a' and λ_a' are other choices, then

$$\varphi = (B_a - \overline{\partial}\lambda_a \wedge \theta_a) - (B_a - \overline{\partial}\lambda_a \wedge \theta_a') = \overline{\partial}\lambda_a' \wedge \theta_a' - \overline{\partial}\lambda_a \wedge \theta_a$$

is a smooth, $\overline{\partial}$-closed form on $\mathcal{M}$ the support of which is disjoint from K. Thus, $\int_\Gamma f\varphi = 0$, and F is well-defined.

We shall show that $F = \tilde{f}$ on $V \cap D$ provided V is small enough. If $a \in V \cap D$, then there is a function $h \in \mathcal{O}(\overline{D})$ with $h(a) = 0$ and such that for small $\varepsilon_0 > 0$, the set $\{\zeta \in G = D \cup V : |h(\zeta)| \leq \varepsilon_0\}$ is disjoint from K. We can choose V so that G is strictly pseudoconvex. Then the domain $G_h = G\backslash\{h = 0\}$ is Stein, so there is a smooth $(n, n-2)$-form θ_a' such that $B_a = \overline{\partial}\theta_a'$ in G_h. Let $\lambda_a' \in \mathcal{C}^\infty(M)$ be one on a neighborhood of $\overline{D}\backslash\{|h| < \varepsilon_0\}$, zero on a neighborhood of $\mathcal{M}\backslash G_h$. Define $B_a' = B_a - \overline{\partial}\lambda_a' \wedge \theta_a'$, and choose $\varepsilon \in (0, \varepsilon_0)$ so that $G \cap \{|h| = \varepsilon\}$ is a smooth real hypersurface transverse to Γ and so that $\lambda_a' = 1$ on a neighborhood of $\overline{D} \cap \{|h| = \varepsilon\}$. As $B_a'|\overline{D}$ is supported on $\overline{D} \cap \{|h| < \varepsilon\}$, we have

$$F(a) = \int_\Gamma f B_a' = \int_{\Gamma \cap \{|h| < \varepsilon\}} f B_a' = \int_{bD_\varepsilon} \tilde{f} B_a'$$

where $D_\varepsilon = D \cap \{|h| < \varepsilon\}$. As $\tilde{f} \in \mathcal{O}(D_\varepsilon)$ and $\overline{\partial}B_a' = \delta_a$, the last integral is $\tilde{f}(a)$. Thus, $F = \tilde{f}$ on $V \cap D$.

To conclude the proof, it is enough to show that F is real-analytic in D, for then F is holomorphic. To do this, we invoke 1.19 to conclude that the form $\theta_a(\zeta)$ above can be chosen to be real-analytic.

We have taken the manifold $\mathcal{M}$ to be a submanifold of $\mathbb{C}^N$, so there is a neighborhood $\widetilde{\mathcal{M}}$ of $\mathcal{M}$ in $\mathbb{C}^N$ that admits a holomorphic retraction, τ, onto $\mathcal{M}$. See [31]. For fixed $a \in D$ and $w \in \mathbb{C}^N$ with $|w| < \varepsilon_1$ for a suitably small ε_1, the mapping τ_w given by $\tau_w(\zeta) = \tau(\zeta - w)$ is biholomorphic from a neighborhood G of $\overline{D}$ in $\mathcal{M}$ into $\mathcal{M}$, and $G \cap \tau_w(G) \supset \overline{D}$. The $(n, n-1)$-form $B_w^a = \tau_w^* B_a - \overline{\partial}\lambda_a \wedge \tau_w^* \theta_a$ satisfies the equation $\overline{\partial}B_w^a = \delta_{\tau(a+w)}$ and is real-analytic in its dependence on w because B_a and θ_a are real-analytic. Its support does not meet K provided ε_1 is sufficiently small. Also, $B_w^a - B_{a+w}^a$ is a smooth, $\overline{\partial}$-closed form on a neighborhood of $\overline{D}$ with support disjoint from K. Consequently, $F(\tau(a+w)) = \int_\Gamma f B_w^a$, which shows that F is real-analytic in a neighborhood of a. Thus, F is the required holomorphic extension of f into D.

We now prove the converse: If K is weakly removable, then $H^{n,n-1}(K) = 0$.

Let $A(D \cup \Gamma)$ be the space of functions continuous on $\Gamma \cup D$, holomorphic on D. Granted that K is weakly removable, the subspace $A(D \cup \Gamma)|\Gamma$ of $\mathcal{C}(\Gamma)$ is closed: It is the intersection of the closed subspaces $\mathcal{F}_\varphi = \{f \in \mathcal{C}(\Gamma) : \int_\Gamma f\varphi = 0\}$ where φ runs through the family of smooth, $\overline{\partial}$-closed $(n, n-1)$-forms φ on $\overline{D}$ with support disjoint from K.

A compact subset X of $\overline{D} \backslash K$ can be covered by finitely many domains of type $V_a = \{\tau(a + w) : |w| < \varepsilon\}$, $a \in X$. For $z = \tau(a + w) \in V_a$ and $f \in A(D \cup \Gamma)$, we have $F(z) = \int_{\Gamma \backslash U_a} f B_w^a$ as above for a neighborhood U_a of K. As $A(D \cup \Gamma)$ is an algebra, it follows that for $k \in \mathbb{N}$,

$$|f(z)|^k \le (\max_{\Gamma \backslash U_a} |f|^k)\|B_w^a\|_{L^1(\Gamma)}.$$

As this holds for all $k \in \mathbb{N}$, it follows that $|f(z)| \le \max_{\Gamma \backslash U_a} |f|$. Since the covering in question is finite, it follows that there is a compact set $\Gamma_X \subset \Gamma$ such that

$$|f(z)| \le \max_{\Gamma_X} |f|$$

for all $z \in X$, all $f \in A(D \cup \Gamma)$[6].

Let now φ be a smooth, $\overline{\partial}$-closed $(n, n-1)$-form on a neighborhood U of K, and let $\lambda \in \mathcal{C}^\infty(\mathcal{M})$, $\lambda = 0$ on a neighborhood of $\mathcal{M} \backslash U$, $\lambda = 1$ on a neighborhood V of K. The map $f \mapsto \int_D f\overline{\partial}\lambda \wedge \varphi$ is a continuous linear functional on $A(D \cup \Gamma)$. As $X = \overline{D} \cup \operatorname{supp} \overline{\partial}\lambda$ is a compact set disjoint from K, there is a measure μ on Γ_X with $\int_D f\overline{\partial}\lambda \wedge \varphi = \int_\Gamma f d\mu$ for all $f \in A(D \cup \Gamma)$. If χ_D denotes the characteristic function of D, then the measure $\sigma = \chi_D \overline{\partial}\lambda \wedge \varphi - \mu$ is orthogonal to $A(D \cup \Gamma)$. As D is strictly pseudoconvex, there is a strictly pseudoconvex domain G' with $\operatorname{supp} \sigma \subset G'$, $K \cap \overline{G}' = \emptyset$ and $\mathcal{O}(D \cup \Gamma)$ dense in $\mathcal{O}(G')|(G' \cap (D \cup \Gamma))$. (To get G', push bD slightly into D near K and slightly out of D near $\operatorname{supp} \sigma$.) The measure σ is orthogonal to $\mathcal{O}(G')$, and so, as G' is Stein, there is an $(n, n-1)$-current φ_0 compactly supported in G' such that $\sigma = \overline{\partial}\varphi_0$ on G'. By a regularity theorem for the $\overline{\partial}$-problem (see, e.g., [16]), we can take φ_0 to be smooth on $D \cap G'$.

On the other hand, there is a strictly pseudoconvex domain G'' with $K \subset G'' \subset D \cup V$. The restriction of $\sigma = \overline{\partial}\varphi_0$ to G'' gives $\overline{\partial}\lambda \wedge \varphi = \overline{\partial}\varphi_0$, i.e., $\overline{\partial}(\lambda\varphi - \varphi_0) = 0$ on G''. (We have implicitly extended $\lambda\varphi$ by zero to have it defined on all of G''.) The domain G'' is Stein, so there is a smooth $(n, n-2)$-form ψ on G'' such that $\lambda\varphi - \varphi_0 = \overline{\partial}\psi$ there. We have that $\lambda = 1$ and $\varphi_0 = 0$ on a neighborhood of K, so it follows that near K, $\overline{\partial}\psi = \varphi$. Thus, $H^{n,n-1}(K) = 0$.

This completes the proof of Theorem 2.3.2. $\square$

To prove Theorem 2.3.1, we notice that 4) implies 2) by 18) and that 2) implies 1) by Theorem 2.1.1.

---

[6] It is worth observing that in this argument we can proceed more abstractly: Given that $A(D \cup \Gamma)|\Gamma$ is a closed subspace of $\mathcal{C}(\Gamma)$, it follows that the restriction map ρ given by $A(D \cup \Gamma) \ni f \xrightarrow{\rho} f|\Gamma$ is a topological isomorphism onto $A(D \cup \Gamma)|\Gamma$. Let χ be its inverse. Given $z \in D$, the Hahn-Banach theorem provides a continuous linear functional $\psi : \mathcal{C}(\Gamma) \to \mathbb{C}$ that on $A(D \cup \Gamma)|\Gamma$ agrees with the map $f|\Gamma \mapsto \chi(f|\Gamma)(z)$. Thus, by the Riesz Representation Theorem, there is a finite measure μ on Γ with compact support such that $f(z) = \int_\Gamma f d\mu$ for all $f \in A(D \cup \Gamma)$. We have $f^k(z) = \int_\Gamma f^k d\mu$ for $k \in \mathbb{N}$. This yields $|f(z)| \le \sup\{|f(\eta)| : \eta \in \operatorname{supp} \mu\}\|\mu\|^{1/k}$ for every k. Thus, $|f(z)| \le \sup\{|f(\eta)| : \eta \in \operatorname{supp} \mu\}$ for all $f \in A(D \cup \Gamma)$.

Conversely, 1) implies 4). Since K is removable, it is weakly removable whence, by Theorem 4, $H^{n,n-1}(K) = 0$. For the other part of the assertion, we prove first that the spaces $H^{n,n-2}(K)$ and $H_c^{n,n-1}(\mathcal{M} \backslash K)$ are topologically isomorphic.

The cohomology mapping

$$\delta : H^{n,n-2}(K) \to H_c^{n,n-1}(\mathcal{M} \backslash K)$$

is defined as follows. (See 1.9.) Let ψ be a smooth $\overline{\partial}$-closed $(n, n-2)$-form on a neighborhood U of K. Let ψ_K be its germ on K, and let $[\psi_K] \in H^{n,n-2}(K)$ be its associated cohomology class. Let $\lambda \in \mathcal{C}^\infty(\mathcal{M})$ have its support a compact subset of U and be identically one on a neighborhood of K. Then $\overline{\partial}\lambda \wedge \psi$, extended by zero outside U is an $(n, n-1)$-form on $\mathcal{M}$ the support of which is a compact subset of $\mathcal{M} \backslash K$. By definition, $\delta[\psi_K] = [\overline{\partial}\lambda \wedge \psi] \in H_c^{n,n-1}(\mathcal{M} \backslash K)$. The map δ defined in this way is well defined, $i.e.$, the cohomology class $[\overline{\partial}\lambda \wedge \psi]$ does not depend on the choice of λ: If λ_1 is another λ, then $\overline{\partial}(\lambda_1 - \lambda) \wedge \psi = \overline{\partial}((\lambda_1 - \lambda)\psi)$, and $(\lambda_1 - \lambda)\psi$ is compactly supported in $\mathcal{M} \backslash K$.

The manifold $\mathcal{M}$ is Stein, so $H_c^{n,n-1}(\mathcal{M}) = 0$. In particular, the space $\overline{\partial}\mathcal{D}^{n,n-2}(\mathcal{M})$ is closed in $\mathcal{D}^{n,n-1}(\mathcal{M})$. We denote by $\mathcal{Z}_c^{n,n-2}$ the space of $\overline{\partial}$-closed forms in $\mathcal{D}^{n,n-2}(\mathcal{M})$, and we then have the continuous, one-to-one map

$$\overline{\partial} : \frac{\mathcal{D}^{n,n-2}}{\mathcal{Z}_c^{n,n-2}}(\mathcal{M}) \to \overline{\partial}\mathcal{D}^{n,n-2}(\mathcal{M}) = \mathcal{Z}_c^{n,n-1}(\mathcal{M}).$$

The open mapping theorem for LF spaces implies that this mapping is a topological isomorphism. (Recall footnote 3.)

As $\mathcal{Z}_c^{n,n-1}(\mathcal{M} \backslash K) \subset \mathcal{Z}_c^{n,n-1}(\mathcal{M})$, we have the continuous mapping

$$\mathcal{Z}_c^{n,n-1}(\mathcal{M} \backslash K) \ni \varphi \mapsto \overline{\partial}^{-1}\varphi = [\psi] \in \frac{\mathcal{D}^{n,n-2}(\mathcal{M})}{\mathcal{Z}_c^{n,n-2}(\mathcal{M})}$$

for some $\psi \in \mathcal{D}^{n,n-2}(\mathcal{M})$. Then $\overline{\partial}\psi = \varphi$ vanishes on a neighborhood of K. It follows that the map $\psi \to \psi_K$ that takes ψ to its germ ψ_K on K has values in $\mathcal{Z}^{n,n-2}(K)$. Since $H_c^{n,n-2}(\mathcal{M}) = 0$ for Stein manifolds of dimension $n \geq 3$, $\{\psi_K\} = 0$ in $H^{n,n-2}(K)$ if $\psi \in \mathcal{Z}_c^{n,n-2}(\mathcal{M})$. In this way we obtain a continuous mapping

$$H_c^{n,n-1}(\mathcal{M} \backslash K) \ni [\varphi] \overset{\overline{\partial}^{-1}}{\to} [\psi] \overset{r}{\to} [\psi_K] \in H^{n,n-2}(K) .$$

It is inverse to δ, for

$$\overline{\partial}\lambda \wedge \psi - \varphi = \overline{\partial}((\lambda - 1)\psi) + (1 - \lambda)\varphi - \varphi = \overline{\partial}((\lambda - 1)\psi)$$

if supp $\lambda \cap$ supp $\varphi = \emptyset$. That is, $\delta[\psi] = [\overline{\partial}\lambda \wedge \psi] = [\varphi]$ in $H_c^{n,n-1}(\mathcal{M} \backslash K)$. Thus, δ is a topological isomorphism.

Let now L be the closure in $\mathcal{D}^{n,n-1}(\mathcal{M} \backslash K)$ of the space $\overline{\partial}\mathcal{D}^{n,n-2}(\mathcal{M} \backslash K)$. If $^\sigma H_c^{n,n-1}(\mathcal{M} \backslash K) \neq 0$, there is a $(0, 1)$-current T on $\mathcal{M} \backslash K$ such that $T|L = 0$ but $\langle T, \varphi \rangle \neq 0$ for some $\varphi \in \mathcal{Z}_c^{n,n-1}(\mathcal{M} \backslash K)$. As $\langle T, \overline{\partial}\psi \rangle = 0$ for $\psi \in \mathcal{D}^{n,n-2}(\mathcal{M} \backslash K)$, the current T is $\overline{\partial}$-closed. The set K is removable, so $H^{0,1}(\mathcal{M} \backslash T) = 0$ by Theorem 1. It follows from the Dolbeault isomorphism that $T = \overline{\partial}S$ for a $(0,0)$-current, $i.e.$, a distribution, S on $\mathcal{M} \backslash K$. But then $\langle T, \varphi \rangle = -\langle S, \overline{\partial}\varphi \rangle = 0$. This contradiction shows that $^\sigma H_c^{n,n-1}(\mathcal{M} \backslash K) = 0$. As δ is a topological isomorphism, it follows that $^\sigma H^{n,n-2}(K) = 0$ as well.

This completes the proof. $\qquad \qquad \square$

2.5   A Geometric Characterization of Weak Removability

In the previous section, we have essentially proved the following geometric characterization of weak removability, which is, in some sense, dual to the characterization of removability in case $n = 2$.

Theorem 2.5.1 *Let D be a relatively compact, strictly pseudoconvex domain with C^2 boundary in a Stein manifold $\mathcal{M}$ of dimension $n \geq 2$. For a compact set $K \subset bD$, the following conditions are equivalent.*

$1)_w$ K is weakly removable.

$5)_w$ For an arbitrary compact set $X \subset \overline{D}\setminus K$, there is a compact set $\Gamma_X \subset \Gamma$ with
$$X \subset \widehat{(\Gamma_X)}_{\mathcal{O}(\overline{D})}.$$

$5)'_w$ $D \subset \widehat{\Gamma}_{\mathcal{O}(\overline{D})}.$

Here $\widehat{\Gamma}_{\mathcal{O}(\overline{D})} = \bigcup\{\widehat{E}_{\mathcal{O}(\overline{D})} : E \subset \Gamma \text{ is compact}\}$. Condition $5)'_w$ is formally weaker than condition $5)_w$: it means that for each $z \in D$, there is a compact set $E_z \subset \Gamma$ such that $z \in (\widehat{E_z})_{\mathcal{O}(\overline{D})}$.

Proof   To see that $5)_w$ implies $1)_w$ repeat word by word the two last paragraphs of the proof that $1)_w$ implies $4)_w$ in Section 2.4 to find that $5)_w$ implies $H^{n,n-1}(K) = 0$, whence K is weakly removable.

For the implication $1)_w$ implies $5)_w$, repeat the first two paragraphs of the proof $1)_w$ implies $4)_w$ implies in Section 2.4, and note at the end that $\mathcal{O}(\overline{D}) \subset A(D \cup \Gamma)$.

To see that $5)'_w$ implies $5)_w$, consider a compact set E in Γ. Since D is strictly pseudoconvex, each point of bD is a peak point for $\mathcal{O}(\overline{D})$. It follows that there is a strictly pseudoconvex domain $D' \supset D$ so near D that $bD' \cap \overline{D} = K$, $\mathcal{O}(\overline{D'})|\overline{D}$ is dense in $\mathcal{O}(\overline{D})$, and so that for a compact set $E' \subset \Gamma' = bD'\setminus K$, we have $E \subset \widehat{E'}_{\mathcal{O}(\overline{D})}$. Thus, for each $a \in \overline{D}\setminus K$ we have that there is a compact set $E'_a \subset \Gamma'$ such that $a \in (\widehat{E'_a})_{\mathcal{O}(\overline{D'})}$. We can assume that $\overline{D}$ and $\overline{D}'$ are $\mathcal{O}(\mathcal{M})$-convex by shrinking $\mathcal{M}$ if necessary, and then we can suppose that $\mathcal{M}$ is a complex submanifold of $\mathbb{C}^N$. In $\mathbb{C}^N$, compact subsets of $\mathcal{M}$ are $\mathcal{O}(\mathcal{M})$-convex exactly when they are polynomially convex.

Again let $\tau : \widetilde{\mathcal{M}} \to \mathcal{M}$ be a holomorphic retraction of a neighborhood $\widetilde{\mathcal{M}}$ of $\mathcal{M}$ onto $\mathcal{M}$. For a compact set $S \subset \mathbb{C}^N$, let $\widehat{S}$ denote the polynomially convex hull of S. Rossi's local maximum principle [64] implies that for a compact set E with $\widehat{E} \subset \widetilde{\mathcal{M}}$, we have $\tau(\widehat{E}) \subset \widehat{\tau(E)}$. Denote by $(E)_\varepsilon$ the closed ε-neighborhood of E, which is compact when E is. We have $(\widehat{E})_\varepsilon \subset \widehat{(E_\varepsilon)}$, so $\tau((\widehat{E})_\varepsilon) \subset \tau(\widehat{(E)_\varepsilon}) \subset \widehat{\tau((E)_\varepsilon)}$. Apply this to $E = E'_a$, denote $\tau((E)_\varepsilon)$ by E''_a, and choose $\varepsilon > 0$ so small that $E''_a \cap \overline{D}$ and $\widehat{E''_a} \cap K$ are empty. Then $\widehat{E''_a}$ contains a neighborhood V_a of a in $\overline{D}$. The local maximum principle implies that V_a is contained in the hull $\widehat{\Gamma_a}$ of the compact set $\Gamma_a = \widehat{E''_a} \cap bD \subset \Gamma$. As $\overline{D}$ is polynomially convex, we have constructed a compact set $\Gamma_a \subset \Gamma$ such that $(\widehat{\Gamma_a})_{\mathcal{O}(\overline{D})}$ contains the neighborhood V of a.

Finally, if $X \subset \overline{D} \backslash K$ is compact, finitely many sets of the form V_a just constructed cover X, and hence there is a compact set $\Gamma_X \subset \Gamma$ with $X \subset \widehat{(\Gamma_x)}_{\mathcal{O}(\overline{D})}$.

The implication $5)_w$ implies $5)'_w$ is immediate, so the proof of Theorem 2.5.1 is complete.

$\square$

3 Removable singularities in general manifolds

In this section we treat removable singularities and weakly removable singularities in general manifolds. In particular, we do not restrict our attention to pseudoconvex or strictly pseudoconvex domains. In this material, we draw in essential ways on the results given in Section 1.

3.1 Removable Sets

For domains in general manifolds, we have the following two results.

Theorem 3.1.1 *Let M be connected noncompact complex manifold of dimension $n \geq 2$ with $H^{n,n-1}(\mathcal{M}) = 0$, let D be a relatively compact domain in $\mathcal{M}$, and let E be a compact subset of $\mathcal{M}$ such that $\Gamma = bD \backslash E$ is locally a Lipschitz graph. Assume in addition that*

1) $D \backslash E = int(\overline{D} \backslash E)$,

2) $\mathcal{M} \backslash (\overline{D} \cup E)$ is connected, and

3) $H^{n,n-1}(E) = {}^{\sigma}H^{n,n-2}(E) = 0$ if $n \geq 3$, and $H^{0,1}(E) = 0$ and $\Omega^2(\mathcal{M})$ is dense in $\Omega^2(E)$ if $n = 2$.

Then if f is a CR-function on Γ, there is $F \in \mathcal{O}(D \backslash E)$ that assumes the boundary values f along Γ in the following sense:

a) If Γ is a hypersurface of class C^k, $k \geq 1$, and if $f \in C^s(\Gamma)$, $0 \leq s \leq k$, then $F \in C^s((D \backslash E) \cup \Gamma)$ and $F|\Gamma = f$.

b) If $f \in L^p_{loc}(\Gamma)$, $p \geq 1$, and if $g_j : \Gamma \to \Gamma_j$ are Lipschitz homeomorphisms of Γ onto hypersurfaces $\Gamma_j \subset D \backslash E$ such that $g_j(z) \to z$ in the locally Lipschitz topology, then

$$\lim_{j \to \infty} \int_S |F(g_j(z)) - f(z)|^p d\sigma_\Gamma = 0$$

for every compact set $S \subset \Gamma$ if $d\sigma_\Gamma$ denotes a volume form on Γ.

c) If Γ is smooth and $g_j : \Gamma \to \Gamma_j$ as above are diffeomorphisms, and f is a distribution on Γ, then

$$\lim_{j\to\infty} \int_{\Gamma_j} F\varphi = \lim_{j\to\infty} \int_{\Gamma} (F \circ g_j)\varphi = \langle f, \varphi \rangle$$

for an arbitrary smooth $(n, n-1)$ form φ with compact support in $\mathcal{M}\backslash E$.

Note that the hypothesis that $H^{n,n-1}(\mathcal{M}) = 0$ is satisfied by every Stein manifold.

The density hypothesis in 3) (in case $n = 2$) means that given a holomorphic 2-form α on a neighborhood of E, α can be approximated on a possibly smaller neighborhood by holomorphic 2-forms defined on $\mathcal{M}$.

For $n \geq 3$, the condition that $H^{n,n-1}(E) = {}^\sigma H^{n,n-2}(E) = 0$ can be replaced by the condition.

3)' $H^{n,n-1}(E) = 0$ and $\mathcal{Z}^{n,n-2}(\mathcal{M})$ is dense in $\mathcal{Z}^{n,n-2}(E)$.

This theorem, with hypothesis 3)' rather than 3) was given by Lupacciolu [53]. For manifolds $\mathcal{M}$ with $H^{n,n-2}(\mathcal{M}) = 0$, e.g., for Stein manifolds, the condition 3)' is equivalent to 3).

Theorem 3.1.2 *Let $\mathcal{M}$ be a connected, noncompact complex manifold of dimension $n \geq 2$. Let D be a relatively compact domain in $\mathcal{M}$. Let E be a compact subset of $\mathcal{M}$ such that $D\backslash E = int(\overline{D}\backslash E)$ and such that $\mathcal{M}\backslash(\overline{D} \cup E)$ is connected. Let $\Gamma = bD\backslash E$ be locally a Lipschitz graph. If $H_\Phi^{0,1}(\mathcal{M}\backslash E) = 0$, then for every CR-function f on Γ, there is $F \in \mathcal{O}(D\backslash E)$ that assumes the boundary values f along Γ in the sense of Theorem 3.1.1.*

Corollary 3.1.3 *Let $\mathcal{M}$ be a connected, noncompact complex manifold of dimension $n \geq 2$ with $H_c^{0,1}(\mathcal{M}) = 0$. Let $E \subset \mathcal{M}$ be a compact subset such that $D\backslash E = int(\overline{D}\backslash E)$ and $\mathcal{M}\backslash(\overline{D}\cup E)$ is connected. Let $\Gamma = bD\backslash E$ be locally a Lipschitz graph. If $H^{0,1}(\mathcal{M}\backslash E) = 0$, then for each CR-function f on Γ, there is $F \in \mathcal{O}(D\backslash E)$ that assumes the boundary values f along Γ in the sense of Theorem 3.1.1.*

3.2   Proofs

Theorem 3.1.1 is a formal consequence of Theorem 3.1.2 because of the following Lemma.

Lemma 3.2.1 *Let E be a compact set in a noncompact, complex n-dimensional manifold $\mathcal{M}$ with $H^{n,n-1}(\mathcal{M}) = 0$. If $H^{n,n-1}(E) = {}^\sigma H^{n,n-2}(E) = 0$ for $n \geq 3$ or if $H^{n,n-1}(E) = 0$ and $\mathcal{Z}^{n,n-2}(\mathcal{M})$ is dense in $\mathcal{Z}^{n,n-2}(E)$ for $n \geq 2$, then $H_\Phi^{0,1}(\mathcal{M}\backslash E) = 0$.*

Proof   As $H^{n,n-1}(E) = 0$, the space $H_\Phi^{0,1}(\mathcal{M}\backslash E)$ is separated by 1.14). By 1.12) it suffices to show that ${}^\sigma H_\Psi^{n,n-1}(\mathcal{M}\backslash E) = 0$. We have $H^{n,n-1}(E) = 0$, so by 1.10) we have the exact sequence

$$H^{n,n-2}(\mathcal{M}) \xrightarrow{r} H^{n,n-2}(E) \xrightarrow{\delta} H_\Psi^{n,n-1}(\mathcal{M}\backslash E) \to 0.$$

Thus, the map δ is surjective. If ${}^\sigma H^{n,n-2}(E) = 0$, then ${}^\sigma H_\Psi^{n,n-1}(\mathcal{M}\backslash E) = 0$ as we wish. If $\mathcal{Z}^{n,n-2}(\mathcal{M})$ is dense in $\mathcal{Z}^{n,n-2}(E)$, then $im\, r$ is dense in $H^{n,n-2}(E)$, so $0 = im\,(\delta \circ r)$ is dense in $H_\Psi^{n,n-1}(\mathcal{M}\backslash E)$, for δ and r are continuous. In this case, ${}^\sigma H_\Psi^{n,n-1}(\mathcal{M}\backslash E)$ is zero. $\square$

Next, Corollary 3.1.3 is also a formal consequence of Theorem 3.1.2: With the hypotheses of the corollary, $H_\Phi^{0,1}(\mathcal{M}\backslash E) = 0$. This is contained in 1.7).

The proof of Theorem 3.1.2 proceeds as follows. (See [15] and [47].)

Given a CR-function $f \in L^1_{loc}(\Gamma)$, we have an associated current $f[\Gamma]^{0,1}$ of bidegree $(0,1)$ on $\mathcal{M}\backslash E$ that acts on smooth compactly supported forms $\varphi \in \mathcal{D}^{2n-1}(\mathcal{M}\backslash E)$ by

$$\langle f[\Gamma]^{0,1}, \varphi \rangle = \int_\Gamma f\varphi^{n,n-1}$$

if $\varphi^{n,n-1}$ denotes the component of φ of bidegree $(n, n-1)$ and if Γ is given the orientation induced as part of bD. If Γ is a class C^k and f is a distribution of order not more than $k - 1$, then by definition $f[\Gamma^{0,1}]$ is the $(0,1)$-current acting on forms φ by

$$\langle f[\Gamma^{0,1}], \varphi \rangle = \langle f, \varphi^{n,n-1}|\Gamma \rangle.$$

The weak tangential Cauchy-Riemann equations have the form

$$\langle f[\Gamma]^{0,1}, \overline{\partial}\psi^{n,n-2} \rangle = 0,$$

which is equivalent to the statement that the current $f[\Gamma]^{0,1}$ is $\overline{\partial}$-closed: f is a CR-function if and only if $f[\Gamma]^{0,1}$ is $\overline{\partial}$-closed.

The space $H_\Phi^{0,1}(\mathcal{M}\backslash E)$ is algebraically isomorphic to $\mathcal{Z}_\Phi'^{0,1}/\overline{\partial}\mathcal{D}_\Phi'^{0,0}$. As this space is zero by hypothesis, there is a $(0,0)$-current, $i.e.$, a distribution h on $\mathcal{M}\backslash E$ with $\overline{\partial}h = -f[\Gamma]^{0,1}$. In $\mathcal{M}\backslash(E \cup \Gamma)$ this h is a holomorphic function since $\overline{\partial}h = 0$ there, and as $\mathcal{M}\backslash(E \cup \overline{D})$ is connected and noncompact, h is identically zero there. Set $F = h|(D\backslash E)$.

If h_1 is a solution of the equation $\overline{\partial}h_1 = -f[\Gamma]^{0,1}$ in a coordinate neighborhood $U \subset \mathcal{M}\backslash E$, then $h - h_1 \in \mathcal{O}(U)$, and the boundary behavior of h is precisely that of h_1: The boundary behavior of h on Γ can be studied locally.

In a coordinate ball in $\mathbb{C}^N$, the solution of the $\overline{\partial}$-problem for a $\overline{\partial}$-closed $(0,1)$-current is given, up to a smooth summand, by convolution with the Bochner-Martinelli kernel. Property b) for the Bochner-Martinelli transform, in an equivalent form, has been given by Kytmanov [43]. Properties a) and c) are proved in [15]. The last one can be proved as well by the method used by Hörmander [38], Chapter III.

3.3   Weakly Removable Sets

For weakly removable sets in general manifolds, there is the following result.

Theorem 3.3.1 *Let D be a relatively compact domain in a complex manifold of dimension $n, n \geq 2$. Let E be a compact subset of $\mathcal{M}$ such that $\Gamma = bD\backslash E = b\overline{D}\backslash E$ is locally a Lipschitz graph. If $H^{n,n-1}(E) = 0$, then every CR-function on Γ that is orthogonal to all smooth $(n, n - 1)$-forms on $\mathcal{M}$ with support disjoint from E that are $\overline{\partial}$-closed in a neighborhood of $\overline{D}$ can be continued holomorphically into $D\backslash E$.*

Note that in this theorem there is not the hypothesis that $\mathcal{M}\backslash(\overline{D}\cup E)$ be connected, which was used essentially in the discussion of removable sets.

Theorem 3.3.1 was proved by Lupacciolu [53]; we give a proof based on Dolbeault cohomology.

Proof We can assume that $\mathcal{M}$ is noncompact.

Denote by $G_1, G_2, \ldots$, the relatively compact components of $\mathcal{M}\backslash(\overline{D}\cup E)$ and by G_o the union of the components of $\mathcal{M}\backslash(\overline{D}\cup E)$ that are not relatively compact. For each $j = 1, 2, \ldots$, let a_j be a point of G_j. Put $\mathcal{M}_m = \mathcal{M}\backslash\{a_1, \ldots .a_m\}$ for $m = 1, 2, \ldots$. Denote by Φ_m the family of closed sets in $\mathcal{M}_m\backslash E$ that have compact closure in $\mathcal{M}_m$. The result 1.14 from the Preliminaries applied to the manifold $\mathcal{M}_m$ implies that $H^{0,1}_{\Phi_m}(\mathcal{M}_m\backslash E)$ is separated. Therefore, as in the proof of Theorem 2.1.2, there is a distribution h_m on $\mathcal{M}_m\backslash E$ with support in the family Φ_m such that $-f[\Gamma]^{0,1} = \overline{\partial}h_m$. By the uniqueness theorem for holomorphic functions, $h_m|(D\backslash E)$ vanishes on $G_o\cup\ldots\cup G_m$. It follows that the holomorphic function $h_m/(D\backslash E)$ has boundary values f on $\Gamma\cap bG_j, j = 0,\ldots, m$. Let $a\in D\backslash E$, and let D_a be the component of $D\backslash E$ that contains a. If $bD_a\subset E$, we define $F|D_a = 0$. If $bD_a\cap\Gamma$ is not empty, then there is $j\geq 0$ such that $bD_a\cap bG_j$ is not empty–recall that $bD\backslash E = b\overline{D}\backslash E$. Set then $F|D_a = h_m|D_a$ for $m\geq j$. By boundary uniqueness for holomorphic functions, we have that $h_{m+1}|D_a = h_m|D_a$, so the function F is well defined on the whole of $D\backslash E$. As $\Gamma = \cup(\Gamma\cap bG_j)$, the boundary values of F on Γ coincide with f. $\square$

3.4   Extension from noncompact hypersurfaces

Let S be an oriented closed, not necessarily compact, locally Lipschitz hypersurface in a complex manifold $\mathcal{M}$. The notion of CR-function on S is well defined, but we do not have in this case any evident domain into which CR-functions should continue holomorphically. Even if S divides $\mathcal{M}$ into two disjoint open sets, $D^{\pm}$, there is no way to distinguish these two domains from each other. Moreover, examples show that we cannot guarantee a holomorphic extension of $CR(S)$ into the whole of either D^+ or D^-. A simple example is the case that $S\subset C^2$ is the surface given by the equation $|w|^2(1+|z|^2) = 1$ and f is the function given by $f(z,w) = \frac{1}{w(w-2)}$. Thus, in general, we have to consider holomorphic continuation into an open set $W = int\,\overline{W}$ in $\mathcal{M}$ that is *attached to* S in the sense that it contains one-sided neighborhoods of every point in S: If $a\in S$, there is a neighborhood $U\ni a$ in $\mathcal{M}$ such that $W\cap U$ contains a connected component of $U\backslash S$, and $S\cap U$ belongs to the boundary of this component. In general W can contain a nonempty open subset of S.

The property that $W = int\,\overline{W}$ is essential. It implies, *e.g.*, that $W\supset U$ whenever $W\supset U\backslash S$. This is reasonable because in the case that a CR-function f on S extends to W and $W\supset U\backslash S$, we want the boundary values of this extension from the two sides of $S\cap U$ to exist and coincide with $f|S\cap U$. Extensions to such W arise naturally in many local problems.

Lemma 3.4.1 *Let S be an orientable closed hypersurface that is locally a Lipschitz graph in a smooth N-dimensional manifold $\mathcal{M}$, and let U and W be open subsets of $\mathcal{M}$ with $W\subset U$ such that W is attached to $S\cap U$. Then there is a hypersurface $S'\subset(S\backslash U)\cup W$*

that is homotopic in $U \cup S$ to S under a homotopy that leaves $S \setminus U$ fixed pointwise and that is closed in $\mathcal{M}$ and smooth in U. We have that $\int_S \varphi = \int_{S'} \varphi$ for every smooth d-closed form $\varphi \in \mathcal{D}^{N-1}(\mathcal{M})$.

Proof As S is orientable, there is a smooth vector field v without critical points in a neighborhood of S that is transverse to S in the sense that $v(a)$ does not belong to the tangent cone[7] $C(a)$ to S at a for any $a \in S$. By the definition of W, there is an open subset W' also attached to $S \cap U$ and such that the intersection of W' with each integral curve of the field v is an arc, either intersecting S or with an endpoint on S. Fix a smooth Riemannian metric on $\mathcal{M}$ and introduce a parameter t of the directed length on these integral arcs with $t = 0$ on S such that the positive direction in t corresponds to the direction of v. Write $S \cap U = S^+ \cup S^o \cup S^-$ where $S^o = S \cap W'$ and $a \in S^{\pm}$ if there is an arbitrarily small neighborhood $U \ni a$ such that W' contains only that component of $U \setminus S$ on which $\pm t > 0$. Choose a continuous function λ on S such that $\lambda = 0$ on $S \setminus U$, $\lambda > 0$ on S^+, $\lambda < 0$ on S^- (S^o is open in S, and $S^+ \cap S^- = \emptyset$), and such that for each $a \in S \cap U$, the set W' contains an integral arc of the field v of length $2|\lambda(a)|$ with a as end point (or initial point).

The ends of the arcs corresponding to $t = \lambda(a)$ constitute a hypersurface S' such that $S' \setminus U = S \setminus U$, and $S' \cap U$ is smooth with appropriate choice of λ. If we make the change of variables $\tau = t - \lambda(a)$ for the integral curve through a, we obtain a smooth function in W' that we can take as a local coordinate. Thus S' is closed in $\mathcal{M}$, is orientable and is homotopic to S in U. Stokes's formula $\int_{S'} \varphi = \int_S \varphi$ follows because of this homotopy along the field $grad\ \tau$. $\qquad\qquad\square$

Thus, we see that holomorphic continuation of $CR(S)$ to a set W attached to S in some sense reduces general CR-functions on S to functions that are holomorphic on a neighborhood of S'. By using approximation theorems, we can even choose S' to be real-analytic. Such a reduction can be very useful in many problems related to CR-functions.

We shall say that a real hypersurface S in a complex manifold $\mathcal{M}$ has the *one-sided extension property* in $\mathcal{M}$ if there is an open set $W \subset \mathcal{M}$ attached to S such that every CR-function on S extends to a holomorphic function in W. An example is that of a compact hypersurface in a noncompact $\mathcal{M}$ with $H_c^{0,1}(\mathcal{M}) = 0$.

The hypersurface S has the one-sided extension property *at a point* $a \in S$ if $S \cap U$ has the one-sided extension property in U for some neighborhood $U \ni a$ in $\mathcal{M}$. A standard example is the case of a point $a \in S$, S of class C^2, at which the Levi form of S is nondegenerate.

Trépreau [82] has given conditions sufficient for the one-sided extension property at a point.

If we take this one-sided extension property the basic one, it is natural to introduce the notion of removable sets as sets that do not hinder such extensions. Thus, we make the following definition:

Definition *Let S be a closed orientable locally Lipschitz hypersurface in a complex manifold $\mathcal{M}$ of dimension $n \geq 2$. A closed set $X \subset S$ is said to be removable with respect*

---

[7]The notion of tangent cone we use is that of Federer [27]: Given a set S in the vector space X, the tangent cone of S at a is the set $C(S, a)$ of vectors v such that for every $\varepsilon > 0$ there are $s \in S$ and $r \in \mathbb{R}$, $r > 0$, such that $|r(s - a) - v| < \varepsilon$.

to S if there is an open set $W \subset \mathcal{M}$ attached to S such that each CR-function on $S \setminus X$ extends to a holomorphic function on W. The set X is weakly removable with respect to S if we have a holomorphic extension to W for each CR-function on S that is orthogonal on S to all $\overline{\partial}$-closed smooth $(n, n-1)$-forms in $\mathcal{M}$ with support that meets the set $S \setminus X$ in a compact set.

Again, there is the issue of how the extended function F is to assume the boundary values f along S. We shall be explicit about this point as required.

It follows from the definition of removable sets that if S has a removable set, then it also has the one-sided extension property. We do not include this condition in the definition because sometimes it is simple to prove the removability of X under some other conditions on S. (See Sections 4.6, 4.7.)

For $\mathcal{M}$ with $H_c^{0,1}(\mathcal{M}) = 0$ and S compact and connected, these new definitions coincide with the definitions given in the introduction, so they are natural generalizations.

There are some simple cohomological conditions similar to those in Sections 3.1 and 3.3 that are sufficient for removability and weak removability with respect to a noncompact hypersurface.

We say that $\mathcal{M} \setminus S$ *extends holomorphically through* S if there is a (not necessarily connected) Riemann domain $(\widetilde{\mathcal{M}}, \pi)$ over $\mathcal{M}$ with holomorphic projection π and an embedding $\iota : \mathcal{M} \setminus S \to \widetilde{\mathcal{M}}$ such that $\pi \circ \iota(z) = z, \pi(\widetilde{\mathcal{M}}) = \mathcal{M}$, such that for all $z \in \mathcal{M}$, the number of points in the fiber $\pi^{-1}(z)$ is not more than two, and, finally, such that every holomorphic function on $\iota(\mathcal{M} \setminus S)$ extends holomorphically into $\widetilde{\mathcal{M}}$. The set of points z for which $\pi^{-1}(z)$ consists of two distinct points is an open set W in $\mathcal{M}$ that is attached to S.

Proposition 3.4.2 *Let S be a locally Lipschitz closed orientable hypersurface in a complex manifold $\mathcal{M}$ of dimension $n \geq 2$. Assume that the group $H_c^{n,n}(\mathcal{M})$ is separated and that $\mathcal{M} \setminus S$ extends holomorphically through S. Then a closed subset $X \subset S$ is weakly removable if $H_c^{n,n-1}(X) = 0$.*

Recall that by 1.13), the group $H_c^{n,n}(\mathcal{M})$ is separated if and only if $H^{0,1}(\mathcal{M})$ is separated. Thus, *e.g.*, the proposition applies when $\mathcal{M}$ is Stein, for in this case $H^{0,1}(\mathcal{M}) = 0$.

Proof The proof essentially repeats the proof of Theorem 2.3.2.

Let f be a CR-function on $S \setminus X$ orthogonal on $S \setminus X$ to every $\overline{\partial}$-closed smooth $(n, n-1)$-form on $\mathcal{M}$ with support compact in $\mathcal{M} \setminus X$. Denote by $f[\Gamma]^{0,1}$ the corresponding $\overline{\partial}$-closed $(0, 1)$-current on $\mathcal{M} \setminus X$. Let L be the closure in $\mathcal{D}'^{0,1}(\mathcal{M} \setminus X)$ of the space $\overline{\partial}\mathcal{D}^{0,0}(\mathcal{M} \setminus X)$. Let $\varphi \in \mathcal{D}^{n,n-1}(\mathcal{M} \setminus X)$ be orthogonal to L. Then $\overline{\partial}\varphi = 0$, that is $\varphi \in \mathcal{Z}_c^{n,n-1}(\mathcal{M} \setminus X)$, whence $\langle f[\Gamma]^{0,1}, \varphi \rangle = 0$.

It follows that $f[\Gamma]^{0,1} \in L$. As $H^{0,1}(\mathcal{M} \setminus X)$ is separated because $H_c^{n,n-1}(X) = 0$–recall 1.17–we have that $L = \overline{\partial}\mathcal{D}'^{0,0}(\mathcal{M} \setminus X)$, and thus $f[\Gamma]^{0,1} = \overline{\partial}h$ for some distribution h on $\mathcal{M} \setminus X$. Then $h|(\mathcal{M} \setminus S)$ is a holomorphic function , and f is the difference of the boundary values of h on S from its two sides. See [15]. As $\mathcal{M} \setminus X$ extends holomorphically through S, there is an open set W attached to S and a Riemann domain $(\widetilde{\mathcal{M}}, \pi)$ as in the definition. Let $\tilde{h}$ be the holomorphic continuation of $h \circ \pi^{-1}$ from $\iota(\mathcal{M} \setminus S)$ to $\widetilde{\mathcal{M}}$. For $z \in W$ denote by $\iota^+(z)$ the point in $\pi^{-1}(z)$ different from $\iota(z)$, and set $F(z) = \tilde{h}(\iota(z)) - \tilde{h}(\iota^+(z))$. Then $F \in \mathcal{O}(W)$, and F or $-F$ has boundary values on $S \setminus X$ that coincide with f.   □

Proposition 3.4.3 *Let S be a locally Lipschitz closed orientable hypersurface in an n-dimensional complex manifold $\mathcal{M}$ with $H_c^{n,n-1}(\mathcal{M}) = 0$, $n \geq 3$. Let X be a closed set in S that is weakly removable and that satisfies $^\sigma H_c^{n,n-2}(X) = 0$. Then X is removable with respect to S.*

The hypothesis on $\mathcal{M}$ is satisfied, *e.g.*, when $\mathcal{M}$ is a Stein manifold.

Proof Let $f \in CR(S \backslash X)$ and $\varphi \in \mathcal{Z}_c^{n,n-1}(\mathcal{M} \setminus X)$. As $H_c^{n,n-1}(\mathcal{M}) = 0$, there is a form $\psi \in \mathcal{D}^{n,n-2}(\mathcal{M})$ such that $\overline{\partial}\psi = \varphi$. In particular, $\overline{\partial}\psi = 0$ in a neighborhood X. As $^\sigma H_c^{n,n-2}(X) = 0$, there are a smaller neighborhood $V \supset X$, compact set $E \subset \mathcal{M}$ and smooth $(n, n-3)$-forms g_j on V such that $\psi = \lim \overline{\partial} g_j$ in $\mathcal{E}^{n,n-2}(V)$ and $V \cap \operatorname{supp} g_j \subset V \cap E$ for all j. Let $\lambda \in \mathcal{D}(V)$, $\lambda = 1$ in a neighborhood of $E \cap X$. Then $\langle f[\Gamma]^{0,1}, \varphi \rangle = \langle f[\Gamma]^{0,1}, (1 - \lambda)\overline{\partial}\psi \rangle = \langle f[\Gamma]^{0,1}, \overline{\partial}((1 - \lambda)\psi) + \overline{\partial}\lambda \wedge \psi \rangle = \langle f[\Gamma]^{0,1}, \overline{\partial}\lambda \wedge \psi \rangle = -\lim \langle f[\Gamma]^{0,1}, \overline{\partial}(\overline{\partial}\lambda \wedge g_j) \rangle = 0$ because $f \in CR(S \backslash X)$ and $\overline{\partial}\lambda \wedge g_j \in \mathcal{D}^{n,n-2}(\mathcal{M} \backslash X)$. Thus, f satisfies the condition of weak removability, and by the condition on X, f extends holomorphically into an open set W attached to S in $\mathcal{M}$. $\qquad\qquad\square$

Corollary 3.4.4 *Let $\mathcal{M}$ and S be as in Proposition 3.4.2, and suppose $H_c^{n,n}(\mathcal{M})$ to be separated. A closed set $X \subset S$ is removable if $H_c^{n,n-1}(X) = {}^\sigma H_c^{n,n-2}(X) = 0$.*

Proof The conditions on $\mathcal{M}$ together with $H_c^{n,n-1}(X) = 0$ imply by Proposition 3.4.2 that X is weakly removable. Hence X is removable by Proposition 3.4.3. $\qquad\qquad\square$

We do not know whether the cohomological conditions in Propositions 3.4.2 and 3.4.3 are necessary for general strictly pseudoconvex noncompact hypersurfaces in Stein manifolds. They are in the case of compact strictly pseudoconvex hypersurfaces. Recall the results of Section 2.

3.5   Unions of removable sets

We start with the following simple corollary of the characterizations of removable and weakly removable sets in strictly pseudoconvex domains given in Section 2.3.

Proposition 3.5.1 *Let D be a relatively compact strictly pseudoconvex domain with boundary of class C^2 in a Stein manifold of dimension n, and let K_1, K_2 be disjoint compact sets in bD. If $n \geq 2$ and K_1, K_2 are weakly removable, then $K_1 \cup K_2$ is weakly removable. If $n \geq 3$ and K_1, K_2 are removable, then $K_1 \cup K_2$ is removable.*

The last statement is not true for $n = 2$. A simple example is given by the set consisting of the two disjoint circles in bB_2 that constitute the boundary of the intersection of the one-dimensional variety with equation $z_1 z_2 = \epsilon$ for small $\epsilon > 0$ with the ball B_2.

The case in which $K_1 \cap K_2$ is not empty is much more complicated.

Proposition 3.5.2 *Let D be a relatively compact domain with locally Lipschitz boundary in a Stein manifold $\mathcal{M}$ of dimension $n \geq 3$, let $K \subset bD$ be a compact set and $\Gamma = bD \setminus K$. Let Γ' be a locally Lipschitz hypersurface that is a closed subset of a neighborhood of $\overline{D}$ and that is foliated by a one-parameter family of complex $(n-1)$-dimensional complex manifolds Γ'_t. Assume that Γ' divides D into two domains $D^{\pm}$ each of which meets Γ and that the sets $bD^{\pm}$ are locally Lipschitz[8]. Suppose that*
(a) $\Gamma \cap U$ has the one-sided extension property in a neighborhood $U \supset \Gamma \cap \Gamma'$,
(b) the sets $K \cap bD^{\pm}$ are removable with respect to $D^{\pm}$, and
(c) $K \cap \Gamma'_t$ is removable in Γ'_t with respect to $\Gamma'_t \cap D$ for each t.
Then K is removable.

Proof Let $f \in CR(\Gamma)$. By (a) we can assume that f is holomorphic in a neighborhood of $\Gamma \cap \Gamma'$. By (c) and a generalization of Hartogs's lemma–see [67], f extends holomorphically into a neighborhood of $\Gamma' \cap D$. By using these extensions, we have two CR-functions $f^{\pm}$ on $bD^{\pm} \setminus K$ respectively such that $f^{\pm} = f$ on $\Gamma \cap bD^{\pm}$. Then by (b), these functions $f^{\pm}$ extend to holomorphic functions $F^{\pm}$ in $D^{\pm}$. By a boundary uniqueness theorem, they constitute a well defined holomorphic function F on D with boundary values f on Γ. $\square$

By using arguments similar to that in the proof of the Mayer-Vietoris sequences, we can obtain a series of sufficient conditions for removability of unions in terms of Dolbeault cohomology.

Lemma 3.5.3 *Let $X = X_1 \cup X_2$ where X_1, X_2 are closed sets in a complex manifold $\mathcal{M}$ of dimension n. Assume that $H^{p,q}(X_j) = 0, j = 1, 2$ for some $p \geq 0, q \geq 2$.*
(a) If $H^{p,p-1}(X_1 \cap X_2) = 0$, then $H^{p,q}(X) = 0$.
(b) If $^{\sigma}H^{p,q-1}(X_1 \cap X_2) = 0$ (for $n \geq 3$) or if $\mathcal{Z}^{p,q}(X_1)$ is dense in $\mathcal{Z}^{p,q}(X_1 \cap X_2)$, then $^{\sigma}H^{p,q}(X) = 0$.

Proof Let $\varphi \in \mathcal{Z}^{p,q}(U)$ for some neighborhood $U \supset X$. Then by hypothesis, there are neighborhoods $U_j \supset X_j$ in U and forms $\psi_j \in \mathcal{E}^{p,q-1}(U_j)$ such that $\overline{\partial}\psi_j = \varphi$ in U_j. It follows that $\psi_1 - \psi_2 \in \mathcal{Z}^{p,q-1}(U_1 \cap U_2)$. If $H^{p,q-1}(X_1 \cap X_2) = 0$, there is a neighborhood $V \supset X_1 \cap X_2$ in $U_1 \cap U_2$ on which there is a form $g \in \mathcal{E}^{p,q-2}(V)$ such that $\psi_1 - \psi_2 = \overline{\partial}g$ on V.

Let λ be a smooth function in a neighborhood of X with $\lambda = 1$ in a neighborhood of $X_1 \setminus V$ and $\lambda = 0$ in a neighborhood of $X_2 \setminus V$. Put

$$\psi = \lambda\psi_1 + (1 - \lambda)\psi_2 + \overline{\partial}\lambda \wedge g.$$

This form is well defined and smooth in a neighborhood of X, and there

$$\overline{\partial}\psi = \varphi + \overline{\partial}\lambda \wedge (\psi_1 - \psi_2) - \overline{\partial}\lambda \wedge \overline{\partial}g = \varphi.$$

Given that $^{\sigma}H^{p,q-1}(X_1 \cap X_2) = 0$, we obtain forms $g^{\nu} \in \mathcal{E}^{p,q-2}(V)$ such that $\psi_1 - \psi_2 = \lim \overline{\partial}g^{\nu}$ on V. Set

$$\psi^{\nu} = \lambda\psi_1 + (1 - \lambda)\psi_2 + \overline{\partial}\lambda \wedge g^{\nu}.$$

---

[8]This hypothesis amounts to a hypothesis that Γ and Γ' be transverse in a suitable sense.

Then the forms ψ^ν are smooth in a neighborhood of X independent of ν, and we have that $\overline{\partial}\psi^\nu = \varphi + \overline{\partial}\lambda \wedge (\psi_1 - \psi_2 - \overline{\partial}g^\nu) \to \varphi$ on this neighborhood.

Finally assuming that the space of germs $\mathcal{Z}^{p,q}(X_1)$ restricts to a dense set in $\mathcal{Z}^{p,q}(X_1 \cap X_2)$, we obtain $h^\nu \in \mathcal{Z}^{p,q}(V_1)$ for some $V_1 \supset X_1$ such that $\psi_1 - \psi_2 = \lim\ h^\nu$ on V. The form $\psi^\nu = \lambda\psi_1 + (1 - \lambda)\psi_2 - \lambda g^\nu$, which is smooth on a neighborhood of X satisfies

$$\overline{\partial}\psi^\nu = \varphi + \overline{\partial}\lambda \wedge (\psi_1 - \psi_2) - \overline{\partial}\lambda \wedge g^\nu \to \varphi$$

in this neighborhood. $\qquad\square$

In the following result we shall use the notion of universally removable set. We say that the compact K in the complex manifold $\mathcal{M}$ is *universally removable* if whenever $D = int\overline{D} \subset \mathcal{M}$ is a relatively compact domain with $\mathcal{M}\backslash\overline{D}$ connected, with $bD \supset K$ and with $bD\backslash K$ a locally Lipschitz hypersurface, the set K is removable for the domain D. There is a parallel notion of *universal weak removability*.

Proposition 3.5.4 *Let $D = int\overline{D}$ be a relatively compact domain in a noncompact complex manifold $\mathcal{M}$ of dimension $n \geq 3$ with $\mathcal{M}\backslash\overline{D}$ connected. Let K_1 and K_2 be compact subsets of bD, and put $K = K_1 \cup K_2$. Assume that $bD \setminus K$ is locally Lipschitz, that $H^{n,n-1}(K_j) = 0, j = 1, 2$, and that $H^{n,n-2}(K_1 \cap K_2) = 0$. Then $H^{n,n-1}(K) = 0$, and K is (universally) weakly removable. Moreover, if $H^{n,n-1}(\mathcal{M}) = 0$ and $^\sigma H^{n,n-2}(K_1 \cap K_2) = 0$ (for $n \geq 3$) or $\mathcal{Z}^{n,n-2}(K_1)$ is dense in $\mathcal{Z}^{n,n-2}(K_1 \cap K_2)$, then $^\sigma H^{n,n-2}(K) = 0$, and K is universally removable.*

Proof The proof follows from Lemma 3.5.3 and Theorems 2.3.2 and 2.3.1. $\qquad\square$

A closed set E in a complex manifold is called *holomorphically convex* if every continuous multiplicative linear functional on the algebra $\mathcal{O}(E)$ of germs of holomorphic functions on E endowed with the inductive limit topology has the form $f \mapsto f(a)$ for some point $a \in E$. The holomorphically convex subsets of a Stein manifold are characterized [33] by the condition that $H^{p,q}(K) = 0$ for all $p \geq 0$, $q \geq 1$. This condition is essentially as difficult to verify as is the condition in the definition. An important subclass of the holomorphically convex sets are the Stein compacta, which we have discused above, *i.e.*, the compact subsets that have neighborhood bases consisting of Stein open sets. A simple example of Björk [10] shows that a holomorphically convex compact set need not be a Stein compactum. (Björk's example is a compact Reinhardt set in $\mathbb{C}^2$.)

Corollary 3.5.5 *If $\mathcal{M}$ is a Stein manifold of dimension at least three, if D is a relatively compact strictly pseudoconvex domain in $\mathcal{M}$ with boundary of class C^2, and if $K_1 \cap K_2$ is holomorphically convex, then $K = K_1 \cup K_2$ is weakly removable if and only if K_1 and K_2 are weakly removable.*

Corollary 3.5.6 *Let $\mathcal{M}$ be a Stein manifold of dimension n, let D be a relatively compact strictly pseudoconvex domain in $\mathcal{M}$ with boundary of class C^2. If $K_1 \cap K_2$ is holomorphically convex (in case $n \geq 4$) or $\mathcal{O}(\overline{D})$-convex if $n \geq 3$, then K is removable if and only if K_1 and K_2 are removable.*

Removability with respect to noncompact hypersurfaces (see Section 3.4) is more flexible.

Lemma 3.5.7 *Let* X_1, X_2 *be closed subsets of the complex manifold* $\mathcal{M}$ *with* $H_c^{p,q}(X_j) = 0, j = 1, 2$ *for some* $p \geq 0, q \geq 1$. *Let* $X_1 \cap X_2$ *coincide with the closure of* $(X_1 \cap X_2)^\circ$, *the set of interior points of* $X_1 \cap X_2$ *in* $X = X_1 \cup X_2$.
(a) If $H_c^{p,q}((X_1 \cap X_2)^\circ) = 0$, *then* $H_c^{p,q}(X) = 0$.
(b) If $^\sigma H_c^{p,q}((X_1 \cap X_2)^\circ) = 0$, *then* $^\sigma H_c^{p,q}(X) = 0$.

Proof Let U be a neighborhood of X and let $\varphi \in \mathcal{Z}^{p,q}(U)$ have the property that $X \cap supp\, \varphi$ is compact. Then by hypothesis, there are neighborhoods $U_j \supset X_j$ in U and forms $\psi_j \in \mathcal{E}^{p,q-1}(U_j)$ such that $\overline{\partial}\psi_j = \varphi$ in U_j and the sets $X_j \cap supp\, \psi_j$ are compact. It follows that $\overline{\partial}(\psi_1 - \psi_2) = 0$ in $U_1 \cap U_2$. Let λ be a smooth function in a neighborhood of $X, \lambda = 1$ in a neighborhood of $X_1 \setminus (X_1 \cap X_2)^\circ$ and $\lambda = 0$ in a neighborhood of $X_2 \setminus (X_1 \cap X_2)^\circ$.

If $H_c^{p,q}((X_1 \cap X_2)^\circ) = 0$, then there is a smooth form ψ_0 on a neighborhood of X such that $\overline{\partial}\psi_0 = \overline{\partial}\lambda \wedge (\psi_1 - \psi_2)$ and $(X_1 \cap X_2)^\circ \cap supp\, \psi_0$ is compact. Set $\psi = \lambda\psi_1 + (1 - \lambda)\psi_2 - \psi_0$. Then ψ is smooth in a neighborhood of X, $X \cap supp\, \psi$ is compact, and $\overline{\partial}\psi = \varphi + \overline{\partial}\lambda \wedge (\psi_1 - \psi_2) - \overline{\partial}\psi_0 = \varphi$.

If $^\sigma H_c^{p,q}((X_1 \cap X_2)^\circ) = 0$, and if φ and ψ_1, ψ_2 are as above, then there are a neighborhood V of $(X_1 \cap X_2)^\circ$, a compact set $E \subset \overline{V}$, and smooth forms g^ν in V such that $\overline{\partial}g^\nu \to \overline{\partial}\lambda \wedge (\psi_1 - \psi_2)$ in V, $supp\, g^\nu \subset E$, and $E \cap (X_1 \cap X_2)^\circ$ is compact. Extending g^ν by zero outside V and setting $\psi^\nu = \lambda\psi_1 + (1 - \lambda)\psi_2 - g^\nu$, we obtain smooth forms on a neighborhood of X that is independent of ν such that $\overline{\partial}\psi^\nu = \varphi + \overline{\partial}\lambda \wedge (\psi_1 - \psi_2) - \overline{\partial}g^\nu \to \varphi$.
This completes the proof. $\square$

From Lemma 3.5.7, Propositions 3.4.2 and 3.4.3, we obtain the following result:

Proposition 3.5.8 *Let* S *be a closed locally Lipschitz orientable hypersurface in a complex manifold* $\mathcal{M}$ *of dimension* n. *Let* X_1, X_2 *be closed sets in* S *such that* $X_1 \cap X_2$ *is the closure of its interior* $(X_1 \cap X_2)^\circ$ *in* $X = X_1 \cup X_2$. *Suppose that* $H_c^{n,n-1}(X_j) = 0, j = 1, 2$ *and that* $H_c^{n,n-1}((X_1 \cap X_2)^\circ) = 0$.
(a) If $n \geq 2$, *then* X *is weakly removable with respect to* S.
(b) If, moreover, $n \geq 3$, $H_c^{n,n-2}(X_j) = 0, j = 1, 2$ *and* $^\sigma H_c^{n,n-2}((X_1 \cap X_2)^\circ) = 0$, *then* X *is removable with respect to* S.

The special classes of singularities studied in Section 4 exhibit some applications of the results of this section.

3.6   Removable Sets for Holomorphic Functions

The cohomological condition studied above for closed sets X can be applied to the problems of removable sets for holomorphic functions. We prove here a simple result similar in spirit to Theorem 2.3.1 on removable sets for CR-functions.

We shall say that a closed subset X of a connected complex manifold $\mathcal{M}$ is *removable for holomorphic functions*, if every holomorphic function f on $\mathcal{M} \setminus X$ extends to a holomorphic function on $\mathcal{M}$.

Proposition 3.6.1 *Let M be a complex manifold of dimension $n \geq 2$ such that the group $H_c^{n,n}(M)$ is separated. A closed subset X of M is removable for holomorphic functions if $^\sigma H_c^{n,n-1}(X) = H_c^{n,n}(X) = 0$.*

Proof From the exact sequence - see 1.9,

$$H_c^{n,n-1}(X) \xrightarrow{\delta} H_c^{n,n}(M \setminus X) \xrightarrow{\iota} H_c^{n,n}(M) \xrightarrow{r} H_c^{n,n}(X) = 0$$

we find that ι is surjective. As $H_c^{n,n}(M)$ is separated, the space $im\ \delta = ker\ \iota = \iota^{-1}(0)$ is closed, for ι is continuous. Since 0 is dense in $H_c^{n,n-1}(X)$ and δ is continuous, its image $\delta(0) = 0$ is dense in the closed set $im\ \delta$. Thus we obtain that $\overline{0} = im\ \delta$ in $H_c^{n,n}(M)$, and thus the mapping $\iota' : {}^\sigma H_c^{n,n}(M \setminus X) \to H_c^{n,n}(M)$ is continuous and one-to-one. By Serre duality, we have that $^\sigma H_c^{n,n}(M \setminus X) \simeq (\mathcal{O}(M \setminus X))^*$, and $H_c^{n,n}(M) \simeq (\mathcal{O}(M)^*)$. Thus, ι' induces the continuous one-to-one mapping $r^* : (\mathcal{O}(M \setminus X))^* \to (\mathcal{O}(M))^*$ dual to the restriction map r, i.e., $\langle r(f), \varphi \rangle = \langle f, r^*(\varphi) \rangle$ for $f \in \mathcal{O}(M), \varphi \in (\mathcal{O}(M \setminus X))^*$. It follows that r^* is a topological isomorphism [68], p. 170]. But then the same is true for r. In particular we have that $r(\mathcal{O}(M)) = \mathcal{O}(M)|(M \setminus X) = \mathcal{O}(M \setminus X)$. $\qquad\square$

Corollary 3.6.2 *If M and X are as in Proposition 3.6.1, then $M \setminus X$ is connected.*

The conditions of the Proposition 3.6.1 are not necessary in general. To see this let X be a closed subset in C^n, $n \geq 2$ that is disjoint from the closed unit ball $\overline{B}_n$, and let $K \subset bB_n$ be the semisphere $bB_n \cap \{\text{Re}\ z_1 \leq 0\}$. Then K is not weakly removable with respect to the ball – see Theorem 2.5.1, so $H^{n,n-1}(K) \neq 0$, by Theorem 2.3.2. It follows that $H_c^{n,n-1}(X \cup K) \neq 0$, but $X \cup K$ is removable if the set X is removable in C^n.

Corollary 3.6.3 *Let X be a pseudoconcave subset of a Stein manifold M of dimension at least two. Then either $H_c^{n,n-1}(X) \neq 0$ or $H_c^{n,n}(X) \neq 0$. Moreover, $H_c^{n,n-1}(X) = 0$ if and only if $M \setminus X$ is a Runge domain in M, and this is equivalent to the condition that $M \setminus X$ be $\mathcal{O}(M)$-convex.*

Recall that X is pseudoconcave if $M \setminus X$ is pseudoconvex. A domain G in M is a Runge domain if $\mathcal{O}(M)$ is dense in $\mathcal{O}(G)$. Finally G is $\mathcal{O}(M)$-convex if $\widehat{K}_{\mathcal{O}(M)} \subset G$ for each compact set $K \subset G$.

Proof The manifold $M \setminus X$ is Stein, so $H_c^{n,n}(M \setminus X)$ is separated. As $H_c^{n,n-1}(M) = 0$, we see from the exact sequence

$$0 \to H_c^{n,n-1}(X) \xrightarrow{\delta} H_c^{n,n}(M \setminus X) \xrightarrow{\iota} H_c^{n,n}(M)$$

that δ is an embedding, so $0 = \delta^{-1}(0)$ is closed in $H_c^{n,n-1}(X)$. That is $^\sigma H_c^{n,n-1}(X) = H_c^{n,n-1}(X)$. Thus, the first part of the Corollary follows from Proposition 3.6.1.

If $H_c^{n,n-1}(X) = 0$, then ι is an embedding. Thus, if $\varphi \in \mathcal{D}^{n,n}(M \setminus X)$ is orthogonal to $\mathcal{O}(M)$, then $\iota[\varphi] = 0$, which implies that $[\varphi] = 0$, i.e., that $\varphi = \overline{\partial}\psi$ for some $\psi \in \mathcal{D}^{n,n-1}(M \setminus X)$. But then φ is orthogonal to $\mathcal{O}(M \setminus X)$ as well, and thus we obtain by the Hahn-Banach theorem that $\mathcal{O}(M)|M \setminus X$ is dense in $\mathcal{O}(M \setminus X)$.

In the opposite direction, if r has dense image and $\iota[\varphi] = 0$, then as φ is orthogonal to $\mathcal{O}(\mathcal{M})$, it is orthogonal to $\mathcal{O}(\mathcal{M} \setminus X)$ as well. The manifold $\mathcal{M} \setminus X$ is Stein, so it follows that $[\varphi] = 0$, and thus that ι is an embedding. Consequently $0 = ker\ \iota = im\ \delta$, from which it follows that $H_c^{n,n-1}(X) = 0$ because δ is an embedding.

If $\mathcal{M} \setminus X$ is $\mathcal{O}(\mathcal{M})$-convex, then the space $\mathcal{O}(\mathcal{M})|(\mathcal{M} \setminus X)$ is dense in $\mathcal{O}(\mathcal{M} \setminus X)$ by the Oka-Weil theorem on Stein manifolds. See [38]. The manifold $\mathcal{M} \setminus X$ is also a Stein manifold, so $\widehat{K}_{\mathcal{O}(\mathcal{M}\setminus X)}$ is compact in $\mathcal{M} \setminus X$ for each compact set $K \subset \mathcal{M} \setminus X$. If $\mathcal{O}(\mathcal{M})|(\mathcal{M} \setminus X)$ is dense in $\mathcal{O}(\mathcal{M} \setminus X)$, it follows that $\widehat{K}_{\mathcal{O}(\mathcal{M}\setminus X)} = \widehat{K}_{\mathcal{O}(\mathcal{M})}$, whence $\mathcal{M} \setminus X$ is $\mathcal{O}(\mathcal{M})$-convex.

This completes the proof. $\square$

The last condition in Corollary 3.6.3 is obviously similar to the geometric condition for weak removability in Section 2.5.

Corollary 3.6.4 *Let X be a closed set of topological dimension m in an n-dimensional complex manifold $\mathcal{M}$ such that the group $H_c^{n,n}(\mathcal{M})$ is separated. If $m < n - 1$, then every function holomorphic on $\mathcal{M} \setminus X$ extends to a function holomorphic on $\mathcal{M}$, i.e., the set X is removable for holomorphic functions.*

Proof By the Dolbeault isomorphism, $H_c^{n,q}(X)$ is isomorphic to $H_c^q(X, \mathbf{\Omega}^n)$, the Cech cohomology with compact supports. As $dim\ X = m$, we have for an open covering $\{U_j\}$ of X a refinement covering $\{V_k\}$ such that $V_{k_o} \cap \ldots \cap V_{k_{m+1}} = \emptyset$ for all choices $k_o, \ldots, k_{m+1}$ of distinct indices. It follows that $H_c^q(X, \mathbf{\Omega}^n) = 0$ if $q \geq m + 1$, thus, for $m < n - 1$, we have that $H_c^{n-1}(X, \mathbf{\Omega}^n) = H_c^n(X, \mathbf{\Omega}^n) = 0$ or, what is the same $H_c^{n,n-1}(X) = H_c^{n,n}(X) = 0$.

This yields the result from Proposition 3.6.1. $\square$

It is to be underlined that in this Corollary, we are dealing with dimension in the topological sense, not the metric sense. For topological dimension theory see [40].

The Corollary implies that any closed one-dimensional subset, for instance a closed subset homeomorphic to the real line, in C^n, $n \geq 3$ is removable, a two-dimensional surface in C^n with $n \geq 4$ is removable, *etc.*

The simplest open question in this direction is whether a closed subset of C^2 homeomorphic to the real line is removable. More generally, is a topologically m-dimensional subset of C^{m+1} removable? In this case, we have $H_c^{n,n}(X) = 0$, but it is not evident that $^\sigma H_c^{n,n-1}(X)$ vanishes.

Related problems for CR-functions are discussed below in Section 4.1.

4 Special classes of singularities

4.1 Low Dimensional singularities

The intrinsic characterizations in Section 3 imply that certain purely topological conditions on a set are sufficient for removability and weak removability. We give here some results involving the dimension of the set. We take *dimension* in the topological sense, not in the metric sense.

We have the following result.

Proposition 4.1.1 *Let M be a noncompact complex manifold of dimension $n \geq 2$, and let K be a compact subset of the boundary of a relatively compact domain $D \subset M$. Assume that $bD \setminus K$ is locally a Lipschitz graph and that $int\overline{D} = D$. Then K is weakly removable if $\dim K < n - 1$. If, in addition $H^{n,n-1}(M) = 0$ and $M \setminus \overline{D}$ is connected, then K is removable if $\dim K < n - 2$.*

Thus, totally disconnected sets are removable for $n \geq 3$ and are weakly removable for $n \geq 2$, arcs[9] are removable for $n \geq 4$, and so on. Note that topologically thin sets can be very massive in the metric sense. For example, there are totally disconnected–and hence zero-dimensional–closed sets of positive volume in M. We see no direct proof of Proposition 1, no proof that does not depend on the intrinsic cohomological conditions from Section 3.

Proof By the Dolbeault isomorphism, the space $H^{p,q}(K)$ is isomorphic to $H^q(K, \Omega^p)$. The cohomology group $H^q(K, \Omega^p)$ is defined to be the direct limit of the groups $H^q(\mathcal{U}, \Omega^p)$ as $\mathcal{U}$ runs through the open coverings of K. (Čech cohomology is discussed in [38] and in [31].) If $\dim K \leq d$, then for an arbitrary covering $\mathcal{U}$ of K, there is a finer locally finite covering $\mathcal{V} = \{V_j\}_{j \in J}$ of K with the property that every collection of $d + 2$ distinct V_j's has empty intersection. It follows that $H^{d+1}(K, \Omega^p) = 0$. Equivalently, $H^{p,d+1}(K) = 0$ Thus, the conclusions of Proposition 4.1.1 follow from Theorems 3.3.1 and 3.1.1. $\square$

In the following corollary, we shall use the terminology that γ *is an arbitrary curve with a totally disconnected set of self-intersections* in bD if $\gamma \subset bD$ is the range of a continuous map $g : I = [0, 1] \to bD$ such that for a totally disconnected subset $\Sigma \subset I$, the set $g(\Sigma)$ is totally disconnected, and g carries $I \setminus \Sigma$ homeomorphically onto $g(I) \setminus g(\Sigma)$.

Corollary 4.1.2 *Let M be a Stein manifold, let D be a strictly pseudoconvex domain in M with boundary of class C^2, and let $\gamma \subset bD$ be an arbitrary curve with a totally disconnected set of self-intersections. If $n \geq 4$ and K is removable, then $K \cup \gamma$ is also removable. If $n \geq 3$ and K is weakly removable, then $K \cup \gamma$ is weakly removable.*

Proof As $\dim \gamma = 1$, we have $H^{p,q}(\gamma') = 0$ for an arbitrary subset $\gamma' \subset \gamma$ and for all $q \geq 2$. For each closed set $E \subset M$, we can represent $E \cup \gamma$ as $E \cup \gamma'$ with $E \cap \overline{\gamma}'$ totally

---

[9]By an *arc* we understand a homeomorph of the closed interval $[0, 1] \subset R$.

disconnected. The curve γ is *locally removable* if $n \geq 4$ in the sense that given a point $p \in \gamma$, there is a neighborhood U of p in $\mathcal{M}$ such that each CR-function on $(U \cap \Gamma)\backslash\gamma$ extends holomorphically into $U \cap D$. This follows from the strict pseudoconvexity of the domain D and Proposition 3.4.3. As D is strictly pseudoconvex, there is a strictly pseudoconvex domain $D' \subset D$ such that $(\gamma \setminus K) \cap \overline{D}' = \emptyset, K \subset bD'$, and every CR-function on $bD \setminus (K \cup \gamma)$ extends holomorphically into $D \setminus D'$. As the removability of K does not depend on the strictly pseudoconvex domain in question–recall the remarks at the end of Section 2.1–we have holomorphic continuation from $bD'\backslash K$ into D'. The resulting continuation is holomorphic in the domain D.

If K is weakly removable, then $H^{n,n-1}(K) = 0$ by Theorem 2.3.2. We have $K \cup \gamma = K \cup \gamma'$ with $K \cap \gamma' = \emptyset$, whence $K \cap \overline{\gamma}'$ is totally disconnected. This implies $H^{n,n-2}(K \cap \overline{\gamma}') = 0$ when $n \geq 3$, and from this, by a Mayer-Vietoris argument of the kind we have used before, $H^{n,n-1}(K \cup \gamma) = 0$: Let φ be a $\overline{\partial}$-closed $(n, n-1)$-form defined and smooth on a neighborhood of $K \cup \gamma$. We must prove φ to be $\overline{\partial}$-exact on a neighborhood of $K \cup \gamma$. There is a neighborhood U_1 of K on which φ is $\overline{\partial}$-exact, say $\varphi = \overline{\partial}\psi_1$ there. There is also a neighborhood U_2 of $\overline{\gamma}'$ on which φ is $\overline{\partial}$-exact, say $\varphi = \overline{\partial}\psi_2$ there. There is then a neighborhood W of $K \cap \overline{\gamma}'$ on which $\psi_1 - \psi_2$ is $\overline{\partial}$-exact, say $\psi_1 - \psi_2 = \overline{\partial}g$ in W. Let λ be a smooth function defined near $K \cup \gamma$ with $\lambda = 1$ near $K\backslash W$, $\lambda = 0$ near $\overline{\gamma}'\backslash W$. The form $\psi = \lambda\psi_1 + (1 - \lambda)\psi_2 + \overline{\partial}\lambda \wedge g$ satisfies $\overline{\partial}\lambda = \varphi$ near $K \cup \gamma$. Thus $H^{n,n-1}(K \cup \gamma) = 0$, and $K \cup \gamma$ is weakly removable as we wish. $\qquad\qquad\square$

We discuss the removability of metrically thin sets below in Section 4.6.

4.2 Holomorphically Convex Singularities

The intrinsic sufficient conditions given in Theorems 3.1.1 and 3.3.1 imply the following result.

Proposition 4.2.1 *Let D be a relatively compact domain in a noncompact complex n-dimensional manifold $\mathcal{M}$, and let $E \subset \mathcal{M}$ be a compact holomorphically convex set such that $E \cap \overline{D} = K \subset bD$ and $D = int\,\overline{D}$. Let $bD\backslash K$ be locally a Lipschitz graph. Then K is is weakly removable if $n \geq 2$. If $n \geq 3$ and $H^{n,n-1}(\mathcal{M}) = 0$, then K is removable.*

The condition that $H^{n,n-1}(\mathcal{M}) = 0$ is satisfied if $\mathcal{M}$ is a Stein manifold and if $\mathcal{M}$ is completely q-convex with $q = 1$. See [4] or [37]. The definition of completely q-convex manifold is given in Section 4.3 below.

For strictly pseudoconvex domains we have intrinsic characterizations of removable and weakly removable sets, which yield the following corollary.

Corollary 4.2.2 *Let D be a relatively compact, strictly pseudoconvex domain with boundary of class C^2 in a two-dimensional Stein manifold. A compact subset K of bD is removable if and only if it is $\mathcal{O}(\overline{D})$-convex, and it is weakly removable, if and only if it is holomorphically convex.*

Among the concrete examples of holomorphically convex sets – actually Stein compacta – are the *minimum sets* of plurisubharmonic functions.

Proposition 4.2.3 *Let D be a relative compact domain in a Stein manifold of dimension n. Let $K \subset bD$ be a compact set such that bD is a strictly pseudoconvex hypersurface in a neighborhood U of K. Assume there is a positive plurisubharmonic function u on $D \cap U$ such that $\limsup_{z \to \zeta} u(\zeta) = 0$ for each $\zeta \in K$. Then K is a Stein compactum. Thus, if $bD \backslash K$ is locally a Lipschitz graph, the set K is weakly removable if $n \geq 2$ and removable if $n \geq 3$.*

A similar result has been given by Alexander [2].

Proof Let V be an arbitrary neighborhood of K with compact closure in U. Since $bD \cap U$ is strictly pseudoconvex, there is an open set $U' \subset U$ such that $K \subset U'$, $bU' \cap V$ is strictly pseudoconvex, and $bV \cap U'$ is a relatively compact subset of $U \cap D$. (The domain U' is obtained by pushing $bD \cap U$ away from D in a neighborhood of K and into D in a neighborhood of $bV \cap bD$.) As $u > 0$ in $U \cap D$, it follows that $u > \epsilon$ on $bV \cap U'$ for some constant $\epsilon > 0$. Set $W = (U' \backslash D) \cup \{z \in V \cap D : u(z) < \epsilon\}$. Then W is pseudoconvex at each point of its boundary and so is a Stein open set. Thus, as claimed, K is a Stein compactum. $\square$

Corollary 4.2.4 *Let D be a relatively compact strictly pseudoconvex domain with C^2 boundary in a Stein manifold of dimension two. Let D be a positive plurisubharmonic function in D such that $\limsup_{z \to \zeta} u(z) = 0$ for each point ζ in a compact set $K \subset bD$. Then K is removable.*

Proof Let $\hat{K}$ be the hull of K with respect to $\mathcal{O}(\overline{D})$. By strict pseudoconvexity, $\hat{K} \cap bD = K$. If $\hat{K} \neq K$, then $\sup_{\hat{K}} u = M > 0$. As u vanishes on K, there is a point $a \in \hat{K} \backslash K$ such that $u(a) = M$. This, however, contradicts the maximum principle for plurisubharmonic functions on hulls [38], [78], so $\hat{K} = K$, and the result follows. $\square$

A special class of sets provided by Proposition 4.2.3 are the maximum modulus sets. A closed subset $E \subset bD$ is a *maximum modulus set* if there is $f \in A(D) = \mathcal{C}(\overline{D}) \cap \mathcal{O}(D)$ with $E = \{z \in bD : |f(z)| = \max_{\overline{D}} |f|\}$.

Corollary 4.2.5 *Let D be a relatively compact, strictly pseudoconvex domain with C^2 boundary in a Stein manifold of dimension n. If $K \subset bD$ is a maximum modulus set then K is removable if $n \geq 3$ and is weakly removable for $n \geq 2$.*

Proof Take $u(z) = -\log\left(|f(z)|/\max_{\overline{D}}|f|\right)$ and apply Proposition 4.2.3. $\square$

It is proved in [81] that a maximum modulus set $K \subset bD$ is removable for $n = 2$ if it satisfies the additional hypothesis that $H^1(K, \mathbb{Z}) = 0$, but this result does not seem accessible by the methods of the present paper.

4.3 *q*-Convex Singularities

We follow here Lupacciolu [55]. A complex manifold $\mathcal{M}$ of dimension n is called *q-complete* or *completely q-convex*, $0 \le q \le n-1$, if there is an exhaustion function ρ of class C^2 the Levi form of which admits everywhere at least $q+1$ positive eigenvalues. (We are using the terminology of [37]; the same manifolds are called $(n-q)$-*complete* in [4] and $(n-q-1)$-*complete* in [55].) A compact set $E \subset \mathcal{M}$ is called *q-complete* if it admits a fundamental sequence of relatively compact q-complete neighborhoods. Andreotti and Grauert [4] proved that $H^{p,r}(\mathcal{M}) = 0$ for all $p \ge 0$ and all $r \ge n-q$ if $\mathcal{M}$ is q-complete. (See also [37].) Thus, if E is a q-complete compact set in a complex manifold, it follows that $H^{p,r}(E) = 0$ for all $p \ge 0$ and all $r \ge n-q$. We are interested mainly in the cases that $r = n-1$ or $n-2$, because these special cases of the Andreotti-Grauert theorem yield sufficient conditions for removability and weak removability. (See [55].)

Proposition 4.3.1 *Let D be a relatively compact domain in a noncompact complex manifold $\mathcal{M}$ of dimension n. Assume that $D = int\overline{D}$. Let $E \subset \mathcal{M} \setminus D$ be a compact set. If E is 1-complete and $n \ge 2$, then $K = E \cap bD$ is weakly removable. If E is 2-complete, if $n \ge 3$, and if $H^{n,n-1}(\mathcal{M}) = 0$, then K is removable.*

Proof The proof follows from the Andreotti-Grauert theorem and Theorems 3.1.1 and 3.3.1.   □

Specific classes of sets provided by Proposition 4.3.1 are considered below in Sections 4.4–4.6.

A compact set $E \subset \mathcal{M}$ is called *q-convex* (in $\mathcal{M}$) if for an arbitrary neighborhood $U \supset E$, there is an exhaustion function ρ of class C^2 the Levi form of which admits everywhere at least $q+1$ positive eigenvalues and such that $\rho < 0$ on E and $\rho > 0$ on $\mathcal{M} \setminus U$. (It follows in particular that both $\mathcal{M}$ and E are q-complete.) Note that for a Stein manifold $\mathcal{M}$, a compact set $E \subset \mathcal{M}$ is $(n-1)$-complete if and only if it is $\mathcal{O}(\mathcal{M})$-convex. (See [38].) Thus, the following result can be considered as a generalization for $n > 2$ of the condition of $\mathcal{O}(\overline{D})$-convexity in Theorem 2.1.1.

Proposition 4.3.2 *Let D be a relatively compact domain in a noncompact complex manifold $\mathcal{M}$ of dimension $n \ge 2$. Assume $D = int\overline{D}$. Let $E \subset \mathcal{M} \setminus D$ be a 1-convex compact set. Then the set $K = E \cap bD$ is removable.*

Proof By the Andreotti-Grauert theorem, $H^{n,n-1}(E) = H^{n,n-1}(\mathcal{M}) = 0$. Moreover, each continuous $\overline{\partial}$-closed $(n, n-2)$-form in a neighborhood of E is uniformly approximated in a (smaller) neighborhood of E by continuous $\overline{\partial}$-closed $(n, n-2)$-forms on $\mathcal{M}$. (See [37], Corollary 12.12.) By using regularization operators-see [16], we thus obtain that $\mathcal{Z}^{n,n-2}(\mathcal{M})$ is dense in $\mathcal{Z}^{n,n-2}(E)$. It follows by Theorem 3.1.1 that K is removable. □

This result was established by Lupacciolu [55] with a different proof.

4.4   Convexity with respect to one-dimensional varieties

Let X be a subset of the complex manifold $\mathcal{M}$, and let K be a subset of X. We say that K is *convex in X with respect to p-dimensional varieties* if for every point $a \in X \setminus K$ there is a purely p-dimensional closed complex analytic subset of a neighborhood of X that contains the point a and is disjoint from K. If X is $\mathcal{O}(\mathcal{M})$-convex and $\mathcal{M}$ is a Stein manifold, we can assume that these varieties are closed subvarieties of the whole $\mathcal{M}$. In this case if $H^2(\mathcal{M}, \mathbb{Z}) = 0$ so that the second Cousin problem is solvable on $\mathcal{M}$, convexity with respect to $(n-1)$-dimensional varieties, $n = \dim \mathcal{M}$, coincides with meromorphic convexity in $\mathcal{M}$, which is rational convexity if $\mathcal{M}$ is properly embedded in C^N.

In this section we are interested especially in the case of convexity with respect to one-dimensional varieties.

Proposition 4.4.1 *Let D be a relatively compact domain in a noncompact complex manifold $\mathcal{M}$ of dimension $n \geq 2$ such that $\overline{D} \setminus K$ contains no compact one-dimensional complex varieties. If $K \subset bD$ is a compact set that is convex in $\overline{D}$ with respect to one-dimensional varieties, and if $bD \setminus K$ is locally a Lipschitz graph, then K is weakly removable.*

A similar result, under the hypothesis of convexity with respect to *nonsingular* curves has been obtained by Lupacciolu (personal communication).

Proof   If $\mathcal{M}$ is a Stein manifold and D is strictly pseudoconvex, the proposition follows from Theorem 2.5.1 by the maximum modulus principle for analytic varieties. See [17] or [31].

Assume first that $n \geq 3$.

Let $a \in D$ and let A be a one-dimensional complex variety in a neighborhood of $\overline{D}$ such that $a \in A$ and $A \cap K = \emptyset$. By the result of the Appendix, we can suppose that A is nonsingular. We can also suppose that for some Stein neighborhood V of $A \cap \overline{D}$ disjoint from K, in $\mathcal{M}$, $A \cap V = f^{-1}(0)$ for a regular holomorphic map $f : V \to \mathbb{C}^{n-1}$. In particular, $A \cap V$ is an ideal-theoretic complete intersection in V. To see this, let V_1 be a neighborhood of A, which by Siu's theorem [76] we may assume to be Stein. By a theorem of Docquier and Grauert – see [31], there is a neighborhood $V_2 \subset V_1$ of A that is equivalent to the zero section of the normal bundle to the embedding $A \hookrightarrow V_1$. As holomorphic vector bundles on open Riemann surfaces are trivial, it follows that a Stein neighborhood of $A \cap \overline{D}$ is holomorphically equivalent to $A' \times \mathbb{B}_{n-1}$ for a neighborhood A' of $A \cap \overline{D}$ in A. The projection of the product onto $\mathbb{B}_{n-1}$ provides the necessary map.

The function $\varphi = 2 \log |f|$ is plurisubharmonic in V so its Levi form is positive semidefinite at each point of $V \setminus A$. This form has at least two zero eigenvalues at each point $b \in V \setminus A$. This is so, for after a unitary transformation in C^{n-1}, we can assume that $f_j(b) = 0, j = 2, \ldots, n-1$. Then $\partial |f|^2|_b = \overline{f_1(b)} df_1|_b$ whence

$$i\partial\overline{\partial}\varphi|_b = i|f_1(b)|^{-2} \sum_{2}^{n-1} df_j \wedge d\overline{f}_j|_b.$$

The space $\{v \in T_b\mathcal{M} : df_j(v) = 0, j = 2, \ldots, n-1\}$ is at least two-dimensional over C. As $i\partial\overline{\partial}\varphi|_b$ is positive semidefinite it follows that the number of nonzero eigenvalues is at most $n-2$.

We let $\epsilon > 0$ be so small that $A_\epsilon = \{z \in \overline{V} : |f(z)| < \epsilon\}$ has no limit points on $bV \cap \overline{D}$. Let $\lambda(t) \geq 0$ be a convex decreasing function on $(\epsilon, +\infty)$ such that $\lim_{t \longrightarrow \epsilon^+} \lambda(t) = +\infty$. If $\psi = \lambda \circ \varphi$, we have

$$i\partial\overline{\partial}\psi = (\lambda' \circ \varphi)i\partial\overline{\partial}\varphi + (\lambda'' \circ \varphi)i\partial\varphi \wedge \overline{\partial}\varphi.$$

This form is positive semidefinite on the kernel of $i\partial\overline{\partial}\varphi$, and it follows that $i\partial\overline{\partial}\psi$ has at least two nonegative eigenvalues at each point of $V \setminus A_\epsilon$. Let ρ be a smooth strictly plurisubharmonic exhaustion function for the Stein manifold V. Then $\rho + \psi$ is a smooth 2-convex exhaustion function for $V \setminus \overline{A}_\epsilon$, that is $V \setminus \overline{A}_\epsilon$ is a 1-complete manifold. It follows by a result of Andreotti and Grauert [4] that $H^{n,n-1}(V \setminus \overline{A}_\epsilon) = 0$. Note that here ϵ is taken sufficiently small.

By Theorem 3.3.1 applied to $A_\epsilon \cap D$ instead of D and $E = \overline{A_{2\epsilon}} \setminus A_\epsilon$ in the Stein manifold V, we find that every CR-function f on Γ that is orthogonal to $\overline{\partial}$-closed $(n, n-1)$-forms α on $\overline{D}$ with $\overline{D} \cap supp\,\alpha \subset A_\epsilon$, extends holomorphically to $D \cap A_\epsilon$. This extension to the point $a \in A$ does not depend on A (see the proof Theorem 2.3.2). Thus, this continuation of f along tubes around one-dimensional varieties defines in D a unique holomorphic function F with boundary values f on Γ.

Consider now the simpler case that $n = 2$. Let $a \in D$, and let A be a one-dimensional irreducible variety in a neighborhood of $\overline{D}$ that contains a and that is disjoint from K. By Siu's theorem used above, there is a domain W in $\mathcal{M}$ that is Stein and that contains the compact set $A \cap \overline{D}$. As A has codimension one in W, the domain $W \setminus A$ is also Stein. Let $W' \supset A \cap \overline{D}$ be a relatively compact subdomain of W. Then the set $X = bW' \cap \overline{D}$ in $W \setminus A$ is compact. Let $\widetilde{X}$ be the holomorphically convex hull of X with respect to the Stein manifold $W \setminus A$. Thus, $\widetilde{X}$ is a compact subset of $\mathcal{M}$ that satisfies $H^{2,1}(\widetilde{X}) = 0$. Moreover, $a \notin \widetilde{X}$. Let Δ be the component of $D \setminus \widetilde{X}$ that contains a, and let Γ_a be the part of the boundary of Δ not in $\widetilde{X}$ so that Γ_a is an open subset of Γ. Theorem 3.3.1 implies that $f|\Gamma_a$ extends holomorphically into Δ, so we again have continuation into a tube around $A \cap D$. The result follows as before.

The proof is complete. $\square$

4.5   Convexity with respect to p-dimensional varieties

In contrast with what one might expect, the convexity of a compact set $K \subset bD$ with respect to p-dimensional varieties in $\mathcal{M}$ with $p \geq 2$ does not guarantee the removability of K. This is known; see [51] and the example given below. It does, however, give continuation into a Riemann domain spread over D, which, as we shall see, can be multisheeted. The notion of one-sided extension property that we use is defined in Section 3.4.

Proposition 4.5.1 *Let $D = int\overline{D}$ be a domain the closure of which is a Stein compactum in the complex manifold $\mathcal{M}$ of dimension $n \geq 3$. Let K be a compact set in bD such that $\Gamma = bD \setminus K$ is connected and locally a Lipschitz graph and has the one-sided extension property. If K is convex in $\overline{D}$ with respect to p-dimensional varieties for some $p \geq 2$, then* *there is a Riemann domain $(\tilde{D}, \pi)$ spread over D such that every CR-function on Γ extends holomorphically into $\tilde{D}$. Moreover, π is one-to-one over $D \setminus \hat{K}_{\mathcal{O}(\overline{D})}$.*

The last property explains what we mean by holomorphic continuation from Γ to $\tilde{D}$: The domain $\tilde{D}$ can be endowed with a boundary $\tilde{\Gamma}$ in such a way that $\pi|(\tilde{D}, \tilde{\Gamma}) \rightarrow (D, \Gamma)$ is a locally biholomorphic mapping of manifolds with boundary, and the continuation assertion in the proposition means continuation of CR-functions of form $f \circ \pi$ from $\tilde{\Gamma}$ into $\tilde{D}$.

Proof As $\overline{D}$ is a Stein compactum, we can assume without loss of generality that $\mathcal{M}$ is a Stein manifold. Because of the one-sided extension property of Γ, there is a domain U attached to Γ that contains one-sided neighborhoods of every point in Γ, such that every CR-function on Γ extends holomorphically into U. (In this connection, recall the discussion in Section 3.4.) Let $(\tilde{U}, \pi)$ be an envelope of holomorphy of U. (Domains in Stein manifolds have envelopes of holomorphy according to [65].) Thus, $(\tilde{U}, \pi)$ is a Riemann domain over $\mathcal{M}$ with projection π, and there is an embedding $\iota : U \rightarrow \tilde{U}$ such that $\pi \circ \iota$ is the identity map on U. Every CR-function on Γ extends holomorphically to $\tilde{U}$.

We show, first, that $D \subset \pi(\tilde{U})$. To do this, let $a \in D$ be an arbitrary point, and let $A_o \ni a$ be a p-dimensional variety in a neighborhood of $\overline{D}$ such that $A_o \cap K = \emptyset$. There is then an irreducible p-dimensional variety A with boundary $bA \subset U$ such that $a \in A \subset A_o$. (See [17] for the notion of variety with boundary.) Then $\iota(bA)$ is an irreducible MC-cycle in $\tilde{U}$ in the sense of Harvey and Lawson [32]. As $\tilde{U}$ is a Stein manifold, it follows from the result of [32] that $\iota(bA)$ is the boundary of an irreducible, relatively compact p-dimensional variety in $\tilde{U}$ with the same boundary bA as A. By a boundary uniqueness theorem for analytic sets (see [32] or [17]), we have $A = \pi(\tilde{A})$. In particular then, $a \in \pi(\tilde{A}) \subset \pi(\tilde{U})$: Note as well that $\pi|\tilde{A} \rightarrow A$ is one-to-one over a neighborhood of bA. The map π is locally biholomorphic on $\tilde{U}$, so it follows that for each positive integer n, the set $\{a \in A : \pi^{-1}(a) \cap \tilde{A}$ consists of k points$\}$ is open. As $\tilde{A}$ is irreducible, we conclude that π is one-to-one on $\tilde{A}$ and so is a homeomorphism. Thus, as $\pi|\tilde{V} \rightarrow V$ is locally biholomorphic, it follows that for some neighborhoods $V \supset A$ and $\tilde{V} \supset \tilde{A}$, $\pi : \tilde{V} \rightarrow V$ is biholomorphic.

The union of all p-dimensional varieties of type $\tilde{A}$ described above fills a subset $\tilde{G} \subset \tilde{U}$, which is open. To see this, let $h : \tilde{U} \rightarrow C^N$ be a proper holomorphic embedding, and let τ be a holomorphic retraction of a neighborhood of $M = h(\tilde{U})$ in C^N onto M. If $A' = h(\tilde{A})$ for a relatively compact p-dimensional variety $\tilde{A}$ in $\tilde{U}$ with $b\tilde{A} \subset \iota(U)$, then the p-dimensional varieties of the form $r(A' + c), c \in C^N, |c| < \epsilon$, fill a neighborhood of A' in C^N, and $b(\tau(A' + c)) \subset h \circ \iota(U)$ if $\epsilon > 0$ is small enough. This means that $\tilde{G}$ is open. Let $\tilde{D}$ be the connected component of $\pi^{-1}(D) \cap \tilde{G}$ that contains $\iota(D \cap U)$. If $D \cap U = \emptyset$, we take $\tilde{D}$ to be the component of $\pi^{-1}(D) \cap \tilde{G}$ that has an open part of the boundary of $\tilde{U}$ in common with the boundary of U in $\tilde{U}$ over Γ. By the construction and the hypothesis, $\pi(\tilde{D}) = D$, so $(\tilde{D}, \pi)$ is a Riemann domain over D. As $\tilde{D} \subset \tilde{U}$, every function holomorphic in U, and, in particular, every CR-function on Γ extends holomorphically into $\tilde{D}$.

We now show that $\pi|\tilde{U}$ is one-to-one over $(D \cup U) \setminus \hat{K}$, where $\hat{K} = \hat{K}_{\mathcal{O}(\mathcal{M})}$. Thus, let $a \in (D \cup U) \setminus \hat{K}$, and let f be a holomorphic function on $\mathcal{M}$ such that $f(a) = 0 \notin f(\hat{K})$. Let $\tilde{A} = \{z \in \tilde{U} : f \circ \pi(z) = 0\}$ and $A = \{f = 0\} \cap int(\overline{D} \cup U)$. Let A_o be the union of the irreducible components of $\tilde{A}$ that intersect ι at each of its points (U). Then $\tilde{A}_o \cap \iota(U) =$

$\tilde{A} \cap \iota(U) = \iota(A \cap U)$, and thus $\pi(\tilde{A}_o) \cap int(\overline{D} \cup U) = A$. As $\pi|\tilde{A}_o \cap \iota(U) \to A \cap U$ is one-to-one and $\tilde{U}$ is Stein, it follows that the projection $\pi|\tilde{A}_o \to A$ is one-to-one as well. If we assume there are distinct points $b_1, b_2 \in \tilde{A}_o$ such that $\pi(b_1) = \pi(b_2)$, then we have an irreducible component $\tilde{A}_1$ of $\tilde{A}_o$ containing b_1. By the definition of $\tilde{A}_o$, the set $\tilde{A}_1 \cap \iota(U)$ is empty, and the manifold $\tilde{U} \setminus \tilde{A}_1 \supset U$ is pseudoconvex in $\tilde{U}$ because $\tilde{A}_1$ has codimension 1. But this contradicts the definition of $\tilde{U}$ as an envelope of holomorphy of U, and the contradiction implies that $\pi|\tilde{U}$ is one-sheeted over $(D \cup U) \setminus \hat{K}$.

The domain $(\tilde{D}, \pi)$ constructed above does not depend essentially on U and $\mathcal{M}$: If $U_1 \subset U$ and $\mathcal{M}_1 \subset \mathcal{M}$ are other domains with the properties described above and ι_1, π_1 are corresponding embedding and projection for $\tilde{U}_1$, then we have an obvious biholomorphism $\varphi : \tilde{D} \to \tilde{D}_1$ such that $\pi_1 \circ \varphi = \pi$. For an arbitrary point $a \in D \setminus \hat{K}_{\mathcal{O}(\overline{D})}$, there is a Stein neighborhood $\mathcal{M}_1 \supset \overline{D}$ such that $a \in D \setminus \hat{K}_{\mathcal{O}(\mathcal{M}_1)}$. It follows from what was proved above that $\pi|\tilde{D}$ is one-sheeted over $D \setminus \hat{K}_{\mathcal{O}(\overline{D})}$. $\qquad\qquad \square$

There arises the natural question, see [54], whether $\tilde{D}$ is really multisheeted over D. The following example shows that it can be, even in the case of strictly pseudoconvex domains in C^n.

Example Let $K = S_1 \times S_2$ where $S_1 = \{|z| = \frac{1}{\sqrt{2}}\}$ and $S_2 = \{|w| = \frac{1}{\sqrt{2}}\}$ are spheres in C_z^m and C_w^m, respectively. Then K lies on the unit sphere bB in C^{2m}, and $bB \setminus K = \Gamma_1 \cup \Gamma_2$ consists of two connected components, $\Gamma_1 \ni (e, 0)$ and $\Gamma_2 \ni (0, e)$, where $e = (1, 0, \dots, 0) \in C^m$. Let γ be a smooth arc in $(C^{2m} \setminus \overline{B}) \cup \{(e, 0), (0, e)\}$ connecting the points $(e, 0)$ and $(0, e)$ and orthogonal to bB at these points. Assume also that if φ is the function on C^{2m} given by $\varphi(z, w) = z_1 - w_1$, then $\gamma_1 = \varphi(\gamma)$ is an arc in the upper half plane $\Pi \subset C$ that connects the points 1 and -1. Assume finally, that the point $2i$ belongs to the relatively compact component of $\Pi \setminus \gamma_1$ in C. As $|z_1 - w_1| \leq |z| + |w| \leq 1 + \frac{1}{\sqrt{2}} < 2$ on $\Gamma_1 \cup \Gamma_2$, we can define a continuous argument of $z_1 - w_1 - 2i$ on $\Gamma_1 \cup \Gamma_2 \cup \gamma$ that takes values in the interval $(-\pi, 0)$ on Γ_1 and takes values in the interval $(\pi, 2\pi)$ on Γ_2. It follows that the function f given by

$$f(z, w) = \sqrt{z_1 - w_1 - 2i} = \sqrt{|z_1 - w_1 - 2i|}e^{iarg(z_1 - w_1 - 2i)},$$

with the argument function as described above, is holomorphic on a neighborhood of $\Gamma_1 \cup \Gamma_2 \cup \gamma$, and $Im f < 0$ on Γ_1, $Im f > 0$ on Γ_2. If $m > 1$, the hull of holomorphy of Γ_1 contains the domain $D_1 = \{|z| < 1, |w| < \frac{1}{\sqrt{2}}\}$, and the hull of holomorphy of Γ_2 contains the domain $D_2 = \{|z| < \frac{1}{\sqrt{2}}, |w| < 1\}$. It follows that the function $f \in \mathcal{O}(\Gamma_1 \cup \Gamma_2 \cup \gamma)$ extends holomorphically into D_1 and into D_2 with different values: $Im f < 0$ on D_1 and $Im f > 0$ on D_2. Thus, the domain of holomorphy of the function f has two different sheets over $\overline{D}_1 \cap \overline{D}_2 = \hat{K}$, the polynomially convex hull of K in C^{2m}.

For an arbitrary neighborhood $V \supset \gamma$ in C^{2m}, there is a strictly pseudoconvex domain D with $B \cup \gamma \subset D \subset B \cup V$, bD of class C^∞. Taking V so small that f is holomorphic on $\Gamma_1 \cup \Gamma_2 \cup \overline{V}$, we obtain a strictly pseudoconvex domain D with smooth boundary such that $K \subset bD$, $\Gamma = bD \setminus K$ is connected, and K is convex in C^{2m} with respect to m-dimensional varieties. But K is not removable, because the hull of holomorphy of Γ is not one-sheeted over $\hat{K}$.

The number $m = \frac{n}{2}$ that apears in the preceding example is essential indeed because of the following result, which extends a result from [81].

Proposition 4.5.2 *Let D and K be as in Proposition 4.5.1, and suppose K convex in $\overline{D}$ with respect to p-dimensional varieties with some $p \geq [\frac{n}{2}] + 1$. Then K is removable.*

Proof Let $U, \tilde{U}, \tilde{D}$ be as in the proof of Proposition 4.5.1, and let $\tilde{f}$ be the holomorphic continuation of the function $f \in \mathcal{O}(U)$ into $\tilde{D}$. Let $b_1, b_2 \in \tilde{D}$ satisfy $\pi(b_1) = \pi(b_2) = a \in D$. Then by the definition of $\tilde{D}$, there are p-dimensional varieties $A_1 \ni a, A_2 \ni a$ in $int(\tilde{D} \cup U)$ and $\tilde{A}_2 \ni b_1, \tilde{A}_2 \ni b_2$ in $\tilde{U}$ such that $\pi|\tilde{A}_1 \rightarrow A_1$ and $\pi|\tilde{A}_2 \rightarrow A_2$ are biholomorphic mappings, and $b\tilde{A}_j \subset \iota(U)$. As $A_1 \cap A_2 \ni a$ is nonempty and $p \geq [\frac{n}{2}] + 1$, the set $A_1 \cap A_2$ contains an irreducible variety $A_o \ni a$ of dimension at least one with $bA_o \subset U$. The function f_j given by $f_j = \pi_*(\tilde{f}|\tilde{A}_j) = \tilde{f} \circ (\pi|A_j)^{-1}$ is holomorphic on A_j. As $b\tilde{A}_j \subset \iota(U)$ and $\pi|\iota(U) \rightarrow U$ is one-to-one, we have $f_1 = f_2$ on bA_o. It follows that $f_1 = f_2$ on the whole of A_o, and, in particular, that $f_1(a) = f_2(a)$. Thus, we obtain that the values of $\tilde{f}$ on different sheets of $\tilde{D}$ over $a \in D$ are the same. Hence $\pi_*(\tilde{f})$ is a well defined holomorphic function, F, on D, which coincides with f in U. By the definition of U, it follows that every CR-function on Γ extends holomorphically into D, and the proof is complete. $\square$

4.6   Metrically Thin Singularities

By using convexity with respect to one-dimensional varieties, we can give further examples of removable and weakly removable singularities.

Proposition 4.6.1 *Let $D = int\overline{D}$ be a relatively compact domain in a noncompact complex manifold $\mathcal{M}$ of dimension $n \geq 2$, and let $K_o \subset bD$ be a compact subset convex in $\overline{D}$ with respect to 1-dimensional varieties. Let K_1 be a closed subset of $bD \setminus K_o$ with zero $(2n - 2)$-dimensional Hausdorff measure, and let $K = K_0 \cup K_1$. Assume $bD \setminus K$ to be locally a Lipschitz graph. If $\overline{D} \setminus K_o$ contains no compact one-dimensional varieties, the set K is weakly removable.*

The Hausdorff measure in question is computed with respect to a smooth Hermitian metric on $\mathcal{M}$.

The proof of this result, in the case $n = 2$, depends on a lemma.

Lemma 4.6.2 *Let $\mathcal{M}$ be a two-dimensional complex manifold, and let $A \subset \mathcal{M}$ be an irreducible, noncompact one-dimensional analytic subvariety of $\mathcal{M}$. There is a Stein neighborhood V of A such that some $h \in \mathcal{O}(V)$ has A as its zero locus.*

Proof The proof of this lemma depends on the fact that for a one-dimensional Stein space $A, H^2(A, \mathbb{Z}) = 0$. (When A is nonsingular, this is a standard result in the topology of open surfaces. The general case can be deduced from this.)

The variety A is irreducible, noncompact and one-dimensional and so is a Stein space. According to a theorem of Siu [76], A has a neighborhood basis $\{W_\alpha\}_{\alpha \in I}$ of Stein neighborhood in $\mathcal{M}$.

Locally A is defined by a single function. Let $\Omega = \{\Omega_j\}_{j \in J}$ be a locally finite open covering of $\mathcal{M}$ such that for each j, there is $f_j \in \mathcal{O}(\Omega_j)$ that defines $A \cap \Omega_j$ ideal-theoretically. Thus on $\Omega_j \cap \Omega_k$, we have $f_j = h_{jk} f_k$ for a zero-free function h_{jk} holomorphic on $\Omega_j \cap \Omega_k$. The functions h_{jk} define an element $[h]$ of the group $H^1(\mathcal{M}, \mathcal{O}^*)$: If $[h] = 0$, we are done; in general $[h]$ is not zero.

For a given $\alpha \in I$ the inclusion $W_\alpha \subset \mathcal{M}$ induces a map

$$\sigma_\alpha : H^1(\mathcal{M}, \mathcal{O}^*) \to H^1(W_\alpha, \mathcal{O}^*) .$$

Also, if $W_\beta \subset W_\alpha$, there is a map

$$\sigma_{\alpha\beta} : H^1(W_\alpha, \mathcal{O}^*) \to H^1(W_\beta, \mathcal{O}^*) .$$

These maps satisfy $\sigma_{\alpha\beta} \sigma_\alpha = \sigma_\beta$. If W_α has the property that $\sigma_\alpha([h]) = 0$, we are done–we can solve the induced Cousin II problem on W_α:

Since the $W'_\alpha s$ are Stein manifolds there are isomorphisms

$$\iota_\alpha : H^1(W_\alpha, \mathcal{O}^*) \to H^2(W_\alpha, \mathbb{Z})$$

that arise from the exact sequence

$$0 \to \mathbb{Z} \to \mathcal{O} \to \mathcal{O}^* \to 0$$

of sheaves. To verify that $\sigma_\alpha([h]) = 0$, it suffices to verify that $\iota\sigma_\alpha([h]) = 0$.

Let $\widetilde{[h]}_\alpha = \iota_\alpha \sigma_\sigma([h])$.

Let

$$\widetilde{\sigma}_{\alpha\beta} : H^2(W_\alpha, \mathbb{Z}) \to H^2(W_\beta, \mathbb{Z})$$

be given by $\widetilde{\sigma}_{\alpha\beta} = \iota_\beta \sigma_{\sigma\beta} t_\sigma^{-1}$, the map induced by the inclusion $W_\beta \subset W_\alpha$. Also, let

$$\tau_\alpha : H^2(W_\alpha, \mathbb{Z}) \to H^2(A, \mathbb{Z})$$

be the map induced by the inclusion $A \subset W_\alpha$. Since $\{W_\alpha\}_{\alpha \in I}$ is a neighborhood basis for A we have that $H^2(A, \mathbb{Z}) = $ direct limit $H^2(W_\alpha, \mathbb{Z})$.

Each $\tau_\alpha = 0$, so the cohomology class $\widetilde{[h]}$ represents the zero element of the direct limit of $H^2(W_\alpha, \mathbb{Z})$. By the definition of the direct limit, this means that $\widetilde{[h]}$ is zero in some $H^2(W_\alpha, \mathbb{Z})$ *i.e.*, that $\iota_\alpha \sigma_\alpha([h]) = 0$ for some α. On W_α and all subsequent W's our Cousin problem is solvable.

The Lemma is proved. $\qquad\qquad\qquad\qquad\qquad\qquad\qquad\qquad\qquad\qquad$ $\square$

Proof of the Proposition Let $a \in \overline{D} \setminus K_o$, and let $A \ni a$ be a one-dimensional variety in a neighborhood of $\overline{D}$ such that $A \cap K_o = \emptyset$, and such that A is the set of common zeros of the functions $f_1, \ldots, f_{n-1}$ holomorphic on a Stein neighborhood V of A with $V \cap K = \emptyset$. (In the case that $n \geq 3$, we invoke again the theorem of Siu [76] and the discussion of the preceding section. For the case $n = 2$, we use the preceding lemma.) Let $f_n \in \mathcal{O}(V)$ be a function such that $A \cap \{f_n = 0\} = \{a\}$. We have a holomorphic mapping $f : z \mapsto [f_1, \ldots, f_n]$ of $V \setminus \{a\}$ into P^{n-1}. It follows from the theorems of Fubini and Sard that $f(K_1 \cap V)$ has zero $(2n - 2)$-dimensional measure in P^{n-1}. In particular, this set has everywhere dense complement. For $b \in P^{n-1}$ close enough to $f(A \setminus \{a\}) = [0, \ldots, 0, 1]$, the one-dimensional variety $f^{-1}(b) \cup \{a\}$ in V has compact intersection with $\overline{D} \cap V$. If we take such a b that is not in $f(K_1)$, we obtain a one-dimensional variety $A' \ni a$ in a neighborhood of $\overline{D}$ such that $A' \cap K = \emptyset$. This means that K is convex in $\overline{D}$ with respect to one-dimensional varieties. The result follows from Proposition 4.4.1. $\square$

Corollary 4.6.3 *If $K \subset bD$ has zero $(2n - 2)$-dimension measure, then K is weakly removable.*

This corollary was proved in [58] and [51] in the case of domains in C^n.

To obtain removability, we need lower metric dimension and additional conditions on bD.

Proposition 4.6.4 *Let $D = int\ \overline{D}$ be a relatively compact domain in a noncompact complex manifold $\mathcal{M}$ of dimension $n \geq 2$. Let $K_o \subset bD$ be a universally removable compact set and let K_1 be a closed subset of $bD \setminus K_o$ with zero $(2n - 3)$-dimensional Hausdorff measure. Assume that $bD \setminus K_o$ is locally a Lipschitz graph and that $(V \cap bD) \setminus K_1$ it has the one-sided extension property for some neighborhood $V \supset K_1$. Then the set $K = K_o \cup K_1$ is removable.*

Proof By Lemma 3.4.1 and Theorem 3.1.1, it is enough to show that K_1 is removable with respect to $bD \setminus K_0$. Thus, let $a \in K_1$, and let (U, z) be a coordinate neighborhood of a such that $z(a) = 0$ and such that the hypersurface $bD \cap U$ is the graph

$$x_1 = h(y_1, z_2, \ldots, z_n)$$

of a Lipschitz function h in a neighborhood of the origin. We assume that the set $K \cap U \cap \{z_3 = \ldots = z_n = 0\}$ has zero length. This can be achieved by making a unitary transformation of local coordinates. Shrinking U lets us suppose that $U = U' \times U'' \subset V \subset C^2 \times C^{n-2}$ with coordinates $z' = (z_1, z_2), z'' = (z_3, \ldots, z_n)$ and the projection $K \cap U \longrightarrow U'$ is proper. This can be done, for the set $K \cap U \cap \{z'' = 0''\}$ has zero length.

By the one-sided extension property there is a domain $W \subset U$ containing one-sided neighborhoods of each point in $bD \cap (U \setminus K)$ such that every CR-function on $(V \cap bD) \setminus K$ extends holomorphically to W. See Lemma 3.4.1. Fix such a function $f \in \mathcal{O}(W)$.

We can take $U' = I \times U_1$ where U_1 is a subset of $R \times \mathbb{C}$ with coordinates y_1, z_2, and I is an interval in the x_1-axis. Let $B \Subset U_1$ be a ball with center at the origin such that $I \times bB \times 0''$ does not intersect K. The projection of W onto $R \times C^{n-1}$ contains $bB \times U''$, so we can choose a smooth function φ on $bB \times U''$ such that its graph $S : x_1 = \varphi(y_1, z_2, z'')$ belongs to W. Let $S_b = S \cap \{z'' = b\}$, and let $\widehat{S}_b$ be its polynomially convex hull. By a theorem of Bedford and Gaveau [9] and Shcherbina [74], $\widehat{S}_b$ is the graph of a Lipschitz function, and $\widehat{S} = \cup \widehat{S}_b$ has the form $x_1 = \hat{\varphi}(y_1, z_2, z'')$ for some Lipschitz function $\hat{\varphi}$ on $\overline{B} \times U''$. As W is open, we can vary φ so that $\widehat{S}$ does not contain the origin. Then there is a domain G between bD and $\widehat{S}$ such that $0 \in bG \subset \widehat{S} \cup W \cup K$, and $bG \setminus \widehat{S}$ is a Lipschitz graph over some domain in $I \times B \times U''$.

The set $K \cap (U' \times 0'')$ has zero length, so its union with an arbitrary polynomially convex compact set is also polynomially convex. Thus $\widehat{S}_0 \cup (K \cap (U' \times 0''))$ is a polynomially convex set on the boundary of the two-dimensional domain $G \cap \{z''\} = 0''$, and the function f is holomorphic in a neighborhood of the rest of the boundary because it belongs to W. It follows from Theorem 3.1.1 that $f(z', 0'')$ extends holomorphically into $G \cap \{z'' = 0\}$.

The same is true for an arbitrary $b'' \in U''$ such that $K \cap (U' \times \{b''\})$has zero length. The set of such b'' has full measure in U'', and if follows from a generalization of Hartogs's lemma [67] that f extends holomorphically into G. Thus we have proved that K_1 is removable in a neighborhood of an arbitrary point, and thus that K_1 is removable. $\square$

Corollary 4.6.5 *If $K \subset bD$ has zero $(2n - 3)$-dimensional measure, then K is removable.*

For strictly pseudoconvex domains in C^n, this was proved in [58] and for domains with connected boundary in Stein manifolds in [57].

Remark *It would be more natural, in the spirit of the present paper, to prove that $H_c^{n,n-1}(K_1) = 0$ in Proposition 4.6.1 and that $H_c^{n,n-1}(K_1) = {}^{\sigma}H_c^{n,n-2}(K_1) = 0$ in Proposition 4.6.4 and then apply Propositions 3.4.2 and 3.4.3. Then one could drop the additional conditions on bD that we used in the proofs above. However, the triviality of these cohomology groups remains open.*

Corollary 4.6.6 *Let D be a relatively compact strictly pseudoconvex domain in a Stein manifold $\mathcal{M}$ of dimension $n \geq 2$. If K_o is a removable compact subset of bD and if K_1 is a closed subset of $bD \setminus K_o$ with zero $(2n - 3)$-dimensional Hausdorff measure, then the set $K = K_1 \cup K_2$ is removable.*

The proof follows from Proposition 4.6.4 and the necessity part of Theorem 2.3.1. $\square$

4.7 Manifolds of codimension three

Let now K be a compact subset of some C^1-submanifold $M \subset bD$, D a domain in the complex manifold $\mathcal{M}$. If the real codimension of M in $\mathcal{M}$ is more than three, then its metric dimension is not more than $2n - 4$, and thus K is removable by Proposition 4.6.4, granted some mild assumptions on bD near K. Thus, the smallest dimension of M for which its removability does not follow from Proposition 4.6.4 is $dim M = 2n - 3$. In this case, there is an obvious obstruction for removability: If f is a function holomorphic on a neighborhood of $\overline{D}$ and the set $D \cap \{f = 0\}$ is not empty and has boundary $M \subset bD$, then M is not removable because $1/f$ does not extend holomorphically into D. In this case M is a closed, maximally complex (MC)-manifold or an MC-cycle in the sense of Harvey and Lawson [32] if it has singularities of zero $(2n - 3)$-dimensional measure. Recall that a $(2p + 1)$-dimensional $\mathcal{C}^1$-manifold M is called *maximally complex* if its complex tangent planes $T_a^c M = T_a M \cap J(T_a M)$ have complex dimension p for all $a \in M$. (Here J denotes the complex structure operator on the tangent bundle $T\mathcal{M}$.) This appears to be the only nonremovable situation for $dim M = 2n - 3$. We formulate a result for a somewhat more general class of objects.

Let $M = M_\sigma \cup M_r$ be a closed subset of a smooth manifold $\mathcal{M}$ where M_σ is a closed set of zero m-dimensional measure (the singular part of M) and M_r is a closed m-dimensional $\mathcal{C}^1$-submanifold in $\mathcal{M} \setminus M_\sigma$, not necessarily connected but everywhere dense in M. Then we call M a *manifold with singularities M_σ*. If $\mathcal{M}$ and M_r are oriented and M has finite m-dimensional Hausdorff measure, then M defines an m-dimensional current $[M]$ of integration on M_r by

$$\langle [M], \varphi \rangle = \int_{M_r} \varphi$$

for $\varphi \in \mathcal{D}^m(\mathcal{M})$. We say that M is an *m-cycle* if this current is closed: $d[M] = 0$, i.e., $\langle [M], d\psi \rangle = 0$ for $\psi \in \mathcal{D}^{m-1}(\mathcal{M})$. As usual, M' is an *m-subcycle* of M if $M' \subset M$ and M' is also an m-cycle in $\mathcal{M}$. For $\mathcal{M}$ complex, a cycle M is called *maximally complex* – for brevity we speak of MC-cycles – if its regular part is maximally complex.

Theorem 4.7.1 *Let $D = int\overline{D}$ be a relatively compact domain in a noncompact complex manifold $\mathcal{M}$ of dimension $n \geq 3$. Let $K_o \subset bD$ be a universally removable compact set, and let $M \subset bD \setminus K_o$ be a $(2n - 3)$-dimensional manifold with singularities M_σ in $\mathcal{M} \setminus K_o$. Assume that for some neighborhood V of M in $\mathcal{M}$, the set $bD \cap V$ is locally a Lipschitz graph and that $(bD \cap V) \setminus M$ has the one-sided extension property. A compact set $K \subset K_o \cup M$ is removable if it contains no maximally complex connected component of $M \setminus M_\sigma$.*

We separate the proof of this result into two parts, each of which is of interest in its own right.

Lemma 4.7.2 *Let $M \subset bD$ be a $\mathcal{C}^1$-submanifold of dimension $2n - 3$, M generic and closed in some open set $V \subset \mathcal{M}$, $dim \mathcal{M} \geq 3$. Assume that $(bD \setminus M) \cap V$ has the one-sided extension property. Then there is an open set W attached to $bD \cap V$ such that every CR-function on $bD \cap V \setminus M$ extends holomorphically into W, i.e., M is removable with respect to $bD \cap V$ in the sense of the Definition of Section 3.4.*

Recall that M is *generic* if $dim T_a^c M \leq max(0, n - codim M)$ at each point $a \in M$.

Proof As $(bD \setminus M) \cap V$ has the one-sided extension property, there is an open set $W_o \subset \mathcal{M}$ attached to $(bD \setminus M) \cap V$ such that every CR-function on $(bD \setminus M) \cap V$ extends holomorphically into W_o. (See Section 3.4.) Fix such a function f that we take to be holomorphic on W_o.

Fix a point $a \in M$. As M is generic and of codimension three, there are neighborhoods $U_o \supset \overline{U} \ni a$ and a C^2-function $u \geq 0$ on U_o such that $M \cap \overline{U} = U_o \cap \{u = 0\}$ and the Levi form of u at each point of U_o has at least three positive eigenvalues. (See [18], Proposition 6.5.). By shrinking U and U_o as necessary and replacing u by $\lambda \circ u$ for a suitable function λ with $\lambda(t) \geq 0, \lambda(0) = 0$ and $\lambda'' > 0$, we can assume that U is a Stein domain, that U_o is 2-complete, and that u is an exhaustion function for U_o. With these assumptions, $M \cap \overline{U}$ is 2-convex in U_o.

Let $u_1 \geq 0$ be a smooth function with compact support in U_o such that $M \cap bU = U_0 \cap \{u_1 = 0\}$. Then for $\epsilon > 0$ sufficiently small, the function $v = u + \epsilon u_1$ is an exhaustion function for U_o as well, its Levi form has at least three positive eigenvalues at each point of U_o, and $M \cap bU = U_o \cap \{v = 0\}$. Thus, we have shown that $M \cap bU$ is also 2-convex in U_o. It follows – see [37], Corollary 12.12 – that every $\bar{\partial}$-closed smooth $(n, n-3)$-form in a neighborhood of $M \cap bU$ is uniformly approximated (on a smaller neighborhood) by continuous $(n, n-3)$-forms on U_o. We also have by a result of Andreotti and Grauert [4] that $H^{n,q}(M \cap \overline{U}) = H^{n,q}(M \cap bU) = 0$ for $q \geq n - 2$. It follows from 1.16) that $H_c^{n,n-1}(M \cap U) = H_c^{n,n-2}(M \cap U) = 0$. We then obtain by 1.18) that $H^{0,1}(U \setminus M) = 0$.

Therefore, every CR-function on $(bD \cap U) \setminus M$ is represented as a jump of boundary values of holomorphic functions on components of $U \setminus bD$. See [15]. Let $f^{\pm}$ be such functions for our function f. As $f \in \mathcal{O}(W_o)$, we have that both of $f^{\pm}$ extend holomorphically into $W_o \cap U$ with the same difference: $f^+ - f^- = f$.

We can assume now that U is a coordinate neighborhood of a with coordinates $z, z(a) = 0$, and that $bD \cap U$ is a graph, say $x_1 = h(y_1, z_2, z'')$ for some Lipschitz function h. Moreover, as the Hausdorff $(2n - 3)$-measure of $M \cap U$ is finite, we can assume, after a linear transformation that $M \cap U \cap \{z'' = 0''\}$ has finite length. Then there is a small $r > 0$ such that the set $M \cap U \cap \{(z', 0'') : y_1^2 + |z_2|^2 = r^2\}$ has zero length. Let S be a graph of a continuous function φ over bB the boundary of the ball $B = \{y_1^2 + |z_2|^2 < r^2, z'' = 0''\}$ belonging to $(W_o \cup M) \cap U$. By a theorem of Shcherbina [74], [75] there is a continuous function $\hat{\varphi}$ on $\overline{B}$ whose graph $\widehat{S} = \{x_1 = \hat{\varphi}(y_1, z_2), z'' = 0''\}$ has S as boundary, and $\widehat{S}$ is foliated into a family of holomorphic discs. As W_0 is open and $M \cap U$ has zero length over bB, we can vary φ in such a way that $0 \notin \widehat{S}$.

Thus, we can assume that $\hat{\varphi}(0, 0) < 0$. The sets $S + (t, 0, 0''), t > 0$, belong then to $int(\overline{U}^+ \cup W_o) \cap U$ where $U^+ = U \cap \{x_1 > h(y_1, z_2, z'')\}$, provided S is close enough to bD. If $r > 0$ is small enough, we have then that $\widehat{S} + (t_1, 0, 0'') \subset U^+$ for some $t_1 > 0$. But then $0 \in \widehat{S} + (t_o, 0, 0'')$ for some t_o in the interval $(0, t_1)$, and we obtain by the Kontinuitätssatz that the function f^+ extends holomorphically through bD at the point a. It follows that f itself extends holomorphically to a one-sided neighborhood of a in $U \setminus bD$ belonging to U^-. As $a \in M$ is arbitrary, the lemma is proved. $\square$

Remark *The proof of Lemma 4.7.2 simplifies essentially if we assume that for an arbitrary point $a \in M$, there is a Stein neighborhood, U such that one of the two components of $U \setminus bD$ extends holomorphically through bD at the point a. Then we obtain at once, from the representation $f = b.v.f^+ - b.v.f^-$ that f extends holomorphically into a one-sided neighborhood of a in $U \setminus bD$. But it can happen that in a neighborhood of a, bD is Levi* *flat, and then we have to use the global property of $bD \backslash M$ as in the proof just given.*

Lemma 4.7.3 *Let $M \subset bD$ be a connected C^1-manifold of dimension $2n - 3, n \geq 2$, that is closed in some open subset $V \subset M$. Assume that $(bD \setminus M) \cap V$ has the one-sided extension property. Let $K \subset M$ be a proper compact subset of M. Then every CR-function on $(bD \cap V) \setminus K$ extends holomorphically into some open set W attached to $bD \cap V$. That is, K is removable with respect to $bD \cap V$.*

Proof Fix a function $f \in CR((bD \cap V) \setminus K)$. It extends into some open set W_o attached to $(bD \cap V) \setminus M$. Let K_1 be the set of points $z \in K$ such that f extends holomorphically into a one-sided neighborhood of z in $V \setminus bD$. Set $K_2 = K \setminus K_1$. We shall show that the compact set K_2 is empty.

Suppose it is not. Then there is a point $a \in K_2$ in the boundary of K_2 with respect to M. Choose a coordinate neighborhood $U \ni a$ as in Lemma 1 such that $bD \cap U$ is a graph $\{x_1 = h(y_1, z_2, z'')\}$ over a convex domain $G \subset R \times \mathbb{C}^{n-1}$. Assume the level sets z'' =const. are transverse to M in U. After a linear change of coordinates, we can assume that $K_2 \cap U \cap \{z'' = 0''\}$ is a nonempty proper subset of the arc $M \cap U \cap \{z'' = 0\}$ and $0 \in K_2$ is not an interior point of $K_2 \cap U \cap \{z'' = 0''\}$ in this arc.

Let $B = B(0, r)$ be a ball in G and let S be a graph of a continuous function φ over bB that is contained in $(W_o \cup K) \cap U$. By a theorem of Shcherbina [74] and [75] there is a continuous function $\hat{\varphi}$ on $\overline{B}$ such that $\hat{\varphi} = \varphi$ on bD and the graph $\widehat{S}$ of $\hat{\varphi}$ is foliated into holomorphic discs lying in planes $z'' = $ const. As W_o is open, we can vary φ in such a way that $0 \notin \widehat{S}$. If $r > 0$ is small enough and the graph of φ is close enough to bD, there is a domain $D' \subset U$ with boundary $bD' \subset \widehat{S} \cup \Gamma'$ where $\Gamma' \subset W_o \cup M$ is a Lipschitz graph over some ball $B(0, r') \subset B, 0 \in \Gamma'$, and D' contains a one-sided neighborhood of 0 in $U \setminus bD$.

By the construction, the set $K_2 \cap U \cap \{z'' = c''\}$ is a proper subset of the arc $M \cap U \cap \{z'' = c''\}$ for each c'' sufficiently near the origin. By Shcherbina's theorem, the sets $\widehat{S} \cap \{z'' = c''\}$ are polynomially convex. As $0 \notin \widehat{S}$ and 0 is a boundary point of $K_2 \cap U \cap \{z'' = c''\}$ in the arc $M \cap U \cap \{z'' = c''\}$, all the sets $(K_2 \cup \widehat{S}) \cap U \cap \{z'' = c''\}$ are polynomially convex when $|c''|$ is small by a theorem of Stolzenberg [79]. By Theorem 3.1.1, the function f, extends holomorphically into $D' \cap \{z'' = c''\}$ in the (z_1, z_2)-variables, for it is holomorphic in $bD' \setminus (K_2 \cup \overline{S})$. It follows from Hartogs's lemma that these extensions constitute a holomorphic extension of f into D'. This contradicts the definition of K_2. Thus, K_2 is empty, and the lemma is proved. $\qquad\qquad\square$

The proof of Theorem 4.7.1 follows easily from these two lemmas:

Fix a function $f \in CR(bD \setminus K)$ and denote by M' the set of points $z \in M$ such that f extends holomorphically into a one-sided neighborhood of z in $M \setminus bD$. By Lemma 4.7.2, the set M' contains all points of $M \setminus M_\sigma$ at which M is generic. If M_1 is a component of $M \setminus M_\sigma$ such that either $M_1 \cap M'$ or $M_1 \cap K$ is not empty, then $M_1 \subset M'$ by Lemma 4.7.3. As f extends holomorphically into an open set W' attached to $(bD \setminus M) \cup M'$, we can vary bD in a neighborhood of M' into a smooth hypersurface $bD' \subset bD \cup W'$ so that f is a CR-function on the boundary of bD' of a domain $D' \subset D \cup W'$ with $D \subset D' \cup W'$. By invoking Trépreau's theorem [82], we can assume that $bD' \setminus (K_o \cup M_\sigma)$ has the one-sided extension property. Then by Proposition 4.6.4, f extends holomorphically into D.

This completes the proof. $\qquad\qquad\qquad\qquad\qquad\qquad\qquad\qquad\qquad\qquad\qquad\qquad\qquad\qquad$ $\square$

Corollary 4.7.4 *A compact manifold $M \subset bD$ with singularities M_σ is removable if $M \setminus M_\sigma$ contains no maximally complex connected components.*

For strictly pseudoconvex domains, this property is essentially necessary:

Proposition 4.7.5 *Let D be a relatively compact strictly pseudoconvex domain with C^2 boundary in a Stein manifold $\mathcal{M}$ of dimension $n \geq 3$. Let $K_o \subset bD$ be an $\mathcal{O}(\overline{D})$-convex compact set and let M be a $(2n-3)$-cycle in $\mathcal{M} \setminus K_o$ such that the closure of each connected component of $M \setminus M_\sigma$ is also a $(2n-3)$-cycle in $\mathcal{M} \setminus K_o$. Then a compact set $K \subset K_o \cup M$ is removable if and only if $K \cap M$ contains no MC-subcycles of M.*

Proof We can assume that $\overline{D}$ is $\mathcal{O}(\mathcal{M})$-convex. Then K_0 is $\mathcal{O}(\mathcal{M})$-convex and so is universally removable. Thus, the sufficiency of the conditions follows from Theorem 4.7.1.

For the necessity, let $K \subset K_o \cup M$ be removable, and $M_1 \subset K \cap M$ be an MC-cycle of dimension $2n-3$. As each point of bD is a peak point for $\mathcal{O}(\overline{D})$, the set $K \cap K_o$ is $\mathcal{O}(\mathcal{M})$-convex.[10] By a generalization of a theorem of Harvey and Lawson – see [17], Theorem 19.6.2 – there is a complex analytic set $A \subset D$ of dimension $n-1$ such that $bA \subset K$. As D is strictly pseudoconvex, there is a strictly pseudoconvex domain $D' \supset D$ such that $bD \cap \overline{D} = K$. The domain $D' \setminus A$ is locally pseudoconvex because $dim\, A = n-1$, and thus it is the domain of holomorphy of a function $f \in \mathcal{O}(D' \setminus A)$. Then f is holomorphic in a neighborhood of $bD \setminus K$, but it does not extend holomorphically into D. This contradiction shows that the removable set K cannot contain any such M_1. $\qquad\qquad$ $\square$

Remark *If $K_o \cup M_\sigma$ is $\mathcal{O}(\overline{D})$-convex, then K is removable if and only if it contains no MC-component of $M \setminus M_\sigma$. (The proof of this assertion is the same as the proof just given.)*

Corollary 4.7.6 *Let D and K_o be as in Proposition 4.7.5, and let the $(2n-3)$-cycle M that is closed in $\mathcal{M} \setminus K_o$ have connected regular part $M \setminus M_\sigma$. Then a compact set $K \subset K_o \cup M$ is removable if and only if $M \not\subset K$. In particular, a $(2n-3)$-cycle $M \subset bD$ that is closed in $\mathcal{M}$ and that has connected regular part is not removable if and only if M is an MC-cycle.*

For a manifold $M \subset bD$ of class C^2, this result was announced by Jöricke [42].

Theorem 4.7.1 and Proposition 4.7.5 are valid for the case $n = 2$ as well if we understand maximally complex 1-cycles in the sense of Harvey and Lawson [32]. We have:

Corollary 4.7.7 *Let D, K_o be as in Proposition 4.7.5, $n = 2$, and let γ be a locally finite union of C^1 or rectifiable curves in $bD \setminus K_o$. A compact set $K \subset K_o \cup \gamma$ is removable if and only if K contains the boundary of no one-dimensional complex analytic variety in D.*

---

[10] The point is that if $E \subset bD$ is $\mathcal{O}(\overline{D})$-convex and $E_o \subset E$ is closed, then E_o is $\mathcal{O}(\overline{D})$-convex: If $\dagger$ denotes $\mathcal{O}(\overline{D})$-hull, then we have $bD \supset E = E^\dagger \supset E_o^\dagger$. However, since each point of bD is a peak point for $\mathcal{O}(\overline{D})$, $E_o^\dagger \cap bD = E_0$, and we have our assertion.

Proof By Theorem 2.1.1, K is removable if and only if K is $\mathcal{O}(\overline{D})$-convex. By a theorem of Stolzenberg [79], the set described in the corollary is $\mathcal{O}(\mathcal{M})$-convex when the curves in question are of class $\mathcal{C}^1$. When they are merely rectifiable, it is necessary to invoke the work of Alexander [1] to obtain the corresponding result.

We can admit as well a set of singularities γ_σ of length zero, because $K_o \cup \gamma_\sigma$ is then polynomially convex, and we can replace K_o by $K_o \cup \gamma_\sigma$. $\square$

4.8 Manifolds of codimension two

Every compact subset of a generic manifold of codimension two in a complex manifold is 1-complete. (See [18], Proposition 6.5.) Thus, the following result is a corollary of Proposition 4.3.1.

Proposition 4.8.1 *Let $D = int\overline{D}$ be a relatively compact domain in a noncompact complex manifold $\mathcal{M}$ of dimension $n \geq 2$. If $K \subset bD$ is a compact subset of a generic $\mathcal{C}^1$-submanifold M of bD that is of codimension two in $\mathcal{M}$, then K is weakly removable.*

The boundary of a smoothly bounded domain M_o in M is a manifold of codimension three, but it is not generic when $n \geq 3$, for it has points at which its CR-dimension is $n - 2$, and accordingly, we are unable to prove conditions like $H_c^{n,n-1}(M_o) = 0$. Thus, the local variant of Proposition 1, in the spirit of Section 3.4, is an open problem as also is the following question: *Let M be a $\mathcal{C}^1$-submanifold of bD, $\dim M = 2n - 2$, and let M^C be the set of points $a \in M$ at which M is not generic, i.e., $T_a M$ is a complex subspace of $T_a\mathcal{M}$. Assume that the set M^C is weakly removable. Is then M itself also weakly removable?*

Manifolds of codimension two in $\mathcal{M}$ divide bD locally, and they may divide it globally. This is one of the reasons that the problem of removability of such manifolds is so complicated. Some of the hypotheses of the following result seem inessential, but we do not know how to dispense with them.

Theorem 4.8.2 *Let $D = int\overline{D}$ be a relatively compact domain in a noncompact complex manifold $\mathcal{M}$ of dimension $n \geq 3$, and let $\mathcal{M}\backslash\overline{D}$ be connected. Let $K_o \subset bD$ be a compact universally removable set, and let M be a closed submanifold of dimension $2n - 2$ of the open subset $bD \setminus K_o$ of bD. Assume $bD\backslash K_o$ to be locally a Lipschitz graph and M to be of class $\mathcal{C}^{2+\epsilon}$ for some $\epsilon > 0$. Let K_1 be a closed subset of M such that:*

a. K_1 contains no components of M.

b. $K_o \cup K_1$ is compact.

c. M is generic and minimal at all points of K_1.

d. bD is a manifold of class $\mathcal{C}^1$ in some neighborhood U of K_1.

e. $(bD \cap U) \setminus (K_o \cup K_1)$ has the one-sided extension property.

Then the set $K = K_o \cup K_1$ is removable.

We recall that M is called *minimal* at a point $a \in M$ (in the sense of Tumanov [83]) if there is no CR-submanifold $M_1 \subset M$ through a such that $\dim M_1 < \dim M$ and CR-$\dim M = CR$-$\dim M$.

Proof Fix a function $f \in CR(bD \setminus K)$ and denote by K_1' the set of points $a \in K_1$ such that f extends holomorphically into a one-sided neighborhood of a with respect to bD. We have to show that $K_1 = K_1'$, for then the removability of K follows from Section 3.4 and the universal removability of K_o. As $bD \cap U \setminus K$ has the one-sided extension property, we can assume by using a suitable deformation of bD as in Section 3.4 that f is holomorphic in a neighborhood U_1 of $(bD \cap U) \setminus K_1''$, where $K_1'' = K_1 \setminus K_1'$. If we assume K_1'' nonempty, we have a point a that belongs to the boundary of K_1'' with respect to M. (Here we use the property that K_1 contains no connected components of M.) Our statement $K_1 = K_1'$ is local, so we assume that $a = 0 \in C^n$ and that near 0, bD is given by the equation $y_n = h(z', z_{n-1}, x_n)$ with some C^1-function h. We have assumed M to be generic at 0, so we can choose coordinates such that near 0, M is defined by an additional equation $y_{n-1} = h_1(z', x_{n-1}, x_n)$. The minimality of M implies, by a theorem of Tumanov [83], that there are small numbers $\tau_o > \tau > 0$ and an open cone V in the plane $z' = x_{n-1} = x_n = 0$ such that every continuous CR-function on $M \cap \{|z| < \tau_o\}$ extends holomorphically into the wedge

$$W_0 = \{z = \zeta + \eta : \zeta \in M, \eta \in V, |z| < \tau\}.$$

Since V is an open cone and bD is of class C^1 near 0, we are assured that $\overline{W}_o$ intersects bD along M only. This extension property is stable with respect to small deformations of class $C^{2+\epsilon}$ of M by [83]. Thus, we fix a point $b \in M \setminus K_1''$, $|b| < \frac{\tau}{2}$ and include M in a one-parameter family of $C^{2+\epsilon}$-manifolds

$$M_t = \{y_n = h_t(z', z_{n-1}, x_n), y_{n-1} = h_{1t}(z', x_{n-1}, x_n), -\delta < t < \delta\}$$

belonging to U_1 for $t \neq 0$. Here $h_{1t} = t$ for $z \in K_1''$, and $h_{1t} = 0$ for z in a neighborhood of b. That is near b, $M_t = M$ for all t. Then, for $\delta > 0$ sufficiently small, $f|M_t, t \neq 0$, extends holomorphically into a wedge

$$W_t = \{z = \zeta + \eta : \zeta \in M_t, \eta \in V_o, |z| < \frac{\tau}{2}\},$$

where V_o is an open cone such that $\overline{V}_o \subset V \cup \{0\}$. As f is holomorphic in $U_1 \supset M_t, t \neq 0$, it follows by a uniqueness theorem that these extensions of f constitute a single holomorphic function in an open set $\cup_{t \neq 0} W_t$. (We use here the property that $M_t = M$ in a neighborhood of b.) By the choice of the function h_{1t}, the set $\cup_{t \neq 0} W_t$ contains a one-sided neighborhood of 0 with respect to some C^1-hypersurface $\Gamma_1 \subset U_1$ homotopic to bD in U_1 as in Section 3.4. This means that f extends holomorphically into a one-sided neighborhood of 0 with respect to bD, whence $a \in K_1'$. This contradiction shows that $K_1 = K_1'$, and the proof is complete. $\square$

Corollary 4.8.3 *Let D be a relatively compact strictly pseudoconvex domain in a Stein manifold of dimension $n \geq 3$. Let K_o be a compact subset of bD, and let K_1 be a closed subset of a $C^{2+\epsilon}$-manifold $M \subset bD \setminus K_o$ of dimension $2n - 2$. Assume that $K = K_o \cup K_1$ is compact, that K_1 contains no component of M, and that M is generic and minimal at all points of K_1. Then the set K is removable if and only if K_o is removable.*

Proof As bD is strictly pseudoconvex and of class C^2, $bD \cap U$ has the one-sided extension property for every open set U. If K_o is removable with respect to D, then it is universally removable. (See Section 2.3.) Thus, the sufficiency of the condition follows from Theorem 4.8.2. The necessity is evident, because subsets of removable sets are removable. $\quad\square$

Corollary 4.8.4 *Let D be as in Corollary 4.8.3, let M be a locally closed $C^{2+\epsilon}$-manifold of dimension $2n - 2$ in bD. Let Σ be the set of points of M at which M is not generic or not minimal. Let $K \subset M$ be a compact set containing no component of M. Then K is removable if and only if $K \cap \Sigma$ is removable.*

Proof Take $K_o = K \cap \Sigma, K_1 = K \setminus K_o$ and apply Corollary 4.8.3. $\quad\square$

Some results on the removable of subsets of submanifolds of codimension two, similar to Corollary 4.8.4 with empty Σ have been announced by B. Jöricke [42]. Her methods seem to be entirely different from ours.

5 Conclusion

In this paper we have dealt with holomorphic functions and their boundary values. It is plain that similar questions could be discussed for sections of holomorphic vector bundles. This subject has been considered by Lupacciolu [52]. It is probable that many of the results we have given above have their analogues in this more general context and that with mainly formal changes our methods would apply. This is not, however, a subject we have pursued in detail. We point out, though, that in the strictly pseudoconvex case, the vector bundle problems are exactly parallel to those for scalar-valued functions as the following simple proposition shows.

Let D be strictly pseudoconvex domain in the Stein manifold $\mathcal{M}$, and let $V \xrightarrow{\pi} \mathcal{M}$ be a holomorphic vector bundle over $\mathcal{M}$. If the compact set $K \subset bD$ is a removable set, and if φ is a holomorphic section of V over a neighborhood W of $bD \setminus K$ in $\mathcal{M}$, then there is a holomorphic section $\tilde{\varphi}$ of V over the set $W \cup D$ that agrees on W with φ.

This follows immediately from standard facts about vector bundles. The main point is that as $\mathcal{M}$ is Stein, there is an exact sequence of vector bundles

$$0 \to V'' \to V' \xrightarrow{\eta} V \to 0$$

with $\mathcal{V}'$ trivial that splits so that there is a bundle map $\sigma : \mathcal{V} \to \mathcal{V}'$ with $\eta \circ \sigma$ the identity on $\mathcal{V}$ [31]. The bundle $\mathcal{V}'$ is trivial and so may be identified with E^q, the direct sum of q copies of the trivial line bundle E on $\mathcal{M}$.

Denote by e a global, zero-free section of E. The composition $\sigma \circ \varphi$ is a section of $\mathcal{V}'$ over W and so, using the identification of $\mathcal{V}'$ with E^q, $\sigma \circ \varphi$ may be expressed in the form

$$\sigma \circ \varphi = (f_1 e, \dots , f_q e)$$

for some functions $f_1, \dots , f_q$ holomorphic on W. As the set K is supposed to be removable, it follows that there are functions $\widetilde{f}_1, \dots , \widetilde{f}_q$ holomorphic on $W \cup D$ that agree on W with $f_1, \dots , f_q$, respectively. Then $\widetilde{\varphi} = \eta(\widetilde{f}_1 e, \dots , \widetilde{f}_q e)$ is a section of $\mathcal{V}$ over $W \cup D$ that agrees with φ on W.

To conclude, we will indicate a few problems that seem to us worthy of further study.

1. In our work, the statements of the results and the techniques we have used have depended explicitly on the ambient complex manifold. However, it is possible to formulate in terms intrinsic to the bounding manifold questions of the kind we have considered. This observation leads to the general problem of studying the removable singularities of CR-functions on abstractly given CR-manifolds. In the case that the CR-manifold in question is known to be realizable as a submanifold of a complex manifold, methods involving the ambient manifold can be brought to bear; in the contrary case, it is necessary to work in terms of the manifold itself. To formulate a specific problem, let $\mathcal{N}$ be a CR-manifold of class $\mathcal{C}^\infty$, and let K be a compact subset of $\mathcal{N}$. What conditions on $\mathcal{N}$ and K suffice to ensure that if u is a distribution on $\mathcal{N} \setminus K$ that satisfies $\overline{\partial}_b u = 0$ in $\mathcal{N} \setminus K$, then there is a distribution $\tilde{u}$ on $\mathcal{N}$ that satisfies $\overline{\partial}_b \tilde{u} = 0$ on $\mathcal{N}$ and that agrees on $\mathcal{N} \setminus K$ with u? (In this direction, see [7], [22] and [23].)

2. One obvious problem that remains is that of finding a geometric characterization of removable sets in strictly pseudoconvex boundaries analogous to the geometric characterization of weakly removable singularities given in Theorem 2.5.1. One would hope for a characterization in terms of convexity with respect to a reasonable class of functions.

3. Are Cantor sets removable? We know that a Cantor set in the boundary of a domain of dimension at least three is removable. Recent work of Lawrence [48] shows that *tame* Cantor sets in three-dimensional strictly pseudoconvex boundaries are removable, but there remains the case of the *wild, i.e.,* nontame, Cantor set in the boundary of a strictly pseudoconvex domain of dimension two, for example in the boundary of the two-dimensional ball. The distinction between tame and wild Cantor sets in this context does not seem natural, though it arises in a perfectly reasonable way in Lawrence's work. (For the notions of wild and tame Cantor sets one can consult [66].)

A related question is to know whether nonrectifiable arcs in the boundary of a two-dimensional strictly pseudoconvex boundary are removable.

4. It would be desirable to have an extension of the result of Jöricke on the removability of totally real discs to higher-dimensional situations. In particular, it would be natural to ask whether if D is a strictly pseudoconvex domain of dimension n and K is a subset of bD that is contained in a smooth, generic submanifold of bD of codimension two in C^n and that is diffeomorphically a closed ball of dimension $2n - 2$, K is necessarily removable. In this connection, it seems to be an open question whether such a set K can contain a compact MC-cycle of dimension $2n - 3$.

5. There is the natural problem of refining our results to corresponding results for various classes of functions. For example, what conditions on a compact set K in bB_n suffice to assure that for every CR-function f of class $L^1(bB_n \setminus K)$, there is a CR-function F on bB_n that is in $L^1(bB_n)$ and that agrees a.e. with f on $bB_n \setminus K$. Instead of considering the space L^1, one can equally well consider other classes of functions, *e.g.*, the L^p classes, the Hölder classes, *etc.* For some work in the direction of this problem, for L^1 boundary values, see the papers of Anderson and Cima [3] and of Kytmanov [43].

Certain problems that fall under this rubric admit more or less evident partial answers. Thus, if K is an arbitrary removable set in the boundary of a bounded domain D with Lipschitz boundary, then given a *bounded* CR-function f on $bD \setminus K$, there is a bounded CR-function f^* on all of bD that agrees a.e. with f on $bD \setminus K$: The set K is assumed to be removable, so there is a function F holomorphic on D that assumes the boundary values f nontangentially a.e. on $bD \setminus K$. The function F is necessarily bounded and so has nontangential boundary values F^* a.e. on bD. The functions f and F^* agree almost everywhere on $bD \setminus K$. [11]

This simple remark merely shows that the natural class of removable singularities for bounded functions contains the class of removable sets. It is probably a difficult problem to characterize the removable sets for bounded CR-functions. In this connection, we recall that the classical Painlevé problem remains open to this day: Characterize the closed subsets of the open unit disc that are removable sets for bounded holomorphic functions.

The question of considering removable singularities for classes of functions is essentially related to the problem of estimating the growth of extensions. Let D be a domain in C^n, and let the set $K \subset bD$ be known to be removable. Given a CR-function or distribution f on $bD \setminus K$, we then know it to extend holomorphically through the domain D. What estimates can be given for the growth of the extended function F near K? An approach to this problem is to consider the current $\vartheta = \overline{\partial} f[\Gamma]^{0,1}$, which is supported on the set K and ask for solutions of the equation $\overline{\partial} u = \vartheta$ *with bounds* on C^n.

6. The question of removability theorems for the boundary values of pluriharmonic functions can be formulated. It seems that at this time little is known in this direction.

Appendix

This appendix is devoted to a point in analytic geometry that is used in an essential way in Section 4.4 above.

Fix an n-dimensional, $n \geq 3$, complex manifold $\mathcal{M}$ and on it a distance function ρ derived from a smooth Hermitian metric on $\mathcal{M}$.

Lemma *Let V be an irreducible, one-dimensional analytic subvariety of $\mathcal{M}$, let $p_0 \in V$, let (Σ, φ) be a normalization of V, and let $s_0 \in \varphi^{-1}(p_0)$. If $K \subset \Sigma$ is a compact set that contains s_0, then given $\varepsilon > 0$, there is a holomorphic map ψ defined, one-to-one and regular on a neighborhood of K that satisfies $\psi(s_0) = p_0$ and $\rho(\psi(\zeta), \varphi(\zeta)) < \varepsilon$ for all*

---

[11] Incidentally, notice that $\sup|F| = ess\ sup_{bD \setminus K}|f| = ess\ sup_{bD}|F^*|$, even though the set $bD \setminus K$ may be of small area.

$\zeta \in K.$

As (Σ, φ) is a normalization and V is irreducible, Σ is a connected noncompact Riemann surface, and φ is a holomorphic map that carries $\Sigma \backslash \varphi^{-1}(sng\ V)$ biholomorphically onto $V \backslash sng\ V$. For normalizations, see [30] or [17].

Proof The variety V is irreducible and noncompact, so it is a Stein space [61]. From this, it follows from a theorem of Siu [76], cf. [69], that we may suppose $\mathcal{M}$ to be a Stein manifold. Then, by the embedding theorem for Stein manifolds, we can suppose that $\mathcal{M}$ is a complex submanifold of $\mathbb{C}^N$ for some sufficiently large N. By a theorem of Docquier and Grauert – see [31], there is a neighborhood Ω of $\mathcal{M}$ in $\mathbb{C}^N$ on which is defined a holomorphic retraction $\pi : \Omega \to \mathcal{M}$ so that π is holomorphic and $\pi(z) = z$ for $z \in \mathcal{M}$.

The motivation for the lemma stems from the result [30], [38] that since $N \geq 3$, proper, regular embeddings of Σ in $\mathbb{C}^N$ are dense in $\mathcal{O}(\Sigma, \mathbb{C}^N)$ and thus that if $h : \Sigma \to \mathbb{C}^N$ is such an embedding that approximates φ well, then $\pi \circ h$, which is defined on K if h is sufficiently near φ, should give the mapping we seek. Some attention to the details is necessary.

We fix a function $f \in \mathcal{O}(\Sigma)$ with three properties: 1°. df vanishes at no point of the set $\varphi^{-1}(V \backslash sng\ V)$, 2°. f maps $\varphi^{-1}(sng\ V)$ injectively into $\mathbb{C}$, and 3°. $f(s_0) = 0$. Such a function f exists since Σ is a Stein manifold.

For $\varepsilon \in \mathbb{C}^N$ and $\zeta \in \Sigma$, we have the vector $f(\zeta)\varepsilon \in \mathbb{C}^N$. We set $\varphi_\varepsilon(\zeta) = \varphi(\zeta) + f(\zeta)\varepsilon$. Our desired function ψ is $\pi \circ \varphi_\varepsilon$ for a suitable choice of ε. Notice first that every φ_ε assumes the value p_0 at s_0.

Fix once and for all a relatively compact neighborhood W_0 of K in Σ. Associated with W_0 is a constant $\delta_0 > 0$ small enough that $\pi \circ \varphi_\varepsilon$ is defined on W_0 for all $\varepsilon \in \mathbb{C}^N$ with $|\varepsilon| < \delta_0$.

We show first that for a $\delta_1 \leq \delta_0$ and for almost all ε with $|\varepsilon| < \delta_1$, $\pi \circ \varphi_\varepsilon$ is regular. To do this, compute, for a fixed ε, the differential with respect to $\zeta \in W_0$ of $\pi \circ \varphi_\varepsilon$: We have

$$d_\zeta(\pi \circ \varphi_\varepsilon)(\zeta) = d\pi(\varphi_\varepsilon(\zeta))(d\varphi(\zeta) + df(\zeta).\varepsilon)$$

where $df(\zeta).\varepsilon$ denotes the vector $(\varepsilon_1 df(\zeta), \dots, \varepsilon_N\ df(\zeta))$. Thus, if $d\varphi(\zeta) \neq 0$, then

$$d_\zeta(\pi \circ \varphi_\varepsilon)(\zeta)|_{\varepsilon=0} = d\pi(\varphi(\zeta))d\varphi(\zeta) = d\varphi(\zeta) \neq 0 .$$

We are using the observation that since π is the identity on $\mathcal{M}$, $d\pi(\varphi(\zeta))$ acts as the identity on vectors tangent to $\mathcal{M}$. As φ takes values in $V \subset \mathcal{M}$, $d\varphi$ is tangent to $\mathcal{M}$.

If, on the other hand, $d\varphi(\zeta) = 0$, then we compute at $\varepsilon = 0$ the differential with respect to ε of the map $\varepsilon \mapsto (\pi \circ \varphi_\varepsilon)(\zeta)$. By the chain rule, it is the linear map $w \mapsto d_\zeta(\pi(\varphi(\zeta)))df(\zeta).w$. As $df(\zeta) \neq 0$, $w \mapsto df(\zeta).w$ has rank N, and, since the map π is of rank n everywhere, it follows that this differential has everywhere rank n.

We have a map $\Phi : W_0 \times \mathbb{B}_N(\delta_0) \to \mathcal{M} \subset \mathbb{C}^N$, *viz,* $\Phi(\zeta, \varepsilon) = \pi(\varphi_\varepsilon(\zeta))$. In $W_0 \times B_N(\delta_0)$, let A' be the analytic set consisting of the points (ζ_0, ε_0) at which the partial map $\zeta \mapsto \Phi(\zeta, \varepsilon_0)$ is not regular. By the first calculation above, A' is disjoint from the discrete set $\{(\zeta, 0) \in W_0 \times \mathbb{B}_N(\delta_0) : d\varphi(\zeta) = 0\}$. By the second calculation, the rank of Φ at points $(\zeta, 0)$ with $d\varphi(\zeta) = 0$ is $N - n$. As $\overline{W}_0$ is compact, it follows that if δ_1 is sufficiently small, then $A' \cap (W_0 \times \mathbb{B}_N(\delta_1))$ has dimension $N - n$. Thus, the projection of A' into $\mathbb{B}_N(\delta_1)$ is of σ-finite $(N - n)$-dimensional measure, whence, for almost all ε with $|\varepsilon| < \delta_1$, $\zeta \to \Phi(\zeta, \varepsilon)$ is regular on W_0.

A similar analysis shows that for most small ε's, $\zeta \to \Phi(\zeta, \varepsilon)$ is one-to-one. Consider in $((W_0 \times W_0 \setminus \{\zeta = \zeta'\}) \times \mathbb{B}_N(\delta_0))$ the set $A'' = \{(\zeta, \zeta', \varepsilon) : \Phi(\zeta, \varepsilon) = \Phi(\zeta', \varepsilon)\}$, a certain analytic set. If $\varphi(\zeta) \neq \varphi(\zeta')$ then the point $(\zeta, \zeta', 0)$ is not in A''. Consider then a pair of distinct points $\zeta, \zeta' \in W_0$ with $\varphi(\zeta) = \varphi(\zeta')$, and consider $d_\varepsilon \Psi(\zeta, \zeta', \varepsilon)|_{\varepsilon = 0}$, the differential of the partial map $\varepsilon \mapsto \Psi(\zeta, \zeta', \varepsilon) = \Phi(\zeta, \varepsilon) - \Phi(\zeta', \varepsilon)$, with respect to ε evaluated at $\varepsilon = 0$. It is the linear map $w \mapsto d\pi(\varphi(\zeta))f(\zeta)w - d\pi(\varphi(\zeta'))f(\zeta')w = d\pi(\varphi(\zeta))(f(\zeta) - f(\zeta'))w$. Condition $2°$ on f implies that $w \mapsto (f(\zeta) - f(\zeta'))w$ has rank N and thus that $d_\varepsilon \Psi(\zeta, \zeta', \varepsilon)|_{\varepsilon = 0}$ has rank n. Thus, near $(\zeta, \zeta', 0)$, A'' has dimension $N - n$. It follows that if $\delta_2 > 0$ is small enough, then for almost all ε with $|\varepsilon| < \delta_2$, the partial map $\zeta \mapsto \Phi(\zeta, \varepsilon)$ is one-to-one.

Thus, for almost all ε with $|\varepsilon| < \min(\delta_1, \delta_2)$, the map φ_ε is both one-to-one and regular on W_0. If ε is small enough, φ_ε will give the desired approximation to φ.

The lemma is proved. $\qquad\qquad\qquad\qquad\qquad\qquad\qquad\qquad\qquad\qquad\qquad\square$

Bibliography

[1] Alexander, H.: Polynomial approximation and hulls in sets of finite linear measure. Amer. J. Math. **93** (1971), 65-74

[2] Alexander, H.: Removable sets for CR functions. To appear in the proceedings of the Mittag-Leffler special year in several complex variables

[3] Anderson, J. T. and Cima, J. A.: Removable singularities for $L^p CR$ functions. Michigan Math. J (To appear)

[4] Andreotti, A. and Grauert, H.: Théorèmes de finitude pour la cohomologie des espaces complexes. Bull. Soc. Math. France **90** (1962), 193–259

[5] Andreotti, A. and Kas, A.: Duality on complex spaces. Ann. Scuola Norm. Sup. Pisa **XXVII** (1973), 187–263

[6] Bănică, C. and Stănăşila, O.: "Algebraic Methods in the Global Theory of Complex Spaces", editor John Wiley, London 1976

[7] Battelli, F.: Singolarità eliminabili per le tracce delle funzioni pluriarmoniche. Boll. U.M.I. (6) **1** (1982), no. 1, 133–146

[8] Bedford, E. and Fornæss, J.-E.: A construction of peak functions on weakly pseudoconvex domains. Ann. Math. (2) **107** (1978), 555–568

[9] Bedford, E. and Gaveau, B.: Envelopes of holomorphy of certain 2-spheres in $\mathbb{C}^2$. Amer. J. Math. **105** (1983), 975–1009

[10] Björk, J. E.: Holomorphic convexity and analytic structures in Banach algebras. Ark. Mat. **9** (1971), 39–54

[11] Bourbaki, N.: "Topological Vector Spaces", Chapters 1–5, Springer-Verlag, Berlin 1987

[12] Bredon, G. E.: "Sheaf Theory", McGraw-Hill, New York 1967

[13] Cartan, H.: Variétés analytiques réeles et variétés analytiques complexes. Bull. Soc. math. France *85* (1957), 77–99

[14] Cassa, A.: Coomologia separata sulle varietà analitiche complesse. Ann. Scuola Norm. Sup. Pisa (3) **25** (1971), 290–323

[15] Chirka, E. M.: Analytic representation of CR-functions. Mat. Sb. **98(4)** (1975), 591–623

[16] Chirka, E. M.: Regularization and $\bar{\partial}$-homotopy on a complex manifold. Dokl. Akad. Nauk SSSR **244** no. 2 (1979), 300–303

[17] Chirka, E. M.: "Complex Analytic Sets", Nauka, Moscow 1985 (Russian). English translation, Kluwer, Dortrecht and Boston, 1989

[18] Chirka. E. M.: Introduction to the geometry of CR-manifolds . Uspekhi Mat. Nauk **46 (277)** no. 1 (1991), 80–164 (Russian). English translation, Russian Math. Surveys, **46** no. 1 (1991), 81–197

[19] deRham, G.: "Variétés Différentiables", Hermann, Paris 1973

[20] Dieudonné, J. "Treatise on Analysis, vol. III", Academic Press, New York 1972

[21] Dieudonné, J. and Schwartz, L.: La dualité dans les espaces $(\mathcal{F})$ et $(\mathcal{LF})$. Ann. Inst. Fourier (Grenoble) **1** (1949), 61–101

[22] Dini, G. and Parrini, C.: Singolarità rimovibili per CR-distribuzioni su domini piatti. Boll. Un. Mat. Ital. **B**(5) 17 (1980), 286–297

[23] Dini, G. and Parrini, C.: Extending CR-distributions. Bull. Sc. Math. (2)(*106*) (1982), 3–18

[24] Dolbeault, P.: Formes differentielles et cohomolgie sur une variété analytique complexe, I. Ann. Math. **(2) 64** (1956), 83–130

[25] Duval, J.: Surfaces convexes dans un bord pseudoconvex. (to appear)

[26] Ehrenpreis, L.: A new proof and extension of Hartog's theorem. Bull. Amer. Math. Soc. **67** (1967), 507–509

[27] Federer, H: "Geometric Measure Theory", Springer-Verlag, Berlin, Heidelberg, New York 1969

[28] Forstnerič, F. and Stout, E. L.: A new class of polynomially convex sets. Ark. Mat. **29** (1) (1991), 51–62

[29] Guenot, J. and Narasimhan, R.: Introduction à la théorie des surfaces de Riemann. L'Enseignment Mathématique (II) **XXI** (1975), 123–328

[30] Gunning, R. C.: "Introduction to Holomorphic Functions of Several Complex Variables", Wadsworth and Brooks/Cole, Belmont 1990

[31] Gunning, R. and Rossi, H.: "Analytic Functions of Several Complex Variables", Prentice-Hall, Englewood Cliffs 1965

[32] Harvey, F. R. and Lawson, B.: On boundaries of complex analytic varieties. I. Ann. Math. **(II) 102** (1975), 233–290

[33] Harvey, F. R. and Wells, R. O.: Compact holomorphically convex subsets of a Stein manifold. Trans. Amer. Math. Soc. **136** (1969), 509–516

[34] Harvey, F. R. and Wells, R. O.: Holomorphic approximation and hyperfunction theory on a C^∞ totally real submanifold of a complex manifold. Math. Ann. **197** (1972), 287–318

[35] Hatziafratis, T.: On certain integrals associated to CR-functions. Trans. Amer. Math. Soc. **314** (1989), 781–802

[36] Henkin, G. and Leiterer, J.: "Theory of Functions on Complex Manifolds", Birkhäuser Verlag, Basel, Boston, Stuttgart 1984

[37] Henkin, G. and Leiterer, J.: "Andreotti-Grauert Theory by Integral Formulas", Birkhäuser Verlag, Basal, Boston, Stuttgart 1988

[38] Hörmander, L.: "An Introduction to Complex Analysis in Several Variables", Van Nostrand, Princeton 1966

[39] Horváth, J.: "Topological Vector Spaces and Distributions", Addison-Wesley, Reading 1966

[40] Hurewicz, W. and Wallman, H.: "Dimension Theory", Princeton University Press, Princeton 1948

[41] Jöricke, B.: Removable singularities for CR-functions. Ark. Mat. **26** (1988), 117–143

[42] Jöricke, B.: Envelopes of holomorphy and CR-invariant subsets of CR-manifolds. (to appear)

[43] Kytmanov, A. M.: Holomorphic continuation of integrable CR-functions from part of the boundary of a domain. Mat. Zametki **48** (2) (1990), 64–70

[44] Kytmanov,A. M.: Holomorphic extension of CR-function with singularities on a hypersurface. Izvestiya Akad. Nauk CCCP **54** No. 6 (1990), 1320–1330 (Russian), English translation Math. USSR Izvestiya **37** (1991), 681–691

[45] Kytmanov, A. M. and Nikitina, T. N.: On the removable singularities of CR-functions given on a generic manifold. (to appear)

[46] Laurent-Thiebaut, C.: Sur l'extension des fonctions CR dans une variété da Stein. Ann. Mat. Pura Appl.(IV) **150** (1988), 1–21

[47] Laurent-Thiebaut, C. and Leiterer, J.: On the Hartogs-Bochner extension phenomenon for differential forms. Math. Ann. **284** (1989), 103–119

[48] Lawrence, M.: Hulls of tame Cantor sets. (to appear)

[49] Lupacciolu, G.: A theorem on holomorphic extension of CR-functions. Pacific J. Math. **124** (1986), 177–191

[50] Lupacciolu, G.: Holomorphic continuation in several complex variables. Pacific J. Math. **128** (1987), 117–125

[51] Lupacciolu, G.: On the removal of singular sets for the tangential Cauchy-Riemann operator. Arkiv för Mat. **28** (1990), 119–130

[52] Lupacciolu, G.: Some global results on extensions of CR-objects in complex manifolds. Trans. Amer. Math. Soc. **321** (1990), 761–774

[53] Lupacciolu, G.: Characterization of removable sets in strongly pseudoconvex boundaries. (to appear)

[54] Lupacciolu, G.: On the envelopes of holomorphy of strictly Levi-convex hypersurfaces. (to appear)

[55] Lupacciolu, G.: Topological properties of q-convex sets. Trans. Amer. Math. Soc. (to appear)

[56] Lupacciolu, G.: Holomorphic and meromorphic q-hulls. (to appear)

[57] Lupacciolu, G.: Approximation and cohomology vanishing properties of low-dimensional compact sets in a Stein manifold. Math. Z. **211** (1992), 523–532

[58] L upacciolu, G. and Stout, E. L.: Removable singularities for $\bar{\partial}_b$. (to appear in the proceedings of the Mittag-Leffler special year on several complex variables)

[59] Lupacciolu, G. and Tomassini, G.: Un teorema di estensione per le CR-functioni. Ann. Mat. Pura App. (IV) **137** (1984), 257–263

[60] Malgrange, B.: Existence et approximation des solutions des équations aux dérivées partielle et des équations de convolution. Ann. Inst. Fourier (Grenoble) **6** (1955), 271–354

[61] Narasimhan, R.: A note on Stein spaces and their normalizations. Ann. Scuola Norm. Sup. Pisa **(3) 16** (1962), 327–333

[62] Narasimhan, R.: "Introduction to Analytic Spaces" in *Springer Lecture Notes*, vol. 25, Springer-Verlag, Berlin, Heidelberg, New York 1966

[63] Rosay, J.-P. and Stout, E. L.: Radó's theorem for CR-functions. Proc. Amer. Math. Soc. **106** (1989), 1017–1026

[64] Rossi, H.: The local maximum modulus principle. Ann. Math. **(2) 72** (1960), 1–11

[65] Rossi, H.: On envelopes of holomorphy. Comm. Pure Appl. Math. **16** (1963), 9–19

[66] Rushing, T. B.: "Topological Embeddings", Academic Press, New York 1973

[67] Sadullaev, A. and Chirka, E. M.: On the continuation of functions with polar singularities. Mat. Sb. **132 (174)** no. 3 (1987), 383–390 (Russian), English translation, Math. USSR Sbornik **60** (1988), 377–384

[68] Schaefer, H. H.: "Topological Vector Spaces", Springer-Verlag, New York, Heidelberg, Berlin 1971

[69] Schneider, M.: Tubenumgebungen Steinscher Räume. Manuscripta Math. **18** (1976), 391–397

[70] Schwartz, L.: "Théorie des Distributions", Hermann, Paris 1966

[71] Serre, J. P.: "Quelques problèmes globaux relatifs aux variétés de Stein" in *Colloque sur les Fonctions de Plusieurs Variables Complex*, Bruxelles, Mars, 1953; Georges Throne, Liege; Masson, Paris, 1953

[72] Serre, J. P.: Un théorème de dualité. Comm. Math. Helv. **29** (1955), 9–26

[73] Shcherbina, N. V.: On fibering into analytic curves of the common boundary of two domains of holomorphy. Izv. Akad. Nauk SSSR **46** no. 5 (1982), 1106–1123 (Russian), English translation, Math. USSR Izvestiya **21** (1983) 399–413

[74] Shcherbina, N. V.: The polynomial hull of a sphere embedded in $\mathbb{C}^2$. Mat. Zametki **49** (1991), 127–134 (Russian), English translation, Math. Notes, **49** (1991), 89–93

[75] Shcherbina, N. V.: On the polynomial hull of a two-dimensional sphere in C^2. Dokl. Akal. Nauk SSSR **306** (no. 6) (1991), 1315–1319

[76] Siu, Y.-T.: Every Stein subvariety admits a Stein neighborhood. Invent. Math. **38** (1976/1977), 89–100

[77] Słodkowski, Z.: Analytic set-valued functions and spectra. Math. Ann. **256** (1981), 363–386

[78] Słodkowski, Z.: Local maximum property and q-plurisubharmonic functions in uniform algebras. J. Math. Anal. Appl. **115** (1986), 105–130

[79] Stolzenberg, G.: Uniform approximation on smooth curves. Acta. Math. **115** (1966), 185–198

[80] Stout, E. L.: Analytic continuation and boundary continuity of functions of several complex variables. Proc. Royal Soc. Edinburgh **89 A** (1981), 63–74

[81] Stout, E. L.: Removable singularities for the boundary values of holomorphic functions. (to appear in the proceedings of the Mittag-Leffler special year in several complex variables)

[82] Trépreau, J.-M.: Sur le prolongement holomorphe des fonctions C-R défines sur une hypersurface réele de class $C^{\mathcal{E}}$ dans $\mathbb{C}^N$. Invent. Math. **83** (1986), 583–592

[83] Tumanov, A. E.: Extension of CR-functions into a wedge. Mat. Sb. **181** (7) (1990), 951–964 (Russian), English translation, Math. USSR Sbornik **70** (1991), 385–398

Research supported in part by National Science Foundation Grant DMS-9001883.

Regularization of closed positive currents of type (1,1) by the flow of a Chern connection

Jean-Pierre Demailly

1 Introduction

Let X be a compact n-dimensional complex manifold and let T be a closed positive current of bidegree $(1,1)$ on X. In general, T cannot be approximated by closed positive currents of class C^∞: a necessary condition for this is that the cohomology class $\{T\}$ is numerically effective in the sense that $\int_Y \{T\}^p \geq 0$ for every p-dimensional subvariety $Y \subset X$. For example, if $E \simeq \mathbb{P}^{n-1}$ is the exceptional divisor of a one-point blow-up $X \longrightarrow X'$, then $T = [E]$ cannot be positively approximated: for every curve $C \subset E$, we have $\int_C \{E\} = \int_C c_1(\mathcal{O}(-1)) < 0$. However, we will see that it is always possible to approximate a closed positive current T of type $(1,1)$ by closed real currents admitting a small negative part, and that this negative part can be estimated in terms of the Lelong numbers of T and the geometry of X.

Let α be a smooth closed $(1,1)$-form representing the same $\partial\overline{\partial}$-cohomology class as T and let ψ be a quasi-psh function on X (that is, a function which is locally the sum of a plurisubharmonic function and a smooth function) such that $T = \alpha + \frac{i}{\pi}\partial\overline{\partial}\psi$. Such a decomposition exists even when X is non-Kähler, since we can always find an open covering (Ω_k) of X such that $T = \frac{i}{\pi}\partial\overline{\partial}\psi_k$ over Ω_k, and construct a global $\psi = \sum \theta_k \psi_k$ by means of a partition of unity (θ_k) (note that $\psi - \psi_k$ is smooth on Ω_k). If ψ_ε is an approximation of ψ, then $T_\varepsilon = \alpha + \frac{i}{\pi}\partial\overline{\partial}\psi_\varepsilon$ is an approximation of T. We are thus led to study a regularization process for quasi-psh functions. In this context, we prove the following result.

Theorem 1.1 *Let T be a closed almost positive $(1,1)$-current and let α be a smooth real $(1,1)$-form in the same $\partial\overline{\partial}$-cohomology class as T, i.e. $T = \alpha + \frac{i}{\pi}\partial\overline{\partial}\psi$ where ψ is an almost psh function. Let γ be a continuous real $(1,1)$-form such that $T \geq \gamma$. Suppose that T_X is equipped with a smooth hermitian metric ω such that the Chern curvature form $\Theta(T_X)$ satisfies*

$$(\Theta(T_X) + u \otimes \mathrm{Id}_{T_X})(\theta \otimes \xi, \theta \otimes \xi) \geq 0 \; \forall \theta, \xi \in T_X \text{ with } \langle \theta, \xi \rangle = 0,$$

for some continuous nonnegative $(1,1)$-form u on X. Then there is a family of closed almost positive $(1,1)$-currents $T_\varepsilon = \alpha + \frac{i}{\pi}\partial\bar{\partial}\psi_\varepsilon$, $\varepsilon \in \,]0, \varepsilon_0[$, such that ψ_ε is smooth over X, increases with ε, and converges to ψ as ε tends to 0 (in particular, T_ε is smooth and converges weakly to T on X), and such that

(i) *$T_\varepsilon \geq \gamma - \lambda_\varepsilon u - \delta_\varepsilon \omega$ where:*

(ii) *$\lambda_\varepsilon(x)$ is an increasing family of continuous functions on X such that $\lim_{\varepsilon \longrightarrow 0} \lambda_\varepsilon(x) = \nu(T, x)$ (Lelong number of T at x) at every point,*

(iii) *δ_ε is an increasing family of positive constants such that $\lim_{\varepsilon \longrightarrow 0} \delta_\varepsilon = 0$.*

More precise results are given in Theorems 4.1 and 6.1. Such approximations can in turn be used to obtain various estimates of intersection theory [4], or asymptotic inequalities for Dolbeault cohomology [2]. They can be also applied to study compact complex manifolds with partially semipositive curvature in the sense of Griffiths (see Section 5); in that case, we prove for instance that every effective divisor is nef (i.e. numerically effective), and that the variety is projective if and only if it is Moishezon.

Our proof uses some ideas already developed in [1], although more general and more precise results will be obtained here. The main idea is to use a convolution kernel constructed by means of the exponential map associated to a Chern connection on T_X. To get precise estimates of the Hessian forms involved, we determine the Taylor expansion of the exponential map at order 3. The third order coefficients can be calculated explicitly in terms of the curvature tensor of the metric. What is perhaps most remarkable is that we have been ultimately able to find the complete Taylor expansion of the exponential convolution kernel (Proposition 3.1); this is indeed possible because we use a modified exponential map which is made fiberwise quasi-holomorphic (see section 2). Finally, we apply Kiselman's singularity attenuation technique [9], [10] in combination with our main estimates to define a partial regularization process for closed $(1, 1)$-currents: in that way, the Lelong numbers can be killed up to any given level (Theorem 6.1).

Further techniques based on Hörmander's L^2 existence theorems [8] are explained in our recent papers [3], [5]; they lead to similar estimates, but with a numerical hypothesis instead of a curvature hypothesis: namely, u should then be a closed real $(1, 1)$-form such that the cohomology class $c_1(\mathcal{O}_{T_X}(1)) + \pi^\star u$ is nef on the total space of the projective bundle $P(T_X^\star) \xrightarrow{\pi} X$ of hyperplanes in T_X. This condition, which is more natural than a curvature hypothesis from the point of view of algebraic geometry, is also more general than the Griffiths semipositivity of $\Theta(T_X) + u \otimes \mathrm{Id}_{T_X}$. However, it is not clear how the above numerical condition can be related to the partial semipositivity hypothesis made in Theorem 1.1 (see the comments after Definition 5.1); for instance, the partial semipositivity hypothesis is void for curves. Therefore, both types of hypotheses seem to have their own domain of applicability. Moreover, the techniques developed here are considerably simpler and in some sense more precise and more explicit, so we felt interesting to explain this simpler method, which is probably easier to extend to currents of higher bidegrees. The main ideas of this work have been worked out during a stay of the author at Bayreuth University in November 1989. The author wishes to thank this Institution for its hospitality.

2 Exponential map associated to the Chern connection

Suppose that the manifold X is equipped with a smooth hermitian metric $\omega = i \sum \omega_{lm} dz_l \wedge d\overline{z}_m$. Denote by D the Chern connection of T_X and by $\Theta(T_X) = \frac{i}{2\pi} D^2$ the curvature tensor. We define an exponential map $\exp : T_X \longrightarrow X$ as follows: if $\zeta \in T_{X,z}$, then $\exp_z(\zeta)$ is the position at time $t = 1$ of the curve $t \mapsto u(t)$ starting at $u(0) = z$ with initial tangent vector $u'(0) = \zeta$ and satisfying the second order differential equation $D(du/dt) = 0$ (parallel translation with respect to D). If ω is Kähler, the Chern connection coincides with the Levi-Civita connection, so $\exp$ is given in that case by the riemannian geodesics; otherwise, $\exp$ differs from the usual riemannian exponential map. For any $x \in X$, fix analytic coordinates $(z_1 \ldots z_n)$ centered at x such that $(\partial/\partial z_l)$ is an orthonormal basis of T_X at x. Consider the Taylor expansion of second order

$$\omega_{lm}(z) = \langle \frac{\partial}{\partial z_l}, \frac{\partial}{\partial z_m} \rangle = \delta_{lm} + \sum_j (a_{jlm} z_j + \overline{a}_{jml} \overline{z}_j) + \sum_{j,k} (b'_{jklm} z_j z_k + \overline{b}'_{jkml} \overline{z}_j \overline{z}_k)$$
$$+ \sum_{j,k} c'_{jklm} z_j \overline{z}_k + O(|z|^3).$$

We may always arrange that the antisymmetry relation $a_{jlm} = -a_{ljm}$ holds; otherwise the change of variables $z_m = z'_m - \frac{1}{4} \sum (a_{jlm} + a_{ljm}) z'_j z'_l$ yields coordinates (z'_l) with this property. If ω is Kähler, the symmetry of $a_{jlm} = \partial \omega_{lm}/\partial z_j$ in j, l implies $a_{jlm} = 0$; in that case b'_{jklm} is also symmetric in j, k, l and a new change of variables $z_m = z'_m - \frac{1}{3} \sum b'_{jklm} z'_j z'_k z'_l$ gives $b'_{jklm} = 0$ likewise. The holomorphic frame of T_X defined by

$$e_l = \frac{\partial}{\partial z_l} - \sum_m (\sum_j a_{jlm} z_j + \sum_{j,k} b'_{jklm} z_j z_k) \frac{\partial}{\partial z_m}$$

satisfies

(2.1) $\qquad \langle e_l, e_m \rangle = \delta_{lm} - \sum_{j,k} c_{jklm} z_j \overline{z}_k + O(|z|^3),$

(2.2) $\qquad \dfrac{\partial}{\partial z_l} = e_l + \sum_m (\sum_j a_{jlm} z_j + \sum_{j,k} b_{jklm} z_j z_k + O(z^3)) e_m$

with $c_{jklm} = -c'_{jklm} - \sum_p a_{jlp} \overline{a}_{kmp}$ and $b_{jklm} = b'_{jklm} + \sum_p a_{jlp} a_{kpm}$. We may of course suppose that $b_{jklm} = b_{kjlm}$. Also, by a modification of the third order terms in (e_l), we can suppose that no term $O(z^3)$ appears in (2.2). The formula $\partial \langle e_l, e_m \rangle = \langle De_l, e_m \rangle$ easily gives the expression of De_l, $D^2 e_l$ and $\Theta(T_X)_x$:

$$De_l = -\sum_{j,k,m} c_{jklm} \overline{z}_k \, dz_j \otimes e_m + O(|z|^2),$$

(2.3) $\qquad \Theta(T_X)_x = \dfrac{i}{2\pi} \sum_{j,k,l,m} c_{jklm} \, dz_j \wedge d\overline{z}_k \otimes e_l^\star \otimes e_m.$

Given a vector field $\zeta = \sum \zeta_l \, \partial/\partial z_l$ in T_X, we denote by (ξ_l) the components of ζ with respect to the basis (e_l), thus $\zeta = \sum \xi_l e_l$. By (2.2) we have

$$(2.4) \qquad \xi_m = \zeta_m + \sum_{j,l} a_{jlm} z_j \zeta_l + \sum_{j,k,l} b_{jklm} z_j z_k \zeta_l .$$

In the Kähler case everything is much simpler, we take $e_l = \partial/\partial z_l$ and $\xi_m = \zeta_m$. In general, the Chern connection D is given by $D\zeta = D(\sum \zeta_l \, \partial/\partial z_l)$ with

$$
\begin{aligned}
D\Big(\frac{\partial}{\partial z_l}\Big) &= -\sum_{j,k,m} c_{jklm} \overline{z}_k \, dz_j \otimes e_m + \sum_{j,m} a_{jlm} dz_j \otimes e_m \\
&\qquad\qquad\qquad + 2 \sum_{j,k,m} b_{jklm} z_k \, dz_j \otimes e_m + O(|z|^2)\, dz \\
&= -\sum_{j,k,m} (c_{jklm}\overline{z}_k - 2b_{jklm} z_k) dz_j \otimes \frac{\partial}{\partial z_m} \\
&\qquad + \sum_{j,m} \Big(a_{jlm} - \sum_{k,p} a_{jlp} a_{kpm} z_k\Big) dz_j \otimes \frac{\partial}{\partial z_m} + O(|z|^2)\, dz ,
\end{aligned}
$$

$$
\begin{aligned}
(2.5)\quad D\zeta &= \sum_m d\zeta_m \otimes \frac{\partial}{\partial z_m} - \sum_{j,k,l,m} (c_{jklm}\overline{z}_k - 2b_{jklm} z_k)\zeta_l dz_j \otimes \frac{\partial}{\partial z_m} \\
&\quad + \sum_{j,l,m} \Big(a_{jlm} - \sum_{k,p} a_{jlp} a_{kpm} z_k\Big)\zeta_l dz_j \otimes \frac{\partial}{\partial z_m} + O(|z|^2)\!\cdot\!\zeta \, dz .
\end{aligned}
$$

Consider a curve $t \mapsto u(t)$. By a substitution of variables $z_j = u_j(t)$, $\zeta_l = du_l/dt$ in formula (2.5), the equation $D(du/dt) = 0$ becomes

$$(2.6) \qquad \frac{d^2 u_m}{dt^2} = \sum_{j,k,l} (c_{jklm}\overline{u}_k(t) - 2b_{jklm} u_k(t))\frac{du_j}{dt}\frac{du_l}{dt} + O(|u(t)|^2)\!\cdot\!\Big(\frac{du}{dt}\Big)^2 ;$$

the contribution of the terms $\sum a_{jl\bullet}\zeta_l dz_j$ is zero by the antisymmetry relation; moreover the remainder term only contains $\mathbb{C}$-quadratic terms in du/dt. The initial condition $u(0) = z$, $u'(0) = \zeta$ gives $u_m(t) = z_m + t\zeta_m + O(t^2|\zeta|^2)$, hence

$$\frac{d^2 u_m}{dt^2} = \sum_{j,k,l} (c_{jklm}(\overline{z}_k + t\overline{\zeta}_k) - 2b_{jklm}(z_k + t\zeta_k))\zeta_j\zeta_l + O(|\zeta|^2(|z| + |\zeta|)^2).$$

Two successive integrations yield

$$
\begin{aligned}
u_m(t) &= z_m + t\zeta_m + \sum_{j,k,l} c_{jklm}\Big(\frac{t^2}{2}\overline{z}_k + \frac{t^3}{6}\overline{\zeta}_k\Big)\zeta_j\zeta_l \\
&\quad - 2\sum_{j,k,l} b_{jklm}\Big(\frac{t^2}{2} z_k + \frac{t^3}{6}\zeta_k\Big)\zeta_j\zeta_l + O(t^2|\zeta|^2(|z| + |\zeta|)^2).
\end{aligned}
$$

An iteration of this procedure (substitution in (2.6) followed by an integration) easily shows that all terms but the first two in the Taylor expansion of $u_m(t)$ contain $\mathbb{C}$-quadratic factors of the form $\zeta_j\zeta_l$. Let us substitute ζ_m by its expression in terms of z, ξ deduced from (2.4). We find that $\exp_z(\zeta) = u(1)$ has a third order expansion

$$(2.7) \quad \exp_z(\zeta)_m = g_m(z,\xi) + \sum_{j,k,l} c_{jklm}\Big(\frac{1}{2}\bar{z}_k + \frac{1}{6}\bar{\xi}_k\Big)\xi_j\xi_l + O(|\xi|^2(|z| + |\xi|)^2)$$

where

$$g_m(z,\xi) = z_m + \xi_m - \sum_{j,l} a_{jlm}z_j\xi_l + \sum_{j,k,l,p} a_{jlp}a_{kpm}z_j z_k\xi_l$$
$$- \sum_{j,k,l} b_{jklm}\Big(z_j z_k\xi_l + z_k\xi_j\xi_l + \frac{1}{3}\xi_j\xi_k\xi_l\Big)$$

is a holomorphic polynomial of degree 3 in z,ξ and where the remainder involves $\mathbb{C}$-quadratic factors $\xi_j\xi_l$ in all terms. In the Kähler case we simply have $\xi_m = \zeta_m$ and $g_m(z,\xi) = z_m + \xi_m$.

The exponential map is unfortunately non holomorphic. However, we can make it quasi-holomorphic with respect to ζ as follows: for z fixed, we consider the formal power series obtained by eliminating all monomials in the Taylor expansion of $\zeta \mapsto \exp_z(\zeta)$ at the origin which are not holomorphic with respect to ζ. This defines in a unique way a jet of infinite order along the zero section of T_X. E. Borel's theorem shows that there is a smooth map

$$T_X \longrightarrow X, \ (z,\zeta) \longmapsto \exp h_z(\zeta)$$

such that its jet at $\zeta = 0$ coincides with the "holomorphic" part of $\zeta \mapsto \exp_z(\zeta)$ (of course, this map is defined only up to an addition of flat C^∞ functions along the zero section of T_X). Moreover, (2.7) implies that

$$(2.8) \qquad \exp h_z(\zeta)_m = g_m(z,\xi) + \frac{1}{2}\sum_{j,k,l} c_{jklm}\bar{z}_k\xi_j\xi_l + O(|\xi|^2(|z| + |\xi|)^2)$$

By including in g_m all holomorphic monomials of partial degree at most 2 in z and N in ξ ($N \geq 2$ being a given integer), we get holomorphic polynomials $h_m(z,\xi)$ of linear part $z_m + \xi_m$ and total degree $N + 2$, such that

$$\exp h_z(\zeta)_m = h_m(z,\xi) + O(\bar{z}, z\bar{z}, \overline{zz}, |z|^3, \xi^{N-1})\xi^2.$$

Here a notation as $O(\bar{z}, z\bar{z}, \overline{zz}, |z|^3, \xi^{N-1})\xi^2$ indicates an arbitrary function in the ideal of C^∞ functions generated by monomials of the form

$$\bar{z}_k\xi_l\xi_m, \ z_i\bar{z}_j\xi_l\xi_m, \ \bar{z}_i\bar{z}_j\xi_l\xi_m, \ z^\alpha\bar{z}^\beta\xi_l\xi_m \ \text{ and } \ \xi^\gamma,$$

for all multi-indices $|\alpha| + |\beta| = 3$ and $|\gamma| = N + 1$ (the notation $|z|^3$ thus stands for an arbitrary monomial of degree 3 in $z, \bar{z}$, so that $O(|z|^3)$ is compatible with the usual Landau notation). By the implicit function theorem applied to the mapping $h = (h_m)_{1 \leq m \leq n}$ we thus get:

Proposition 2.1 *Let ω be a smooth hermitian metric on X. There exists a C^∞ map*

$$T_X \longrightarrow X, \ (x,\zeta) \longmapsto \exp h_x(\zeta), \ \zeta \in T_{X,x}$$

with the following properties:

(i) *For every $x \in X$, $\mathrm{exph}_x(0) = x$ and $d_\zeta \, \mathrm{exph}_x(0) = \mathrm{Id}_{T_{X,x}}$.*

(ii) *For every $x \in X$, the map $\zeta \mapsto \mathrm{exph}_x(\zeta)$ has a holomorphic Taylor expansion at $\zeta = 0$. Moreover, with respect to ω, there are local normal coordinates $(z_1, \ldots, z_n)$ on X centered at x and holomorphic normal coordinates (ξ_j) on the fibers of T_X near x such that*

$$\mathrm{exph}_z(\zeta) = h_x(z, \rho_x(z, \xi)),$$

where $h_x(z, \xi)$ is a holomorphic polynomial map of degree 2 in z and of degree N in ξ, and where $\rho_x : \mathbb{C}^n \times \mathbb{C}^n \longrightarrow \mathbb{C}^n$ is a smooth map such that

$$
\begin{aligned}
h_{x,m}(z,\xi) &= z_m + \xi_m - \sum_{j,l} a_{jlm} z_j \xi_l + \sum_{j,k,l,p} a_{jlp} a_{kpm} z_j z_k \xi_l \\
&\quad - \sum_{j,k,l} b_{jklm}\left(z_j z_k \xi_l + z_k \xi_j \xi_l + \frac{1}{3}\xi_j \xi_k \xi_l\right) \\
&\quad + O((|z| + |\xi|)^4),
\end{aligned}
$$

(2.9)

$$
\begin{aligned}
\rho_{x,m}(z,\xi) &= \xi_m + \sum_{2 \le |\alpha| \le N}\left(\sum_k d_{\alpha k m} \xi^\alpha \overline{z}_k + \sum_{j,k} e_{\alpha j k m} \xi^\alpha z_j \overline{z}_k\right) \\
&\quad + O(\overline{z}^2, |z|^3, \xi^{N-1}) \xi^2.
\end{aligned}
$$

(iii) *For $\alpha = (0, \ldots, 1_j, \ldots, 1_l, \ldots, 0)$ of degree 2, we have $d_{\alpha k m} = \frac{1}{2} c_{jklm}$ where (c_{jklm}) is the curvature tensor of ω at x.*

Of course, if the hermitian metric ω is real analytic, all the above expansions are convergent, hence $\mathrm{exph}_z(\zeta)$ is real analytic and holomorphic in ζ in a neighborhood of the zero section of T_X. By taking $N = \infty$, we obtain a real analytic map $\rho(x, \xi)$ which is holomorphic in ξ, so the above remainder term becomes $O(\overline{z}^2, |z|^3)\xi^2$.

3   Regularization of quasi-psh functions

We now come to the main idea. Select a cut-off function $\chi : \mathbb{R} \longrightarrow \mathbb{R}$ of class C^∞ such that

$$\chi(t) > 0 \text{ for } t < 1, \; \chi(t) = 0 \text{ for } t \ge 1, \; \int_{v \in \mathbb{C}^n} \chi(|v|^2)\, d\lambda(v) = 1.$$

If ψ is a quasi-psh function on X, we set

(3.10) $$\psi_\varepsilon(z) = \frac{1}{\varepsilon^{2n}} \int_{\zeta \in T_{X,z}} \psi(\mathrm{exph}_z(\zeta)) \, \chi\!\left(\frac{|\zeta|^2}{\varepsilon^2}\right) d\lambda(\zeta), \; \varepsilon > 0.$$

Here $d\lambda$ denotes the Lebesgue measure on $\mathbb{C}^n$, resp. on the hermitian space $(T_{X,z}, \omega(z))$. For $w \in \mathbb{C}$ with $|w| = \varepsilon$, we have $\psi_\varepsilon(z) = \Psi(z, w)$ with

$$(3.11) \qquad \Psi(z,w) = \int_{\zeta \in T_{X,x}} \psi(\,\mathrm{exph}_z(w\zeta))\,\chi(|\zeta|^2)\,d\lambda(\zeta).$$

The change of variable $y = \mathrm{exph}_z(w\zeta)$ expresses $w\zeta$ as a smooth function of y, z in a neighborhood of the diagonal in $X \times X$. Hence Ψ is smooth over $X \times \{0 < |w| < \varepsilon_0\}$ for some $\varepsilon_0 > 0$. We are going to compute $\partial\overline{\partial}\Psi$ over this set and estimate its negative part when $|w|$ is small. For this, we fix a point $x \in X$ and use the coordinates (z, ξ) on T_X introduced in section 2 for simplicity, we omit the index x in the notation of h_x and ρ_x. By (2.1), we have

$$(3.12)$$
$$|\zeta|^2 = \sum_m |\xi_m|^2 - \sum_{j,k,l,m} c_{jklm}\, z_j \overline{z}_k \xi_l \overline{\xi}_m + O(|z|^3)|\xi|^2,$$

$$(3.13)$$
$$d\lambda(\zeta) = \frac{1}{2^n n!}(i\partial\overline{\partial}|\zeta|^2)^n$$
$$= (1 - \sum_{j,k,l} c_{jkll}\, z_j \overline{z}_k + O(|z|^3))\frac{i}{2}d\xi_1 \wedge d\overline{\xi}_1 \wedge \ldots \wedge \frac{i}{2}d\xi_n \wedge d\overline{\xi}_n.$$

In (3.11), we make the change of variables $s = w^{-1}\rho(z, w\xi)$, hence we can write $\mathrm{exph}_z(w\zeta) = h(z, ws)$. By (2.9) we get

$$(3.14)$$
$$s_m = \xi_m + \sum_{2 \leq |\alpha| \leq N} \Big(\sum_k d_{\alpha km} w^{|\alpha|-1} \xi^\alpha \overline{z}_k + \sum_{j,k} e_{\alpha jkm} w^{|\alpha|-1} \xi^\alpha z_j \overline{z}_k \Big)$$
$$+ O(\overline{z}^2, |z|^3, w^{N-1}\xi^{N-1})w\xi^2.$$

Therefore

$$(3.15)$$
$$\xi_m = s_m - \sum_{2 \leq |\alpha| \leq N} \Big(\sum_k d_{\alpha km} w^{|\alpha|-1} s^\alpha \overline{z}_k + \sum_{j,k} e_{\alpha jkm} w^{|\alpha|-1} s^\alpha z_j \overline{z}_k \Big)$$
$$+ O(\overline{z}^2, |z|^3, w^{N-1}s^{N-1})ws^2$$

and $\xi = s + O(w^N s^{N+1})$ for $z = 0$. After a substitution in (3.11), (3.12), (3.13) we get

$$(3.16) \qquad \Psi(z,w) = \int_{\mathbb{C}^n} \psi(h(z, ws))\,\chi(A(z, w, s))\,B(z, w, s)\,d\lambda(s)$$

where

$$A(z,w,s) = \sum_m |s_m|^2 - \sum_{j,k,l,m} c_{jklm} z_j \overline{z}_k s_l \overline{s}_m$$

$$- 2\operatorname{Re} \sum_{\alpha,k,m} d_{\alpha km} w^{|\alpha|-1} s^\alpha \overline{s}_m \overline{z}_k$$

$$- 2\operatorname{Re} \sum_{\alpha,j,k,m} e_{\alpha jkm} w^{|\alpha|-1} s^\alpha \overline{s}_m z_j \overline{z}_k$$

$$+ \sum_{\alpha,\beta,j,k,m} d_{\alpha km} \overline{d_{\beta jm}} \, w^{|\alpha|-1} \overline{w}^{|\beta|-1} s^\alpha \overline{s}^\beta z_j \overline{z}_k$$

$$+ O(z^2, \overline{z}^2, |z|^3, |w|^{N-1} |s|^{N-1}) |w| |s|^3,$$

$$B(z,w,s) = 1 - \sum_{j,k,l} c_{jkll} z_j \overline{z}_k$$

$$- 2\operatorname{Re} \sum_{\alpha,k,m} d_{\alpha km} w^{|\alpha|-1} \alpha_m s^{\alpha-1_m} \overline{z}_k$$

$$- 2\operatorname{Re} \sum_{\alpha,j,k,m} e_{\alpha jkm} w^{|\alpha|-1} \alpha_m s^{\alpha-1_m} z_j \overline{z}_k$$

$$+ \sum_{\alpha,\beta,j,k,l,m} d_{\alpha km} \overline{d_{\beta jl}} \, w^{|\alpha|-1} \overline{w}^{|\beta|-1} \alpha_m \beta_l s^{\alpha-1_m} \overline{s}^{\beta-1_l} z_j \overline{z}_k$$

$$+ O(z^2, \overline{z}^2, |z|^3, |w|^{N-1} |s|^{N-1}) |w| |s| \,;$$

here $(1_m)_{1 \leq m \leq n}$ denotes the standard basis of $\mathbb{Z}^n$, hence $s^{1_m} = s_m$.

Proposition 3.1 *For any integer $N \geq 2$ and any $(\theta, \eta) \in T_{X,x} \times \mathbb{C}$, the Hessian form of Ψ at $(x,w) \in X \times \mathbb{C}$ satisfies an estimate*

$$\partial \overline{\partial} \Psi_{(x,w)} [\theta, \eta]^2 = \int_{\zeta \in T_{X,x}} \partial \overline{\partial} \psi \cdot (\tau \wedge \overline{\tau} + |w|^2 V)_{\mathrm{exph}_x(w\zeta)} \, \chi(|\zeta|^2) \, d\lambda(\zeta)$$

$$+ O(|w|^{N-1}) [\theta, \eta]^2$$

where τ (resp. V) is a vector field (resp. a $(1,1)$-vector field) depending smoothly on the parameters x, w and linearly (resp. quadratically) on θ, η, and where the notation $[\tau]^2$ stands for the $(1,1)$-vector $\tau \wedge \overline{\tau} \in \Lambda^{1,1} T_X$. The vector fields τ, V are given at $y = \mathrm{exph}_x(w\zeta)$ by

$$\tau_y = \partial \operatorname{exph}_{(x,w\zeta)} (\theta^h + \eta \zeta^v + |w|^2 \Xi_y^v),$$

$$V_y = \partial \operatorname{exph}_{(x,w\zeta)} (U^v - |w|^2 \Xi^v \wedge \overline{\Xi^v})_y,$$

where $\theta^h, \zeta^v \in T(T_X)_{(x,w\zeta)}$ are respectively the horizontal lifting of θ with respect to the connection D and the vertical tangent vector associated to ζ, and where Ξ, U can be expressed in the coordinates of Proposition 2.1 (ii) by

$$\Xi_y = \sum_{\alpha,j,l,m} \frac{1}{\chi(|\zeta|^2)} \frac{\partial}{\partial \overline{\zeta}_l}(\chi_1(|\zeta|^2)\overline{\zeta}^{\alpha-1_m})\overline{d_{\alpha j l}} \frac{\alpha_m}{|\alpha|} \overline{w}^{|\alpha|-2}\theta_j \frac{\partial}{\partial z_m},$$

$$U_y = \sum_{l,m} \frac{1}{2}(U_{m,l} + \overline{U_{l,m}})\frac{\partial}{\partial z_m} \wedge \frac{\partial}{\partial \overline{z}_l},$$

$$U_{m,l} = -\frac{\chi_1(|\zeta|^2)}{\chi(|\zeta|^2)}\Big\{ \sum_{j,k} c_{jklm}\theta_j\overline{\theta}_k$$

$$+ 2\sum_{\alpha,j,k} e_{\alpha jkm} w^{|\alpha|-1}\frac{\alpha_l}{|\alpha|}\zeta^{\alpha-1_l}\theta_j\overline{\theta}_k$$

$$+ 2\sum_{\alpha,k} d_{\alpha km}(|\alpha|-1)w^{|\alpha|-2}\frac{\alpha_l}{|\alpha|}\zeta^{\alpha-1_l}\eta\overline{\theta}_k\Big\}$$

$$+ \sum_{\alpha,\beta,j,k} d_{\alpha km}\overline{d_{\beta j l}}\, w^{|\alpha|-1}\overline{w}^{|\beta|-1}\zeta^\alpha\overline{\zeta}^\beta\theta_j\overline{\theta}_k.$$

Here (c_{jklm}) is the curvature tensor of the given hermitian metric ω on X, and χ_1 denotes the primitive $\chi_1(t) = \int_{+\infty}^t \chi(u)\,du$ of χ such that $\chi_1(t) = 0$ for $t \geq 1$. Moreover, $\alpha, \beta \in \mathbb{N}^n$ run over all multi-indices such that $2 \leq |\alpha|, |\beta| \leq N$. The formula is exact when ω is real analytic and $N = \infty$.

Proof A brute force differentiation of (3.16) at $z = 0$ gives

$$\partial\overline{\partial}\Psi_{(x,w)}[\theta,\eta]^2 = \int_{\mathbb{C}^n} \partial\overline{\partial}(\psi \circ h)_{(0,ws)}[\theta,\eta s]^2\, \chi(A(0,w,s))\, B(0,w,s)\,d\lambda(s)$$

$$(3.17) \qquad\qquad - \int_{\mathbb{C}^n} \overline{\partial}(\psi \circ h)_{(0,ws)}[\theta,\eta s]\, E_{(w,s)}[\theta,\eta]\,d\lambda(s)$$

$$- \int_{\mathbb{C}^n} \partial(\psi \circ h)_{(0,ws)}[\theta,\eta s]\, \overline{E_{(w,s)}[\theta,\eta]}\,d\lambda(s)$$

$$(3.18) \qquad\qquad - \int_{\mathbb{C}^n} \psi \circ h(0,ws)\, F_{(w,s)}[\theta,\eta]^2\,d\lambda(s)$$

where

$$E_{(w,s)} = -\partial_{(z,w)}(\chi(A(z,w,s))B(z,w,s)),$$

$$F_{(w,s)} = -\partial\overline{\partial}_{(z,w)}(\chi(A(z,w,s))B(z,w,s))$$

at $z = 0$, $|w| < \varepsilon_0$. Clearly, terms $w^p z\overline{z}$ in A or B play no role in $E_{(w,s)}$, while terms $w^p\overline{z}$ contribute either linearly as $dw \wedge d\overline{z}$ differentials of $F_{(w,s)}$ or quadratically as $dz \wedge d\overline{z}$ differentials. This gives

$$E_{(w,s)}[\theta,\eta] = \chi'(|s|^2) \sum_{\alpha,j,l} \overline{d_{\alpha jl}}\, \overline{w}^{|\alpha|-1} \overline{s}^{\alpha} s_l \theta_j$$

$$+ \chi(|s|^2) \sum_{\alpha,j,l} \overline{d_{\alpha jl}}\, \overline{w}^{|\alpha|-1} \alpha_l \overline{s}^{\alpha - 1_l} \theta_j + O(|w|^{N-1}|s|^N)[\theta,\eta],$$

$$F_{(w,s)}[\theta,\eta]^2 = \chi'(|s|^2) \sum_{j,k,l,m} c_{jklm} s_l \overline{s}_m \theta_j \overline{\theta}_k + \chi(|s|^2) \sum_{j,k,l} c_{jkll} \theta_j \overline{\theta}_k$$

$$+ 2\,\mathrm{Re}\,\{\chi'(|s|^2) \sum_{\alpha,j,k,m} e_{\alpha jkm} w^{|\alpha|-1} s^{\alpha} \overline{s}_m \theta_j \overline{\theta}_k$$

$$+ \chi(|s|^2) \sum_{\alpha,j,k,m} e_{\alpha jkm} w^{|\alpha|-1} \alpha_m s^{\alpha - 1_m} \theta_j \overline{\theta}_k$$

$$+ \chi'(|s|^2) \sum_{\alpha,k,m} d_{\alpha km}(|\alpha|-1) w^{|\alpha|-2} s^{\alpha} \overline{s}_m \eta \overline{\theta}_k$$

$$+ \chi(|s|^2) \sum_{\alpha,k,m} d_{\alpha jkm}(|\alpha|-1) w^{|\alpha|-2} \alpha_m s^{\alpha - 1_m} \eta \overline{\theta}_k \}$$

$$- \chi''(|s|^2) \sum_{\alpha,\beta,j,k,l,m} d_{\alpha km}\overline{d_{\beta jl}}\, w^{|\alpha|-1}\overline{w}^{|\beta|-1} s^{\alpha}\overline{s}^{\beta} s_l \overline{s}_m \theta_j \overline{\theta}_k$$

$$- \chi'(|s|^2) \sum_{\alpha,\beta,j,k,m} d_{\alpha km}\overline{d_{\beta jm}}\, w^{|\alpha|-1}\overline{w}^{|\beta|-1} s^{\alpha}\overline{s}^{\beta} \theta_j \overline{\theta}_k$$

$$- \chi'(|s|^2) \sum_{\alpha,\beta,j,k,l,m} d_{\alpha km}\overline{d_{\beta jl}}\, w^{|\alpha|-1}\overline{w}^{|\beta|-1} s^{\alpha}\overline{s}_m \beta_l \overline{s}^{\beta - 1_l} \theta_j \overline{\theta}_k$$

$$- \chi'(|s|^2) \sum_{\alpha,\beta,j,k,l,m} d_{\alpha km}\overline{d_{\beta jl}}\, w^{|\alpha|-1}\overline{w}^{|\beta|-1} \alpha_m s^{\alpha - 1_m} \overline{s}^{\beta} s_l \theta_j \overline{\theta}_k$$

$$- \chi(|s|^2) \sum_{\alpha,\beta,j,k,l,m} d_{\alpha km}\overline{d_{\beta jl}}\, w^{|\alpha|-1}\overline{w}^{|\beta|-1} \alpha_m \beta_l s^{\alpha - 1_m} \overline{s}^{\beta - 1_l} \theta_j \overline{\theta}_k$$

$$+ O(|w|^{N-2}|s|^N)[\theta,\eta]^2.$$

We find

$$E_{(w,s)}[\theta,\eta] = \sum_{l,m} \frac{\partial^2}{\partial \overline{s}_l \partial s_m}\left(\chi_1(|s|^2) \sum_{\alpha,j} \overline{d_{\alpha jl}}\, \overline{w}^{|\alpha|-1} \frac{\alpha_m}{|\alpha|} \overline{s}^{\alpha - 1_m} \theta_j\right)$$

$$+ O(|w|^{N-1}|s|^N)[\theta,\eta],$$

$$F_{(w,s)}[\theta,\eta]^2 = \sum_{l,m} \frac{\partial^2}{\partial \bar{s}_l \partial s_m}\left(\chi_1(|s|^2) \sum_{j,k} c_{jklm}\theta_j \bar{\theta}_k\right)$$

$$+ 2\operatorname{Re}\left\{ \sum_{l,m} \frac{\partial^2}{\partial \bar{s}_l \partial s_m}\left(\chi_1(|s|^2) \sum_{\alpha,j,k} e_{\alpha jkm} w^{|\alpha|-1} \frac{\alpha_l}{|\alpha|} s^{\alpha-1_l}\theta_j \bar{\theta}_k\right) \right.$$

$$+ \sum_{l,m} \frac{\partial^2}{\partial s_l \partial \bar{s}_m}\left(\chi_1(|s|^2) \sum_{\alpha,k} d_{\alpha km}(|\alpha|-1) w^{|\alpha|-2}\frac{\alpha_l}{|\alpha|}\bar{s}^{\alpha-1_l}\eta\bar{\theta}_k\right)\Big\}$$

$$- \sum_{l,m} \frac{\partial^2}{\partial \bar{s}_l \partial s_m}\left(\chi(|s|^2) \sum_{\alpha,\beta,j,k} d_{\alpha km}\overline{d_{\beta jl}}\, w^{|\alpha|-1}\overline{w}^{|\beta|-1} s^{\alpha}\bar{s}^{\beta}\theta_j \bar{\theta}_k\right)$$

$$+ O(|w|^{N-2}|s|^N)[\theta,\eta]^2.$$

In all these expansions, the remainder terms $O(\bullet)$ involve uniform constants when the origin x of coordinates belongs to a compact subset of a coordinate patch. By the mean value properties of plurisubharmonic functions, we have

$$\int_{|s|<1} |\psi(x+ws)|\, d\lambda(s) \leq C(1+|\log|w||)$$

locally uniformly in x. An integration by parts with compact supports yields

$$\int_{|s|<1} \partial(\psi \circ h)_{(0,ws)}\, O(|w|)\, d\lambda(s) = \int_{|s|<1} \psi \circ h(0,ws)\, d\lambda(s) = O(\log|w|),$$

hence the remainder terms $O(|w|^{N-1})$ in $E_{(w,s)}$ and $O(|w|^{N-2})$ in $F_{(w,s)}$ give contributions of order at most $O(|w|^{N-2}\log|w|)$ in $\partial\bar{\partial}\Psi$ as $|w|$ tends to 0. An integration by parts in (3.17) and (3.18) gives

$$\partial\bar{\partial}\Psi_{(x,w)}[\theta,\eta]^2 = \int_{\mathbb{C}^n} \partial\bar{\partial}(\psi \circ h)_{(0,ws)} \cdot \{(\theta,\eta s) \wedge \overline{(\theta,\eta s)} + |w|^2(0,\Xi) \wedge \overline{(\theta,\eta s)}$$

$$+ |w|^2(\theta,\eta s) \wedge \overline{(0,\Xi)} + |w|^2(0,U)\}\,\chi(A(0,w,s))\,B(0,w,s)\,d\lambda(s)$$

$$+ O(|w|^{N-2}\log|w|)[\theta,\eta]^2$$

with

$$\Xi = \sum_{\alpha,j,l,m} \frac{1}{\chi(|s|^2)}\frac{\partial}{\partial \bar{s}_l}\left(\chi_1(|s|^2)\bar{s}^{\alpha-1_m}\right)\overline{d_{\alpha jl}}\,\frac{\alpha_m}{|\alpha|}\,\overline{w}^{|\alpha|-2}\theta_j\,\frac{\partial}{\partial z_m},$$

$$U = \sum_{l,m} \frac{1}{2}(U_{m,l} + \overline{U_{l,m}})\,\frac{\partial}{\partial z_m} \wedge \frac{\partial}{\partial \bar{z}_l},$$

$$
U_{m,l} = -\frac{\chi_1(|s|^2)}{\chi(|s|^2)} \Big\{ \sum_{j,k} c_{jklm} \theta_j \bar{\theta}_k
$$

$$
+ 2 \sum_{\alpha,j,k} e_{\alpha jkm} w^{|\alpha|-1} \frac{\alpha_l}{|\alpha|} s^{\alpha-1_l} \theta_j \bar{\theta}_k
$$

$$
+ 2 \sum_{\alpha,k} d_{\alpha km} (|\alpha|-1) w^{|\alpha|-2} \frac{\alpha_l}{|\alpha|} s^{\alpha-1_l} \eta \bar{\theta}_k \Big\}
$$

$$
+ \sum_{\alpha,\beta,j,k} d_{\alpha km} \overline{d_{\beta jl}}\, w^{|\alpha|-1} \overline{w}^{|\beta|-1} s^\alpha \bar{s}^\beta \theta_j \bar{\theta}_k .
$$

The choice $\chi(t) = \frac{C}{(1-t)^2} \exp\left(\frac{1}{t-1}\right)$ for $t < 1$ gives $\chi_1(t) = -C \exp\left(\frac{1}{t-1}\right)$, so $\chi_1(t)/\chi(t) = (1-t)^2$ is smooth and bounded, and our vector fields Ξ, U are smooth. We can write

$$
(\theta, \eta s) \wedge \overline{(\theta, \eta s)} + |w|^2 (0, \Xi) \wedge \overline{(\theta, \eta s)} + |w|^2 (\theta, \eta s) \wedge \overline{(0, \Xi)} + |w|^2 (0, U)
$$
$$
= (\theta, \eta s + |w|^2 \Xi) \wedge \overline{(\theta, \eta s + |w|^2 \Xi)} + (0, U - |w|^2 \Xi \wedge \bar{\Xi}),
$$

therefore (3.19) implies the formula in Proposition 3.1 with

$$
\tau = dh_{(0,ws)}(\theta, \eta s + |w|^2 \Xi),
$$
$$
V = dh_{(0,ws)}(0, U - |w|^2 \Xi \wedge \bar{\Xi}).
$$

Since $\mathrm{exph}_z(\zeta) = h(z, \rho(z,\xi))$, $\rho(0,\xi) = \xi + O(\xi^{N+1})$ and $\partial \rho_{(0,\xi)} = d\xi + O(\xi^N) d\xi$ by (2.9), we infer that the $(1,0)$-differential of exph at $(x, \zeta) \in T_X$ is

$$
\partial \mathrm{exph}_{(x,\zeta)} = dh_{(0,\xi)} + O(|\xi|^N) d\xi
$$

modulo the identification of the tangent spaces $T(T_X)_{(x,\zeta)}$ and $T(T_{\mathbb{C}^n})_{(0,\xi)}$ given by the coordinates (z, ξ) on T_X. However, these coordinates are precisely those which realize the splitting $T(T_X)_{(x,\zeta)} = (T_{X,x})^h \oplus (T_{X,x})^v$ with respect to the connection D. Since $s = \xi + O(w^N \xi^{N+1})$ and $\xi = \zeta$ at $z = 0$, we get

$$
\tau = \partial \mathrm{exph}_{(x,w\zeta)}(\theta^h + \eta \zeta^v + |w|^2 \Xi^v) + O(|w|^N |\zeta|^N),
$$
$$
V = \partial \mathrm{exph}_{(x,w\zeta)}(U^v - |w|^2 \Xi^v \wedge \overline{\Xi^v}) + O(|w|^N |\zeta|^N).
$$

We can drop the terms $O(|w|^N)$ in τ and V because

$$
\int_{|\zeta|<1} \partial\bar{\partial}\psi(\mathrm{exph}_x(w\zeta))\, d\lambda(\zeta) = \frac{1}{|w|^{2n}} \int_{|\zeta|<|w|} \partial\bar{\partial}\psi(\mathrm{exph}_x(\zeta))\, d\lambda(\zeta)
$$

$$
(3.19) \hspace{5.5cm} = O(|w|^{-2})
$$

by the usual estimates on Lelong numbers (see e.g. [12]), thus

$$
\int_{|\zeta|<1} \partial\bar{\partial}\psi(\mathrm{exph}_x(w\zeta))\, O(|w|^N)\, d\lambda(\zeta) = O(|w|^{N-2}).
$$

After substituting ζ to s in the formal expression of Ξ and U, we get precisely the formulas given in Proposition 3.1. It remains to explain why the remainder term $O(|w|^{N-2}\log|w|)$ is in fact of the type $O(|w|^{N-1})$. To see this, we increase N by two units and estimate the additional terms in the expansions, due to the contribution of all multi-indices α with $|\alpha| = N+1$ or $N+2$. It is easily seen that the additional terms in Ξ and U are $O(|w|^{N-1})$, so they are $O(|w|^{N+1})$ in τ and $|w|^2 V$. The contribution of these terms to $\partial\bar\partial\Psi_{(x,w)}$ is thus of the form

$$\int_{|\zeta|<1} \partial\bar\partial\psi(\,\mathrm{exph}_x(w\zeta))\, O(|w|^{N+1})\, d\lambda(\zeta) = O(|w|^{N-1}).$$

$\square$

4 Approximation theorem and estimates

A straightforward consequence of our computations is the following approximation theorem for closed positive currents.

Theorem 4.1 *Let X be a compact n-dimensional manifold equipped with a smooth hermitian metric ω. Fix a smooth semipositive $(1,1)$-form u on X such that the Chern curvature tensor $\Theta(T_X) \in \mathrm{Herm}(T_X \otimes T_X)$ of (T_X, ω) satisfies*

$$(\Theta(T_X) + u \otimes \mathrm{Id}_{T_X})(\theta \otimes \xi, \theta \otimes \xi) \geq 0$$

for all $\theta, \xi \in T_X$ such that $\langle\theta, \xi\rangle = 0$. Let $T = \alpha + \frac{i}{\pi}\partial\bar\partial\psi$ be a closed real current where α is a smooth closed real $(1,1)$-form and ψ is quasi-psh. Suppose that $T \geq \gamma$ for some real $(1,1)$-form γ with continuous coefficients. As w tends to 0 and x runs over X, there is a uniform lower bound

$$\alpha_x[\theta]^2 + \frac{i}{\pi}\partial\bar\partial\Psi_{(x,w)}[\theta,\eta]^2 \geq \gamma_x[\theta]^2 - \lambda(x,|w|)\, u_x[\theta]^2 - \delta(|w|)\,|\theta|^2 - \frac{1}{\pi}K(|\theta||\eta| + |\eta|^2)$$

where $K > 0$ is a sufficiently large constant, $\delta(t)$ a continuous increasing function with $\lim_{t\to 0}\delta(t) = 0$, and

$$\lambda(x,t) = t\,\frac{\partial}{\partial t}(\Psi(x,t) + Kt^2).$$

The above derivative $\lambda(x,t)$ is a nonnegative continuous function on $X \times\,]0, \varepsilon_0[$ which is increasing in t and such that

$$\lim_{t\to 0} \lambda(x,t) = \nu(T,x) \ \text{(Lelong number of } T \text{ at } x\text{)}.$$

In particular, the currents $T_\varepsilon = \alpha + \frac{i}{\pi}\partial\bar\partial\Psi(\bullet, \varepsilon)$ are smooth closed real currents converging weakly to T as ε tends to 0, such that

$$T_\varepsilon \geq \gamma - \lambda(\bullet, \varepsilon)\, u - \delta(\varepsilon)\,\omega.$$

Proof It suffices to prove the estimate for $|w| < \varepsilon(\delta)$, with $\delta > 0$ fixed in place of $\delta(|w|)$. Also, the estimate is local on X. If we change ψ into $\psi + \psi_0$ with a smooth function ψ_0, then α is changed into $\alpha - \frac{i}{\pi}\partial\overline{\partial}\psi_0$ and Ψ into $\Psi + \Psi_0$, where Ψ_0 is a smooth function on $X \times \mathbb{C}$ such that $\Psi_0(z,w) = \psi_0(z) + O(|w|^2)$. It follows that the estimate remains unchanged up to a term $O(1)|\eta|^2$. We can thus work on a small coordinate open set $\Omega \subset X$ and choose ψ_0 such that $\gamma - (\alpha - \frac{i}{\pi}\partial\overline{\partial}\psi_0)$ is positive definite and small at x, say equal to $\frac{\delta}{4}\omega_x$. After shrinking Ω and making the change $\psi \mapsto \psi + \psi_0$, we may in fact suppose that $T = \alpha + \frac{i}{\pi}\partial\overline{\partial}\psi$ on $\Omega_{x,\delta} \subset \Omega$, where α satisfies $\gamma_x - \alpha_x = \frac{\delta}{4}\omega_x$ and $\gamma - \frac{\delta}{2}\omega \le \alpha \le \gamma$ on $\Omega_{x,\delta}$. In particular, $\frac{i}{\pi}\partial\overline{\partial}\psi \ge \gamma - \alpha$ and ψ is plurisubharmonic on $\Omega_{x,\delta}$. All we have to show is that

$$(4.20) \qquad \frac{i}{\pi}\partial\overline{\partial}\Psi_{(x,w)}[\theta,\eta]^2 \ge -\lambda(x,|w|)\,u_x[\theta]^2 - \frac{\delta}{2}|\theta|^2 - \frac{1}{\pi}K(|\theta||\eta| + |\eta|^2)$$

for $|w| < w_0(\delta)$ small. For this, we apply Proposition 3.1 at order $N = 2$; order 2 is enough for our purposes since we will neglect all terms in $\partial\overline{\partial}\Psi$ which converge to 0 with w, especially all $O(|w|)$ terms. By (3.19), we can neglect all terms of the form $\partial\overline{\partial}\psi(\mathrm{exph}_x(w\zeta))O(|w|^3)$ under the integral sign. Up to such terms, $\partial\overline{\partial}\psi \cdot (\tau \wedge \overline{\tau} + |w|^2 V)_{\mathrm{exph}_x(w\zeta)}\chi(|\zeta|^2)$ is equal to

$$-|w|^2\chi_1(|\zeta|^2)\,\mathrm{Re}\sum_{l,m}\frac{\partial^2\psi}{\partial\overline{z}_l\partial z_m}\Big\{\frac{\chi(|\zeta|^2)}{-|w|^2\chi_1(|\zeta|^2)}\overline{\tau}_l\tau_m + \sum_{j,k}c_{jklm}\theta_j\overline{\theta}_k$$

$$+ 2\sum_{|\alpha|=2,k}d_{\alpha km}(|\alpha|-1)w^{|\alpha|-2}\frac{\alpha_l}{|\alpha|}\zeta^{\alpha-1_l}\eta\overline{\theta}_k\Big\}$$

$$\ge -|w|^2\chi_1(|\zeta|^2)\sum_{l,m}\frac{\partial^2\psi}{\partial\overline{z}_l\partial z_m}\Big\{\frac{1}{|w|^2}\overline{\tau}_l\tau_m + \sum_{j,k}c_{jklm}(\theta_j\overline{\theta}_k + \frac{1}{2}\zeta_j\eta\overline{\theta}_k + \frac{1}{2}\overline{\zeta}_k\theta_j\overline{\eta})\Big\}$$

$$= -|w|^2\chi_1(|\zeta|^2)\sum_{l,m}\frac{\partial^2\psi}{\partial\overline{z}_l\partial z_m}\Big\{\frac{1}{|w|^2}\overline{\tau}_l\tau_m + \sum_{j,k}c_{jklm}\tau_j\overline{\tau}_k$$

$$- \sum_{j,k}c_{jklm}(\frac{1}{2}\zeta_j\eta\overline{\theta}_k + \frac{1}{2}\overline{\zeta}_k\theta_j\overline{\eta} + \zeta_j\overline{\zeta}_k\eta\overline{\eta})\Big\}$$

in view of Proposition 2.1 (iii) and of the inequality $0 \le -\chi_1 \le \chi$; in the last equality, we have also used the fact that $\tau = \theta + \eta\zeta + O(|w|)$.

By (3.19), the mixed terms $\theta_j\overline{\eta}$, $\eta\overline{\theta}_k$ and the terms $\eta\overline{\eta}$ give rise to contributions bounded below by $-K'(|\theta||\eta| + |\eta|^2)$. Hence we get the estimate

$$\frac{i}{\pi}\partial\overline{\partial}\Psi_{(x,w)}[\theta,\eta]^2 \ge \frac{1}{\pi}|w|^2\int_{\mathbb{C}^n}-\chi_1(|\zeta|^2)$$

$$\sum_{j,k,l,m}\frac{\partial^2\psi}{\partial\overline{z}_l\partial z_m}(\mathrm{exph}_x(w\zeta))\,(c_{jklm} + \frac{1}{|w|^2}\delta_{jm}\delta_{kl})\tau_j\overline{\tau}_k\,d\lambda(\zeta)$$

$$(4.21) \qquad\qquad -K'(|\theta||\eta| + |\eta|^2).$$

We need a lemma.

Lemma 4.2 *Suppose that the curvature assumption of Theorem 4.1 is satisfied. Then for every $\varepsilon > 0$, there is a constant $M_\varepsilon > 0$ such that*

$$\sum_{j,k,l,m} \frac{1}{2\pi}(c_{jklm} + M_\varepsilon \delta_{jm}\delta_{kl})\tau_j\overline{\tau}_k\xi_l\overline{\xi}_m + \sum_{j,k,l} u_{jk}\tau_j\overline{\tau}_k\xi_l\overline{\xi}_l + \varepsilon|\tau|^2|\xi|^2 \geq 0$$

for all tangent vectors τ, ξ.

Proof Let us consider the hermitian form H on $T_X \otimes T_X$ defined by

$$H(\tau \otimes \xi, \tau \otimes \xi) = \sum_{j,k,l,m} \frac{1}{2\pi}c_{jklm}\tau_j\overline{\tau}_k\xi_l\overline{\xi}_m + \sum_{j,k,l} u_{jk}\tau_j\overline{\tau}_k\xi_l\overline{\xi}_l + \varepsilon|\tau|^2|\xi|^2.$$

Let μ be the infimum of $H(\tau \otimes \xi, \tau \otimes \xi)$ on the compact set $\{|\tau| = 1\} \times \{|\xi| = 1\}$. By our curvature assumption we have $H(\tau \otimes \xi, \tau \otimes \xi) \geq \varepsilon$ when $\tau \perp \xi = 0$ and $|\tau| = |\xi| = 1$, therefore $H(\tau \otimes \xi, \tau \otimes \xi) \geq 0$ on some neighborhood $|\langle \tau, \xi \rangle| < r_\varepsilon$ of that set. It follows that

$$H(\tau \otimes \xi, \tau \otimes \xi) + \frac{|\mu|}{r_\varepsilon^2}|\langle \tau, \xi \rangle|^2 \geq 0$$

for all $|\tau| = |\xi| = 1$. Lemma 4.2 follows with $M_\varepsilon = |\mu|/r_\varepsilon^2$. $\qquad\square$

Let us apply the inequality of Lemma 4.2 to each vector ξ in a basis of eigenvectors of $\partial\overline{\partial}\psi$, multiply by the corresponding (nonnegative) eigenvalue and take the sum. We get

$$\frac{1}{2\pi}\sum_{j,k,l,m}\frac{\partial^2\psi}{\partial\overline{z}_l\partial z_m}(c_{jklm} + M_\varepsilon\delta_{jm}\delta_{kl})\tau_j\overline{\tau}_k + \sum_l \frac{\partial^2\psi}{\partial z_l\partial\overline{z}_l}(u[\tau]^2 + \varepsilon|\tau|^2) \geq 0.$$

Combining this with (4.21) for $|w|^2 < 1/M_\varepsilon$, we infer

$$\frac{i}{\pi}\partial\overline{\partial}\Psi_{(x,w)}[\theta,\eta]^2$$

$$\geq -2|w|^2\int_{\mathbb{C}^n} -\chi_1(|\zeta|^2)\sum_l \frac{\partial^2\psi}{\partial z_l\partial\overline{z}_l}(\,\mathrm{exph}_x(w\zeta))\,(u_x[\tau]^2 + \varepsilon|\tau|^2)d\lambda(\zeta)$$

$$- K'(|\theta||\eta| + |\eta|^2)$$

$$\geq -\{2|w|^2\int_{\mathbb{C}^n} -\chi_1(|\zeta|^2)\sum_l \frac{\partial^2\psi}{\partial z_l\partial\overline{z}_l}(\,\mathrm{exph}_x(w\zeta))\,d\lambda(\zeta)\}(u_x[\theta]^2 + \varepsilon|\theta|^2)$$

$$- K''(|\theta||\eta| + |\eta|^2)$$

by (3.19) again, in combination with the equality $\tau = \theta + \eta\zeta + O(|w|)$. The change of variables $\zeta \mapsto s$ defined by $\mathrm{exph}_x(w\zeta) = x + ws$ yields $\zeta = s + O(w^2 s^3)$ by Proposition 2.1 (ii), hence choosing $\varepsilon \ll \delta$ small enough we get

$$\frac{i}{\pi}\partial\overline{\partial}\Psi_{(x,w)}[\theta,\eta]^2 \geq -\lambda_\Omega(x,|w|)\,u_x[\theta]^2 - \frac{\delta}{3}|\theta|^2 - K(|\theta||\eta| + |\eta|^2)$$

where

$$\lambda_\Omega(x, |w|) = 2|w|^2 \int_{\mathbb{C}^n} -\chi_1(|s|^2) \sum_l \frac{\partial^2 \psi}{\partial z_l \partial \overline{z}_l}(x + ws)\, d\lambda(s).$$

By definition, the Lelong number $\nu(\psi, x)$ is $\lim_{r \to 0} \nu(\psi, x, r)$ where

$$\nu(\psi, x, r) = \frac{1}{\pi^{n-1} r^{2n-2}/(n-1)!} \int_{B(x,r)} \frac{2}{\pi} \sum_l \frac{\partial^2 \psi}{\partial z_l \partial \overline{z}_l}\, d\lambda(z).$$

Therefore we have

$$\nu(\psi, x, |w|r) = \frac{1}{\pi^{n-1} r^{2n-2}/(n-1)!} |w|^2 \int_{|s|<r} \frac{2}{\pi} \sum_l \frac{\partial^2 \psi}{\partial z_l \partial \overline{z}_l}(x + ws)\, d\lambda(s).$$

As $-\chi_1(|s|^2) = 2 \int_{|s|}^{+\infty} \chi(r^2)\, r\, dr$, the Fubini formula gives

$$\lambda_\Omega(x, |w|) = 4|w|^2 \int_0^{+\infty} \left\{ \int_{|s|<r} \sum_l \frac{\partial^2 \psi}{\partial z_l \partial \overline{z}_l}(x + ws)\, d\lambda(s) \right\} \chi(r^2)\, r\, dr$$

$$= \frac{2\pi^n}{(n-1)!} \int_0^1 \nu(\psi, x, |w|r)\, \chi(r^2)\, r^{2n-1}\, dr,$$

$$\lambda_\Omega(x, t) = \int_{\mathbb{C}^n} \nu(\psi, x, t|s|)\, \chi(|s|^2)\, d\lambda(s).$$

Hence $\lambda_\Omega(x, t)$ is smooth in (x, t), increasing in t and $\lim_{t \to 0} \lambda_\Omega(x, t) = \nu(\psi, x) = \nu(T, x)$, as desired.

The expected estimate (4.20) thus holds on Ω with $\lambda_\Omega(x, t)$ in place of $\lambda(x, t)$; we only have to show that $\lambda_\Omega(x, t) - \lambda(x, t)$ is small. By (4.20), $\Psi(x, w) + K|w|^2$ is plurisubharmonic in w, and so is a convex function of $\log|w|$; since $\Psi(x, t)$ is bounded above as t tends to 0, the sum $\Psi(x, t) + Kt^2$ which is convex in $\log t$ must be increasing in t. Therefore

$$\lambda(x, t) = \frac{\partial}{\partial \log t}(\Psi(x, t) + Kt^2)$$

is a nonnegative increasing function of t. When we put $\theta = 0$, Proposition 3.1 gives $\Xi = V = 0$ and $\tau = \partial \exp h_{(x, w\zeta)}(\eta\zeta^v)$, thus

$$\frac{\partial^2 \Psi}{\partial w \partial \overline{w}}(x, w) = \int_{\zeta \in T_{X,x}} \partial\overline{\partial}\psi_{\exp h_x(w\zeta)}[\zeta]^2\, \chi(|\zeta|^2)\, d\lambda(\zeta) + O(|w|^{N-1})$$

$$= \int_{\mathbb{C}^n} \partial\overline{\partial}\psi_{x+ws}[s]^2\, \chi(|s|^2)\, d\lambda(s) + O(1),$$

again by the change of variable $\exp h_x(w\zeta) = x + ws$, for which $\zeta = s + O(w^2 s^3)$. Since $\partial^2/\partial w \partial \overline{w} = (t^{-1}\partial/\partial t) \circ (t\,\partial/\partial t)$ for a function of w depending only on $t = |w|$, a multiplication by t followed by an integration implies

$$(4.22) \quad t\frac{\partial \Psi(x,t)}{\partial t} = t\frac{\partial}{\partial t}\int_{\mathbb{C}^n} \psi(x+ts)\,\chi(|s|^2)\,d\lambda(s) + O(t^2)$$

$$= \int_{\mathbb{C}^n} \nu(\psi,x,t|s|)\,\chi(|s|^2)\,d\lambda(s) + O(t^2) = \lambda_\Omega(x,t) + O(t^2)\,;$$

here, we used the well known fact that $\nu(\psi,x,t)$ is equal to the derivative $\partial/\partial\log t$ of the mean value of ψ on the sphere $S(x,t)$. Hence $\lambda_\Omega(x,t) - \lambda(x,t) = O(t^2)$ and the first estimate in Theorem 4.1 follows. Now, it is clear that ψ_ε converges to ψ in L^1_{loc}, so T_ε converges weakly to T; note also that $\psi_\varepsilon + K\varepsilon^2$ is increasing in ε by the above arguments. The proof is complete. $\qquad\square$

Remark 4.3 *An integration of the first equality in (4.22) also shows that*

$$\Psi(x,t) = \int_{|s|<1} \psi(x+ts)\,\chi(|s|^2)\,d\lambda(s) + O(t^2)$$

relatively to the system of normal coordinates at x defined in section 2.

Remark 4.4 *The estimates obtained in Theorem 4.1 can be slightly improved by setting*

$$\widetilde{\Psi}(x,w) = \Psi(x,w) + |w|, \ \widetilde{\lambda}(x,t) = t\frac{\partial}{\partial t}(\widetilde{\Psi}(x,t)).$$

Indeed $\widetilde{\lambda}(x,t) = \lambda(x,t) + t - 2Kt^2$ is increasing in t and larger than $\lambda(x,t)$ for t small, hence $\widetilde{\Psi}(x,w)$ is convex and increasing in $\log|w|$, while

$$\partial\overline{\partial}\widetilde{\Psi}_{(x,w)}[\theta,\eta]^2 = \partial\overline{\partial}\Psi_{(x,w)}[\theta,\eta]^2 + \frac{|\eta|^2}{4|w|}.$$

Since $K|\theta||\eta| \leq 2K^2|w||\theta|^2 + |\eta|^2/(8|w|)$, we get for $|w| < \varepsilon_0$ small enough a lower bound of the form

$$(4.23) \quad \alpha_x[\theta]^2 + \frac{i}{\pi}\partial\overline{\partial}\widetilde{\Psi}_{(x,w)}[\theta,\eta]^2 \geq \gamma_x[\theta]^2 - \widetilde{\lambda}(x,|w|)\,u_x[\theta]^2 - \widetilde{\delta}(|w|)\,|\theta|^2,$$

where $\lim_{t\to 0}\widetilde{\lambda}(x,t) = \nu(T,x)$ and $\lim_{t\to 0}\widetilde{\delta}(t) = 0$, $\widetilde{\delta}$ being continuous and increasing.

5 Compact manifolds with partially semipositive curvature

In case the curvature of the ambient manifold X satisfies suitable semipositivity assumptions, our smoothing theorem 4.1 yields interesting geometric consequences.

Definition 5.1 *We say that X has partially semipositive curvature (in the sense of Griffiths) if there exists a hermitian metric ω on X such that the associated Chern curvature tensor $\Theta(T_X)$ satisfies*

$$\Theta(T_X)(\theta\otimes\xi,\theta\otimes\xi) \geq 0$$

for all $\theta,\xi \in T_X$ with $\langle\theta,\xi\rangle = 0$.

The standard Griffiths semipositivity assumption for T_X would require that the above inequality is satisfied for *all pairs* $\theta, \xi \in T_X$. This is the case if X is a compact complex homogeneous manifold, since the tangent bundle is then generated by global sections; a fortiori, such manifolds satisfy the property in Def. 5.1. However, the partial semipositivity condition is substantially weaker. In fact, every complex curve has partially semipositive curvature, but a curve has a metric of semipositive curvature if and only if it is of genus 0 or 1. This observation gives rise immediately to higher dimensional examples; in fact let $X = \mathbb{P}^1 \times C$ where C is a curve of genus $g \geq 2$, equipped with its unique metric of constant curvature -1. The Fubini-Study metric on $\mathbb{P}^1$ has curvature $+2$, hence we have

$$\langle \Theta(T_X)(\theta \otimes \xi), \theta \otimes \xi \rangle = 2|\theta_1|^2|\xi_1|^2 - |\theta_2|^2|\xi_2|^2,$$

where the subscripts $1, 2$ denote respectively the first and second projection of a tangent vector; clearly the difference is nonnegative when $\langle \theta_1, \xi_1 \rangle + \langle \theta_2, \xi_2 \rangle = 0$, thus $\mathbb{P}^1 \times C$ has partially semipositive curvature. It would be interesting to have a more algebraic understanding of the notion of partial semipositivity. This is certainly a very difficult problem; recall in this context the deep unsolved conjecture of Griffiths that ampleness is equivalent to Griffiths positivity.

The following results were already obtained in [5] under the assumption that T_X is nef (i.e. $\mathcal{O}_{T_X}(1)$ is nef over $P(T_X^\star)$). It is well known that Griffiths semipositivity implies nefness, and that the converse is not true; however, if Griffiths' conjecture holds, nefness would be equivalent to the fact that the curvature tensor $\Theta(T_X)$ can be made larger than $-\varepsilon$ for every $\varepsilon > 0$ (see [6]). The above examples show that partial semipositivity covers different situations.

Proposition 5.2 *Let (X, ω) be a compact hermitian manifold with partially semipositive curvature.*

(i) *For any closed real $(1,1)$-current T on X such that $T \geq \gamma$ for some continuous real $(1,1)$-form γ on X, there is a family T_ε of smooth approximations in the same $\partial\bar{\partial}$-cohomology class as T and converging weakly to T as ε tends to 0, such that $T_\varepsilon \geq \gamma - \delta_\varepsilon \omega$ with $\lim_{\varepsilon \to 0} \delta_\varepsilon = 0$.*

(ii) *Every closed positive $(1,1)$-current T is numerically effective in the sense of [5]; in particular, if X is Kähler, the De Rham cohomology class $\{T\}$ satisfies $\int_Y \{T\}^p \geq 0$ for every p-dimensional analytic subset $Y \subset X$.*

Proof (i) follows immediately from Theorem 4.1 by taking $u = 0$.

To get (ii), we apply (i) with $\gamma = 0$. We then obtain $T_\varepsilon \geq -\delta_\varepsilon \omega$. Let ω_0 be a Kähler metric on X (ω_0 need not be related to ω). Multiplying ω_0 by a suitable constant we get $\omega_0 \geq \omega$ and so $T_\varepsilon + \delta_\varepsilon \omega_0 \geq 0$. As $\{T\} = \{T_\varepsilon\}$, this implies

$$\int_Y (\{T\} + \delta_\varepsilon\{\omega_0\})^p = \int_Y (T_\varepsilon + \delta_\varepsilon\omega_0)^p \geq 0,$$

thus $\int_Y \{T\}^p \geq 0$ in the limit. $\qquad\qquad\qquad\qquad\qquad\qquad\qquad\qquad\qquad\qquad\square$

Proposition 5.2 implies that a compact complex manifold with partially semipositive curvature must be minimal in the sense that it does not contain any divisor D which can be blown down to some lower dimensional variety. Indeed, in the latter case, there is always a curve $C \subset D$ such that $D \cdot C = \int_C \{D\} < 0$.

Proposition 5.3 *Let X be a compact complex manifold with partially semipositive curvature. Then X is Kähler if and only if X is in the Fujiki class $\mathcal{C}$ ($=$ class of complex varieties which are bimeromorphic to Kähler manifolds), and X is projective if and only if X is Moishezon.*

Proof L et X be a manifold in the Fujiki class $\mathcal{C}$ such that T_X has partially semipositive curvature. We know by Fujiki [7] that X has a smooth Kähler modification $\mu : \widetilde{X} \longrightarrow X$. Let β be a Kähler metric on $\widetilde{X}$ and let $T = \mu_\star \beta$ be the direct image current. If ω is a hermitian metric on X, we have $\beta \geq c\,\mu^\star\omega$ on $\widetilde{X}$ for some small constant $c > 0$, thus by taking the direct image we get $T = \mu_\star\beta \geq c\omega$ on X. Now Proposition 5.2 (i) produces a smooth approximation T_ε of T such that $T_\varepsilon \geq c\omega - \delta_\varepsilon\omega \geq \frac{c}{2}\omega$ for ε small enough. Thus T_ε is a Kähler metric and X is Kähler. In particular, if X is Moishezon, we can take β to be the curvature form of an ample line bundle $\mathcal{O}(D)$ over $\widetilde{X}$. Then T and T_ε are cohomologous to $\mu_\star([D]) = [D']$ where $D' = \mu(D)$ is the image of D in X. This implies that $\mathcal{O}(D')$ is ample and therefore X is projective algebraic (we could also have applied the well-known result of Moishezon that every Kähler Moishezon manifold is projective algebraic). The converse implications are trivial. $\qquad\qquad\square$

6 Singularity attenuation process for closed (1,1)-currents

If T is a closed positive or almost positive current on a complex manifold X, we denote by $E_c(T)$ the c-upperlevel set of Lelong numbers:

$$E_c(T) = \{x \in X \; ; \; \nu(T, x) \geq c\}, \; c > 0.$$

A well-known theorem of [13] asserts that $E_c(T)$ is a closed analytic subset of X. A combination of Theorem 4.1 with Kiselman's singularity attenuation technique yields the following partial regularization process. The sets $E_c(T)$ appear precisely to be the obstructions to smoothing when no loss of positivity is admitted.

Theorem 6.1 *Let T be a closed almost positive $(1, 1)$-current and let α be a smooth real $(1, 1)$-form in the same $\partial\bar{\partial}$-cohomology class as T, i.e. $T = \alpha + \frac{i}{\pi}\partial\bar{\partial}\psi$ where ψ is an almost psh function. Let γ be a continuous real $(1, 1)$-form such that $T \geq \gamma$. Suppose that T_X is equipped with a smooth hermitian metric ω such that the Chern curvature form satisfies*

$$(\Theta(T_X) + u \otimes \mathrm{Id}_{T_X})(\theta \otimes \xi, \theta \otimes \xi) \geq 0 \; \forall\theta, \xi \in T_X \text{ with } \langle\theta, \xi\rangle = 0,$$

for some continuous nonnegative $(1,1)$-form u on X. Then for every $c > 0$, there is a family of closed almost positive $(1,1)$-currents $T_{c,\varepsilon} = \alpha + \frac{i}{\pi}\partial\bar{\partial}\psi_{c,\varepsilon}$, $\varepsilon \in {]}0, \varepsilon_0[$, such that $\psi_{c,\varepsilon}$ is smooth on $X \setminus E_c(T)$, increasing with respect to ε, and converges to ψ as ε tends to 0 (in particular, the current $T_{c,\varepsilon}$ is smooth on $X \setminus E_c(T)$ and converges weakly to T on X), and such that

(i) $T_{c,\varepsilon} \geq \gamma - \min\{\lambda_\varepsilon, c\}u - \delta_\varepsilon\omega$ *where:*

(ii) $\lambda_\varepsilon(x)$ *is an increasing family of continuous functions on X such that* $\lim_{\varepsilon\longrightarrow 0}\lambda_\varepsilon(x) = \nu(T,x)$ *at every point,*

(iii) δ_ε *is an increasing family of positive constants such that* $\lim_{\varepsilon\longrightarrow 0}\delta_\varepsilon = 0$,

(iv) $\nu(T_{c,\varepsilon}, x) = (\nu(T,x) - c)_+$ *at every point $x \in X$.*

Proof Following an idea of Kiselman [10], we let $\psi_{c,\varepsilon}$ be the Legendre transform

$$\psi_{c,\varepsilon}(x) = \inf_{|w|<1}\left(\widetilde{\Psi}(x, \varepsilon w) + \frac{\varepsilon}{1 - |w|^2} - c\log|w|\right),$$

where $\widetilde{\Psi}(x, w) = \Psi(x, w) + |w|$ is the function defined in Remark 4.4. It is clear that $\psi_{c,\varepsilon}$ is increasing in ε and that

$$\lim_{\varepsilon\longrightarrow 0}\psi_{c,\varepsilon}(x) = \widetilde{\Psi}(x, 0_+) = \Psi(x, 0_+) = \psi(x).$$

Moreover, as $\widetilde{\Psi}(x, w)$ is convex and increasing in $t = \log|w|$, the function

$$t \longmapsto \Phi_{c,\varepsilon}(x, t) := \widetilde{\Psi}(x, \varepsilon t) + \frac{\varepsilon}{1 - t^2} - c\log t$$

is strictly convex in $\log t$ and tends to $+\infty$ as t tends to 1. It follows that the infimum is attained for $t = t_0(x) \in [0, 1[$ given either by the zero of the $\partial/\partial\log t$ derivative:

$$\widetilde{\lambda}(x, \varepsilon t) + \frac{2\varepsilon t^2}{(1 - t^2)^2} - c = 0$$

when $\nu(T, x) = \lim_{t\longrightarrow 0}\widetilde{\lambda}(x, t) < c$, or by $t_0(x) = 0$ when $\nu(T, x) \geq c$ (in which case the above strictly convex function is increasing on the whole interval $[0, 1[$). Since the $\partial/\partial\log t$ derivative is itself strictly increasing in t, the implicit function theorem shows that $t_0(x)$ depends smoothly on x on $X \setminus E_c(T) = \{\nu(T, x) < c\}$, hence $\psi_{c,\varepsilon}(x) = \Phi_{c,\varepsilon}(x, t_0(x))$ is actually smooth on $X \setminus E_c(T)$.

Now, fix a point $x \in X \setminus E_c(T)$ and $t_1 > t_0(x)$. For all z in a neighborhood V of x we still have $t_0(z) < t_1$, thus

$$\psi_{c,\varepsilon}(z) = \inf_{|w|<t_1}\left(\widetilde{\Psi}(z, \varepsilon w) + \frac{\varepsilon}{1 - |w|^2} - c\log|w|\right) \text{ on } V.$$

By (4.23), all functions involved in that infimum have a complex Hessian in (z, w) bounded below by

$$\gamma_z - \alpha_z - \widetilde{\lambda}(z, \varepsilon t_1)u_z - \widetilde{\delta}(\varepsilon t_1)\omega_z.$$

Hence $\partial\overline{\partial}\psi_{c,\varepsilon}$ has the same lower bound (this is an easy consequence of Kiselman's principle [9] that an infimum $\inf_w u(z,w)$ of plurisubharmonic functions depending only on $\mathrm{Re}\,w$ is plurisubharmonic in z). By taking t_1 arbitrarily close to $t_0(x)$ and by shrinking V, the lower bound comes arbitrary close to

$$\gamma_x - \alpha_x - \widetilde{\lambda}(x,\varepsilon t_0(x))\,u_x - \widetilde{\delta}(\varepsilon t_0(x))\omega_x \geq \gamma_x - \alpha_x - \min\{\widetilde{\lambda}(x,\varepsilon),c\}\,u_x - \widetilde{\delta}(\varepsilon)\omega_x,$$

because $\widetilde{\lambda}(x,\varepsilon t_0(x)) = c - 2\varepsilon t_0(x)^2/(1-t_0(x)^2)^2 \leq c$ and $\widetilde{\lambda}(x,t)$, $\widetilde{\delta}(t)$ are increasing in t. Therefore we get

$$\alpha + \frac{i}{\pi}\partial\overline{\partial}\psi_{c,\varepsilon} \geq \gamma - \min\{\widetilde{\lambda}(\bullet,\varepsilon),c\}\,u - \widetilde{\delta}(\varepsilon)\omega$$

on $X \setminus E_c(T)$. However, as the lower bound is a continuous $(1,1)$-form and as $\psi_{c,\varepsilon}$ is quasi-psh, it is immediate to check that the lower bound extends to X by continuity. Hence the properties of Theorem 6.1 (i), (ii), (iii) are proved.

Remark 4.3 finally shows that $\psi_{c,\varepsilon}$ differs locally from Kiselman's usual Legendre transform only by bounded terms, thus

$$\nu(\psi_{c,\varepsilon},x) = (\nu(\psi,x) - c)_+$$

at every point $x \in X$ by Kiselman's results [9], [10]. Therefore the property of Theorem 6.1 (iv) also holds. $\qquad\square$

Bibliography

[1] Demailly, J.-P.: *Estimations L^2 pour l'opérateur $\overline{\partial}$ d'un fibré vectoriel holomorphe semi-positif au dessus d'une variété kählérienne complète*. Ann. Sci. Ec. Norm. Sup. **15** (1982), 457-511.

[2] Demailly, J.-P.: *Singular hermitian metrics on positive line bundles* in Proceedings of the Bayreuth conference "Complex algebraic varieties", April 2-6, 1990, edited by K. Hulek, T. Peternell, M. Schneider, F. Schreyer, Lecture Notes in Math. n° 1507, Springer-Verlag, (1992).

[3] Demailly, J.-P.: *A numerical criterion for very ample line bundles*. Prépublication n° 153, Institut Fourier, Univ. Grenoble I, September (1991), to appear in J. Differential Geom.

[4] Demailly, J.-P.: *Monge-Ampère operators, Lelong numbers and intersection theory*, Prépub. Inst. Fourier n° 173, Mai (1991), 70p, to appear in the Proceedings "Complex Analysis and Geometry", CIRM, Trento (Italy), edited by V. Ancona and A. Silva.

[5] Demailly, J.-P.: *Regularization of closed positive currents and Intersection Theory*. J. Alg. Geom. **1** (1992) 361–409.

[6] Demailly, J.P., Peternell, Th., Schneider, M.: *Compact complex manifolds with numerically effective tangent bundles*, Prépub. Inst. Fourier n° 184, Octobre (1991), 48p.

[7] A. Fujiki.: *Closedness of the Douady spaces of compact Kähler spaces*. Publ. R.I.M.S., Kyoto Univ. **14** (1978) 1–52.

[8] Hörmander, L.: "An introduction to Complex Analysis in several variables", North-Holland Math. libr., vol.7, Amsterdam, London 1966, third edition, 1991.

[9] Kiselman, C.O.: *The partial Legendre transformation for plurisubharmonic functions*. Invent. Math. **49** (1978) 137–148.

[10] Kiselman, C.O.: *Densité des fonctions plurisousharmoniques*. Bull. Soc. Math. France **107** (1979) 295–304.

[11] Lelong, P.: *Intégration sur un ensemble analytique complexe*. Bull. Soc. Math. France **85** (1957) 239–262.

[12] Lelong, P.: "Plurisubharmonic functions and positive differential forms", Gordon and Breach, New-York, and Dunod, Paris 1969

[13] Siu, Y.T.: *Analyticity of sets associated to Lelong numbers and the extension of closed positive currents*. Invent. Math. **27** (1974) 53–156.

Pseudoconvex domains of semiregular type

Klas Diederich and Gregor Herbort

Dedicated to Professor Pierre Dolbeault on the occasion of his retirement

1 Introduction and main results

In this article we develop the geometric tools needed for obtaining more precise analytic information than known so–far on a relatively large class of bounded pseudoconvex domains $\Omega \subset \mathbb{C}^n$ with C^∞–smooth boundary of finite type. The article is a continuation of [9], where, among other things, the following was shown:

Theorem 1.1 *On the class $\mathcal{C}$ of bounded smooth real–analytic pseudoconvex domains $\Omega \subset \mathbb{C}^3$, $0 \in \partial\Omega$, among the following six local biholomorphic boundary–invariants*
- *i) the 1–type $\tau(\partial\Omega, 0)$,*
- *ii) the multitype $\mathcal{M}(\partial\Omega, 0)$ (in the sense of D. Catlin),*
- *iii) the order of pseudoconvex extendability $N(\partial\Omega, 0)$,*
- *iv) the exponent of pseudoconvex extendability $e(\partial\Omega, 0)$,*
- *v) the order of subellipticity $\epsilon_1(\partial\Omega, 0)$,*
- *vi) the growth exponent for the Bergman kernel $g(\partial\Omega, 0)$,*

neither $\epsilon_1(\partial\Omega, 0)$ nor $g(\partial\Omega, 0)$ is a function of the respective five other invariants.

(For precise definitions of these invariants the reader is refered to [9].)

On the other hand, it is known, that certain inequalities between the invariants mentioned in Theorem 1.1 always hold. In fact, in [9] the following result was proved in this respect:

Theorem 1.2 *Let $\Omega \subset\subset \mathbb{C}^n$ be a pseudoconvex domain, such that $\partial\Omega$ is C^∞–smooth near a point $0 \in \partial\Omega$. Suppose, $\tau(\partial\Omega, 0) < \infty$ and denote the multitype of $\partial\Omega$ at 0 by $\mathcal{M}(\partial\Omega, 0) = (m_1, \ldots, m_n)$. Furthermore, let p be the rank of the Levi–form $\mathcal{L}_{\partial\Omega}(0, \cdot)$. Then the following estimates hold:*

$$(1.1) \qquad 2\left(\sum_{i=1}^{p+2} \frac{1}{m_i} + \frac{n-p-2}{N(\partial\Omega, 0)}\right) \leq 2e(\partial\Omega, 0) \leq g(\partial\Omega, 0) \leq 2\left(\sum_{i=1}^{n} \frac{1}{m_i}\right)$$

Remark *As the examples used for proving Theorem 1.1 show, strict inequalities can occur for certain domains at all places in (1.1).*

These two results stress, what was already known before to some extent: the geometric invariants $\tau(\partial\Omega, 0)$, $\mathcal{M}(\partial\Omega, 0)$, $N(\partial\Omega, 0)$ and $e(\partial\Omega, 0)$ carry a lot of information on the analysis of pseudoconvex domains of finite type. However, they do not even suffice for a precise description of the basic analytic invariants.

The main point of this article is to show, that this difficulty disappears under the following natural, since biholomorphically invariant, additional condition:

Definition 1.3 *A domain Ω as in Theorem 1.2 is called of semiregular type at the point 0, if it satisfies the condition*

$$(1.2) \qquad\qquad m_i = \Delta_{n-i+1}(\partial\Omega, 0)$$

for all $i = 2, \ldots, n$. Here, for $1 \leq q \leq n - 1$, we mean by $\Delta_q(\partial\Omega, 0)$ the q–type (in the sense of d'Angelo) of $\partial\Omega$ at 0.

Remarks *a) It is known from examples (for details see [10]), that in (1.2) strict inequality can occur for any i, $3 \leq i \leq n$. However, for $i = 2$, one always has equality. This shows, in particular, that in dimension 2 all domains of finite type are of semiregular type. (In fact, everything, we will prove in this article for domains of semiregular type, is known for $n = 2$.)*

b) It is important to observe, that the condition (1.2) is not open, as can easily be seen from the following example:
We put

$$\Omega := \left\{ z \in \mathbb{C}^4 \ : \ r(z) := \operatorname{Re} z_1 + |z_2|^6 + |z_3|^6 + |z_4|^6 + |z_2|^2 |z_3|^2 |z_4|^2 < 0 \right\}$$

Then $\partial\Omega$ is of semiregular type at 0 with $\mathcal{M}(\partial\Omega, 0) = (1, 6, 6, 6)$, but at all points $q \in \partial\Omega$ of the form $q = (-a^6, a, 0, 0)$ one has $\mathcal{M}(\partial\Omega, q) = (1, 2, 4, 4)$, whereas $\tau(\partial\Omega, q) = 6$.

c) Our terminology is perhaps a little bit misleading, since it might suggest, that smooth pseudoconvex domains Ω of regular type at a point $z^0 \in \partial\Omega$ in the sense, that for all $1 \leq q \leq n - 1$ the type $\Delta_q(\partial\Omega, z^0)$ is in the appropriate sense realized by the order of contact of a smooth complex manifold of dimension q passing through z^0, is also of semiregular type at z^0. However, this implication does not hold as the following example shows (for more details see [10]):

$$\Omega := \left\{ z \in \mathbb{C}^4 \ : \ r(z) := \operatorname{Re} z_1 + |z_2|^8 + |z_3|^8 + |z_4|^8 + |z_2|^2 |z_3|^2 |z_4|^2 < 0 \right\}$$

For this domain the 1–type at all points $q \in \partial\Omega$ is realized by the order of contact of smooth complex curves passing through q and also the 2–type at all such points is (in the correct sense) realized by the order of contact of 2–dimensional complex manifolds passing through q. Nevertheless one has $\mathcal{M}(\partial\Omega, 0) = (1, 6, 6, 6)$, whereas $\tau(\partial\Omega, 0) = 8$.

The consideration of the class of domains of semiregular type is motivated by the following recent results in the literature. McNeal [17] and, independently, H. Boas and E. Straube [2] showed:

Theorem 1.4 *If the domain Ω as in Theorem 1.2 is linear convex near 0, then $\tau(\partial\Omega, 0)$ is equal to the maximum of the orders of contact between $\partial\Omega$ and all complex lines through the origin.*

Remark *Boas and Straube prove this result, in fact, for a somewhat larger class of domains.*

J. Y. Yu extended the result to the multitype by proving in [23]:

Theorem 1.5 *If the domain Ω as in Theorem 1.2 is linear convex near 0, then one can obtain a realizing coordinate system for the multitype $\mathcal{M}(\partial\Omega, 0) = (m_1, \ldots, m_n)$ by a linear coordinate transformation. In the new coordinates the entries $m_2, \ldots, m_n$ are the orders of contact between the respective coordinate axis and $\partial\Omega$ at 0. Furthermore, one has for each $i \in \{2, \ldots, n\}$ the relation $m_i = \Delta_{n-i+1}(\partial\Omega, 0)$. In other words, the domain Ω is near 0 of semiregular type.*

These results were basic for obtaining important new information on the Bergman kernel and metric on convex domains in the recent articles by J. McNeal [18] and J. McNeal and E. M. Stein [19]. However, in his proofs, McNeal uses the linear convexity also in other important ways. In this article, we want to deduce interesting analytic information just from the semiregularity condition for the type. At first, however, we will show the following geometric facts:

Theorem 1.6 *Let $\Omega \subset\subset \mathbb{C}^n$ be a pseudoconvex domain, $z^0 \in \partial\Omega$. Suppose, that $\partial\Omega$ is C^∞–smooth near z^0 and that it is of semiregular type at z^0. Then one has:*

$$(1.3) \qquad N(\partial\Omega, z^0) = \tau(\partial\Omega, z^0)$$

and

$$(1.4) \qquad e(\partial\Omega, z^0) = \sum_{i=1}^{n} \frac{1}{m_i}$$

where the m_i are the entries of the multitype $\mathcal{M}(\partial\Omega, z^0)$.

Together with Theorem 1.2, one obtains immediately the following analytic information:

Corollary 1.7 *Let Ω and z^0 be as in Theorem 1.6. Then the growth order of the Bergman kernel function of Ω at z^0 is*

$$(1.5) \qquad g(\partial\Omega, z^0) = 2\sum_{i=1}^{n} \frac{1}{m_i}$$

Also the growth order $h(\partial\Omega, z^0)$ of the Bergman metric can be described explicitly from the geometric information on domains of semiregular type. Namely, we will show:

Theorem 1.8 *Let Ω and $z^0 \in \partial\Omega$ be as in Theorem 1.6. Then one has*

$$(1.6) \qquad h(\partial\Omega, z^0) = \frac{1}{\tau(\partial\Omega, z^0)}$$

Remarks *This result can be better understood in the following context:*

a) It is known to hold in $\mathbb{C}^2$ (see D. Catlin [6]);

b) In [16] J. McNeal showed, that, on all pseudoconvex domains Ω with C^∞–smooth boundary of finite type near a point $z^0 \in \partial\Omega$, the growth order of the Bergman metric satisfies the estimate

$$(1.7) \qquad\qquad h\left(\partial\Omega, z^0\right) \geq \epsilon_1\left(\partial\Omega, z^0\right)$$

c) D. Catlin showed in [3] the general estimate

$$\epsilon_1\left(\partial\Omega, z^0\right) \leq \frac{1}{\tau(\partial\Omega, z^0)}$$

d) In [9], for each positive integer m, a pseudoconvex real–analytic domain $\Omega_m \subset\subset \mathbb{C}^n$ with $0 \in \partial\Omega_m$ was constructed such that always $h(\partial\Omega_m, 0) \geq \frac{1}{16}$, whereas $\tau(\partial\Omega_m, 0) = 2m$. This shows, that, in general, one cannot hope, to have equivalence between the two sides of (1.7). Furthermore, because of (1.6), the Ω_m of [9] are not of semiregular type at 0 for all $m > 8$.

The main result behind Theorems 1.6 and 1.8 is the following characterization of domains of semiregular type, which gives, at the same time, precise relations between different geometric invariants (for all definitions see [9]):

Theorem 1.9 *Let $\Omega \subset\subset \mathbb{C}^n$ be a bounded pseudoconvex domain, the boundary of which is C^∞–smooth and of finite type near the point $z^0 \in \partial\Omega$. Then the following properties are equivalent:*

i) $\partial\Omega$ is of semiregular type at z^0;

ii) There is a distinguished weight $\mu = (1, \mu_2, \ldots, \mu_n)$ (in the sense of [5]) and a defining function r of Ω near z^0, which, in the coordinates (z) belonging to μ, has the form

$$(1.8) \qquad r(z) = \operatorname{Re} z_1 + P(z') + R_1(z') + (Im\, z_1)R_2(z') + R_3(z)$$

where

a) P is a real–valued plurisubharmonic polynomial (in which pluriharmonic terms do not appear) in the variables $z' := (z_2, \ldots, z_n)$ of weighted degree 1 with respect to $(\mu_2, \ldots, \mu_n)$,

b) $|R_1(z')| \leq C(\sum_{j=2}^n |z_j|^{\mu_j})^{1+\delta}$ for suitable positive constants C, δ,

c) $|R_2(z')| \leq C|z'|$, $|R_3(z)| \leq C(Im\, z_1)^2$,

d) If we put $r_0(z) := \operatorname{Re} z_1 + P(z')$ and $\Omega_0 := \{r_0 < 0\}$, then 0 is a point of finite type of Ω_0;

iii) The multitype $\mathcal{M}(\partial\Omega, z^0)$ is an admissible n–tuple of orders of extendability for $\partial\Omega$ at z^0;

iv) There is an n–tuple, which is at the same time an admissible weight and an admissible n–tupel of orders of extendability for $\partial\Omega$ at z^o.

Remarks *a) The fact, that i) implies iii), obviously immediately gives (1.4) of Theorem 1.6, and, therefore, also Corollary 1.7.*

b) Notice, that according to the definition of an admissible n–tupel of orders of extendability, iii) can be interpreted as a bumping theorem with precise orders for domains of semiregular type.

As we will prove in section 6, Theorem 1.9 also has the following important consequence:

Theorem 1.10 *Suppose the domain $\Omega \subset\subset \mathbb{C}^n$ is pseudoconvex with smooth C^∞–boundary and such that $\partial\Omega$ is of finite and semiregular type at a point $z^0 \in \partial\Omega$. Suppose, furthermore, that $\overline\Omega$ admits a Stein neighborhood basis. Then z^0 is a peak point for the algebra $A^0(\Omega)$. The function $h_{z^0} \in A^0(\Omega)$ peaking at z^0 can, in fact, be chosen to be Hölder continuous on $\overline\Omega$.*

Remarks *a) The hypothesis on $\overline\Omega$ admitting a Stein neighborhood basis is only used for globalizing the Hölder continuous local peak function constructed at first only near z^0. In case $\partial\Omega$ is everywhere of finite type, the existence of the Stein neighborhood basis follows automatically as shown in [4].*

b) In a forthcoming article ([11]) the authors will deduce further analytic properties of domains of semiregular type from Theorem 1.10.

This article is organized in the following way: In section 2 we show at first a bumping theorem for domains defined by weighted homogeneous polynomials of finite type. This is the basis for the proof of our geometric Theorem 1.9, which will be given afterwards in section 3. Theorem 1.9 is then the main tool in the derivation of the analytic results of Theorems 1.6 and 1.8 together with Corollary 1.7. The rest of the article is devoted to showing the existence of peak functions as claimed in Theorem 1.10. We construct in section 5 at first local holomorphic Hölder continuous peak functions by using the technical Lemma 5.2 the rather tedious proof of which is postponed to section 7. For the convenience of the readers we give in section 6 also the arguments allowing to pass from local to global peak functions.

2 A bumping for weighted homogeneous domains

One of the basic ingredients for proving the main results of this article is the following bumping result:

Proposition 2.1 *Let $P(z')$ be a plurisubharmonic polynomial without pluriharmonic terms and suppose, there are positive numbers $\mu_2,\dots,\mu_n$, such that P is homogeneous of degree 1 with respect to the weight $(\mu_2,\dots,\mu_n)$. Suppose, furthermore, that the origin $0 \in \mathbb{C}^n$ is a point of finite type of the domain*

$$\Omega_0 := \{\, z : \ r_0(z) := \operatorname{Re} z_1 + P(z') < 0 \,\}$$

Then there is a (real–valued) C^3–function $\widetilde P$ on $\mathbb{C}^{n-1}$, homogeneous of degree 1 with respect to $(\mu_2,\dots,\mu_n)$, with the following properties
i) $\widetilde P < P$ except at the origin;

ii) For $\epsilon > 0$ sufficiently small, the function

$$(2.9) \qquad \widetilde{P}_\epsilon(z') := \widetilde{P}(z') - 2\epsilon \sum_{j=2}^{n} |z_j|^{\mu_j}$$

is plurisubharmonic on $\mathbb{C}^{n-1}$.

Remarks *a) One might ask, whether, in this proposition, it is necessary to pass at first from the given plurisubharmonic function P as in the proposition to the new function $\widetilde{P}$ of the same kind and with i), before it is possible to subtract the term $2\epsilon \sum_{j=2}^{n} |z_j|^{\mu_j}$ in the described way. The answer is, in general, yes, as the following example shows: Let $n = 2$ and $P(z_2) := |z_2|^4 + 2\,\mathrm{Re}\,a z_2 \bar{z}_2^3$. It is easy to see, that P satisfies all the requirements of the proposition whenever $|a| \leq \frac{2}{3}$. However, if one choses $a = -\frac{2}{3}$, one has $\frac{\partial^2 P}{\partial z_2 \partial \bar{z}_2}|\mathbb{R} \equiv 0$. Therefore, subtracting $2\epsilon \sum_{j=2}^{n} |z_j|^{\mu_j}$ would destroy plurisubharmonicity.*
b) The proposition was proved by A. Noell in [20] for the special case that $\mu_2 = \cdots = \mu_n$.

Proof of Proposition 2.1: A) We subdivide the proof into two major steps. Namely, we assume, at first, that $4 \leq \mu_2 \leq \cdots \leq \mu_n$. (The other case, where the Levi form of P has positive rank, will be considered in part E) of this proof).

We choose a radius $R > 0$, such that

$$(2.10) \qquad S_0 := \{ (-P(z'), z') : |z'| \leq 2 \} \subset \Gamma := \partial\Omega_0 \cap B(0; R)$$

Since, for $t > 0$, the scaling

$$(2.11) \qquad F_t(z) := (t\,z_1, F_t'(z')) := \left(tz_1, t^{1/\mu_2} z_2, \ldots, t^{1/\mu_n} z_n \right)$$

maps the boundary $\partial\Omega_0$ bijectively to itself, we get from the fact, that the condition of finite type is biholomorphically invariant and open (Theorem 4.11 of [1]), that $\partial\Omega_0$ is everywhere of finite type, since this is the case near 0. Therefore, due to its upper semicontinuity (see the Main Theorem 2.2 of [5]), the multitype has only finitely many different values, say $\mathcal{M}^1 < \mathcal{M}^2 < \cdots < \mathcal{M}^L$, on $\partial\Omega_0$. For simplicity of notation we assume, that $\mathcal{M}^1 = (1, 2, \ldots, 2)$ (allthough, for the given $\partial\Omega_0$, this might not occur, in which case we just add it to the list of the $\mathcal{M}^j$). The set M of weakly pseudoconvex points on $\partial\Omega_0$ can now be stratified as

$$M = M_1 \cup M_2 \cup \cdots \cup M_{L-1}$$

with $M_i = \{ z \in \partial\Omega_0 : \mathcal{M}(\partial\Omega_0, z) = \mathcal{M}^{L-i+1} \}$. Notice, that each set M_i is invariant under the scaling maps (2.11). Because of the upper semicontinuity of the multitype, the set M_1 is closed, and, for each $j \geq 2$, the set M_j is a closed subset of $\partial\Omega_0 \setminus \bigcup_{i<j} M_i$. Furthermore, according to Propositions 2.1, 3.6 and 4.4 of [5], for each point $q \in M_j$ there is an open neighborhood U and a real CR–submanifold $S_q \subset \partial\Omega_0$ of holomorphic dimension 0 with $M_j \cap U \subset S_q$. The sets S_q can be chosen to be of the form

$$S_q = \{ z \in U : r_0(z) = r_{p+2}(z) = \cdots = r_n(z) = 0 \}$$

where $p = rank(\mathcal{L}_{\partial\Omega_0}(q))$ and the functions $r_{p+2},\dots,r_n$ belong to a special boundary system for $\partial\Omega_0$ near q (in the sense of Catlin). In particular, they are C^∞ on U and, as a little technical fact, which we will prove as Lemma 2.4 at the end of this section, shows, they can be chosen to be independent of z_1. Hence, the projection $S'_q := \{w' : (-P(w'), w') \in S_q\}$ is a real submanifold of an open subset of $\mathbb{C}^{n-1}$. We, now, put

$$\sigma(w') := \sum_{j=2}^{n} |w_j|^{\mu_j}$$

and $\Sigma := \{w' \in \mathbb{C}^{n-1} : \sigma(w') = 1\}$. Furthermore, for each $i \in \{1,\dots,L-1\}$, we define $M''_i := \{w' \in \mathbb{C}^{n-1} : (-P(w'), w') \in M_i\}$ and $M'_i := M''_i \cap \Sigma$. The sets M''_i now, obviously, have the following property:

(H) For each $w' \in M''_i$ there is an open neighborhood $U' \subset \mathbb{C}^{n-1}$ and a real submanifold $S' \subset U'$ with $M''_i \cap U' \subset S'$ such that the Levi form $\mathcal{L}_P$ is strictly positive at all points $v' \in M''_i \cap U'$, when applied to a holomorphic tangent vector to S'.

With this we are in a situation, where we can apply a modified version of the methods developed in [7] (in particular Lemmas 3 and 4). We show

Lemma 2.2 *For each $k = 1,\dots,L-1$ there is a constant $c_k > 0$, an open neighborhood U'_k of $M'_1 \cup \cdots \cup M'_k$ and a nonpositive function $g_k \in C^3(\mathbb{C}^{n-1})$, homogeneous of degree 1 with respect to $(\mu_2,\dots,\mu_n)$, such that $g_k < 0$ on U'_k and, for a sufficiently small $\epsilon > 0$,*

$$(2.12) \qquad \partial\overline{\partial}(P + \epsilon g_k) \geq \epsilon c_k \partial\overline{\partial}|z'|^2$$

on U'_k.

B) The proof of this lemma will, of course, be given by induction on k. For this, we need, however, at first the following

Lemma 2.3 *Let $k \in \{1,\dots,L-1\}$ and $\widetilde{M} \subset M'_k$ be compact. Furthermore, let $\widehat{h} \in C^3(\mathbb{C}^{n-1})$ be a function which is homogeneous of degree 1 with respect to $(\mu_2,\dots,\mu_n)$. Then there is a constant $b_k > 0$, an open neighborhood $\widetilde{W}$ of $\widetilde{M}$ and a function $h \in C^3(\mathbb{C}^{n-1})$ which is also homogeneous of degree 1 with respect to $(\mu_2,\dots,\mu_n)$ and such that*
i) h vanishes of at least second order along $\widetilde{M'_k} \cap \widetilde{W}$ and its Leviform $\mathcal{L}_h(z',\cdot)$ is positive semidefinite at all $z' \in \widetilde{M'_k} \cap \widetilde{W}$;
ii) for sufficiently small $\epsilon > 0$ one has

$$(2.13) \qquad \partial\overline{\partial}\left(P + \epsilon\left(\widehat{h} + h\right)\right) \geq \epsilon b_k \partial\overline{\partial}|z'|^2$$

on $\widetilde{W}$.

Proof 1) Choose $q' \in \widetilde{M}$. We use property (H). Thus, there is an open neighborhood $U = U_{q'} \subset \mathbb{C}^{n-1}$ of q' and a real C^∞-submanifold $S' = \{w' \in U : h_1(w') = \cdots = h_d(w') = 0\}$ with $M''_k \cap U \subset S'$ and such that $\mathcal{L}_P(w',\cdot)$ is positive definite on $T^{1,0}_{w'}S'$ for all $w' \in S'$. After shrinking U, we may assume, that the function $dist^2(\cdot, S')$ is C^∞ on U and we may assume, that the defining functions h_j have been chosen such that

$$dist^2(\cdot, S') = h_1^2 + \cdots + h_d^2$$

We now put for $z' \in \mathbb{C}^{n-1}$

$$(2.14) \qquad \varphi(z') := \left(\frac{z_2}{\sigma(z')^{1/\mu_2}}, \ldots, \frac{z_n}{\sigma(z')^{1/\mu_n}} \right)$$

and

$$(2.15) \qquad G_{q'}(z') := dist^2(\varphi(z'), S')$$

By direct calculation one obtains for $X' \in \mathbb{C}^{n-1}$

$$(2.16) \quad \partial\bar{\partial} G_{q'}\left(X', \overline{X}'\right)(q') = 2\sum_{i=1}^{d} |(\partial h_i(q'), X') - 2\,\mathrm{Re}\,R_i(q')(\partial\sigma(q'), X')|^2$$

with

$$R_i(q') := \sum_{l=2}^{n} \frac{\partial h_i}{\partial w_l}(q') \frac{q_l}{\mu_l}$$

If now $t > 0$ is close to 1, we get for F_t' as defined in (2.11), that $F_t'(q') \in M_k'' \cap U$. Therefore, the function $t \longmapsto h_i(F_t'(q'))$ vanishes identically for t close to 1. By differentiating it at $t = 1$ with respect to t, we obtain

$$(2.17) \qquad \mathrm{Re}\,R_i(q') = 0$$

This gives, together with (2.16) and property (H), after possibly shrinking U, that there is a positive constant b_k' and a small $\epsilon_0 > 0$, such that for all $0 < \epsilon' < \epsilon_0$ the following estimate holds on $U \cap \widetilde{M}$:

$$(2.18) \qquad \partial\bar{\partial}(P + \epsilon'\sigma G_q') \geq \epsilon' b_k' \partial\bar{\partial}|z'|^2$$

Next, we choose a positive constant K such that $-\partial\bar{\partial}\widehat{h} < K\partial\bar{\partial}|z'|^2$ on Σ. Then we get for $0 < \epsilon < \epsilon_1 := \frac{b_k'}{2K}\epsilon_0$, if we put $\epsilon' := 2K\epsilon/b_k'$ in (2.18):

$$\partial\bar{\partial}\left(P + \frac{2K}{b_k'}\epsilon\sigma G_{q'} \right) > 2K\epsilon\partial\bar{\partial}|z'|^2 > K\epsilon\partial\bar{\partial}|z'|^2 - \epsilon\partial\bar{\partial}\widehat{h}$$

The function

$$g_{q'} := \frac{2K}{b_k'}\sigma G_{q'}$$

is in $C^3(\varphi^{-1}(U))$, and one has for all $0 < \epsilon < \epsilon_1$

$$(2.19) \qquad \partial\bar{\partial}\left(P + \epsilon\left(g_{q'} + \widehat{h}\right) \right) \geq \epsilon K\partial\bar{\partial}|z'|^2$$

on $U \cap \widetilde{M}$.

Since $\widetilde{M}$ is compact, we can choose finitely many points $q'_{(1)}, \ldots, q'_{(l)} \in \widetilde{M}$ together with pairs of open neighborhoods $U''_i \subset\subset U'_i$ of $q'_{(i)}$ and cut–off–functions $\chi_i \in C_0^\infty(U'_i)$, such that

$$\widetilde{M} \subset \bigcup_{i=1}^{l} U''_i =: U''$$

and $\chi_1 + \cdots + \chi_l \equiv 1$ on U'' and

$$\partial\bar\partial\left(P + \epsilon\left(g_{q'_i} + \widehat{h}\right)\right) \geq \epsilon K \partial\bar\partial |z'|^2$$

on each U'_i, if $0 < \epsilon < \epsilon_2$ for a sufficiently small $0 < \epsilon_2 << \epsilon_1$. With these choices, the function

$$h := \sum_{i=1}^{l} \chi_i(\varphi(z')) g_{q'_i}(z')$$

has the following properties:

a') h vanishes on $M'_k \cap U''$, since $M'_k \cap U'_i \subset S_{q'_{(i)}}$ for all i; furthermore, the Leviform of h is positive semidefinite on $M'_k \cap U''$;

b') On $\widetilde{M}$ the following estimate holds

$$\partial\bar\partial\left(P + \epsilon\left(h + \widehat{h}\right)\right) \geq \epsilon K \partial\bar\partial |z'|^2$$

for sufficiently small $\epsilon > 0$.

The fact about the Levi form of h claimed in a') follows directly from the construction of h and 2.16. Again, since $\widetilde{M}$ is compact, this implies properties a) and b) of Lemma 2.3. $\square$

C) We, now, come to the proof of Lemma 2.2 by induction over k. We start with $k = 1$. The set $\widetilde{M} := M'_1$ is compact. Let h be the function obtained from Lemma 2.3 for $\widehat{h} := 0$. Then the function

$$H := h - \frac{b_1}{2}\sigma$$

is negative near $\widetilde{M}$, since $h|\widetilde{M} \equiv 0$ and $\sigma|\widetilde{M} \equiv 1$, and if we replace h in ii) of Lemma 2.3 by H, the inequality is still satisfied near $\widetilde{M}$. We choose open neighborhoods $U''_1 \subset\subset U'$ of $\widetilde{M}$ and a non–negative cut–off function $\psi \in C_0^\infty(U')$ with $\psi|U''_1 \equiv 1$. Then the function

$$g_1(z') := \psi(\varphi(z'))H(z')$$

is of class $C^3(\mathbb{C}^{n-1})$ and satisfies the requirements of Lemma 2.2 on $U'_1 := U''_1 \cap \varphi^{-1}(U''_1)$.

Next, we assume, that the Lemma has been proved for some $k < L - 1$. Then there is an open neighborhood U'_k of $M'_1 \cup \ldots \cup M'_k$ and a non–positive function g_k which is homogeneous of degree 1 with respect to the weight $(\mu_2, \ldots, \mu_n)$ and such that $g_k < 0$ on U'_k and (2.12) of Lemma 2.2 is satisfied on U'_k for any sufficiently small $\epsilon > 0$. We choose an open neighborhood $U''_k \subset\subset U'_k$ of $M'_1 \cup \ldots \cup M'_k$ and put $\widetilde{M} := M'_{k+1} \cap \complement U''_k$. Since the set $\widetilde{M}$ is compact, Lemma 2.3, applied with $\widehat{h} := g_k$, gives us an open neighborhood $\widetilde{W}'_{k+1}$ of $\widetilde{M}$ with $(M'_1 \cup \ldots \cup M'_k) \cap \widetilde{W}'_{k+1} = \emptyset$ and a function $g'_{k+1} \in C^3(\mathbb{C}^{n-1})$ homogeneous of degree 1 with respect to $(\mu_2, \ldots, \mu_n)$, vanishing of at least second order along $\widetilde{M}_{k+1} := \widetilde{W}'_{k+1} \cap M'_{k+1}$, having semipositive Leviform on $\widetilde{M}_{k+1}$, and such that, for a suitably small positive constant b_{k+1} its Leviform satisfies the estimate

$$(2.20) \qquad \partial\overline{\partial}\big(P + \epsilon(g_k + g'_{k+1})\big) \geq \epsilon b_{k+1} \partial\overline{\partial}|z'|^2$$

on $\widetilde{W}'_{k+1}$. Let now $\widetilde{W}''_{k+1} \subset\subset \widetilde{W}'_{k+1}$ be an open neighborhood of $\widetilde{M}$ and $\psi_1 \in C_0^\infty(\widetilde{W}'_{k+1})$ a non–negative cut–off function with $\psi_1|\widetilde{W}''_{k+1} \equiv 1$. We then define

$$h_{k+1}(z') := g_k(z') + \psi_1(\varphi(z'))g'_{k+1}(z')$$

and show, that, for sufficiently small $\epsilon > 0$, we get on $M'_1 \cup \ldots \cup M'_{k+1}$ the estimate

$$(2.21) \qquad \partial\overline{\partial}(P + \epsilon h_{k+1}) \geq \epsilon c''_{k+1} \partial\overline{\partial}|z'|^2$$

with a small positive constant c''_{k+1}. For this, for a given $q' \in M'_1 \cup \ldots \cup M'_{k+1}$ we will distinguish between three cases:

1st case: $q' \in T_1 := U'_k \cap \complement \widetilde{W}'_{k+1}$. Here we can estimate

$$\partial\overline{\partial}(P + \epsilon h_{k+1})(q') = \partial\overline{\partial}(P + \epsilon g_k)(q') \geq \epsilon c_k \partial\overline{\partial}|z'|^2$$

2nd case: $q' \in T_2 := \widetilde{W}''_{k+1}$. Then we have $\psi_1 \circ \varphi \equiv 1$ near q', such that we get:

$$\partial\overline{\partial}(P + \epsilon h_{k+1})(q') = \partial\overline{\partial}\big(P + \epsilon(g_k + g'_{k+1})\big)(q') \geq \epsilon b_{k+1} \partial\overline{\partial}|z'|^2$$

3rd case: $q' \in T_3 := M'_{k+1} \cap (\widetilde{W}'_{k+1} \cap \complement \widetilde{W}''_{k+1})$. Since $\widetilde{M} \subset \widetilde{W}''_{k+1}$, we get from this

$$q' \in M'_{k+1} \cap \complement\widetilde{M} = M'_{k+1} \cap \complement(M'_{k+1} \cap \complement U''_k) = M'_{k+1} \cap U''_k \subset U''_k$$

And since g'_{k+1} vanishes of at least second order in q', this gives

$$
\begin{aligned}
\partial\overline{\partial}(P + \epsilon h_{k+1})(q') \ &= \ \partial\overline{\partial}(P + \epsilon g_k)(q') + \epsilon\psi_1(q')\,\partial\overline{\partial}g'_{k+1}(q') \\
(2.22) \qquad &\geq \ \partial\overline{\partial}(P + \epsilon g_k)(q') \\
&\geq \ \epsilon c'_{k+1}\partial\overline{\partial}|z'|^2
\end{aligned}
$$

From this the estimate (2.21) now follows, since $M'_1 \cup \ldots \cup M'_{k+1} \subset T_1 \cup T_2 \cup T_3$.

Observe next, that we have $h_{k+1} \leq 0$ on $M'_1 \cup \ldots \cup M'_{k+1}$. This becomes clear from the definition of h_{k+1} together with the following facts:

1) $g_k \leq 0$ everywhere and $g_k < 0$ on $U'_k \supset\supset U''_k$ according to the inductive hypothesis;

2) g'_{k+1} vanishes on $M'_{k+1} \cap \widetilde{W}'_{k+1}$;

3) $\psi_1 \in C_0^\infty(\widetilde{W}'_{k+1})$.

(For the complete details, distinguish between the cases $z' \notin \widetilde{W}'_{k+1}$, $z' \in M'_{k+1} \cap \widetilde{W}'_{k+1}$ and $z' \notin \widetilde{W}'_{k+1}$.)

Therefore, with a small positive constant c''_{k+1}

$$H_{k+1} := h_{k+1} - \frac{1}{2}c''_{k+1}\sigma$$

is a C^3–smooth function, homogeneous of degree 1, negative on an open neighborhood U''_{k+1} of $M'_1 \cup \ldots \cup M'_{k+1}$ and satisfying there the estimate

$$\partial\overline{\partial}(P + \epsilon H_{k+1}) \geq \frac{\epsilon}{2}c''_{k+1}\partial\overline{\partial}|z|^2$$

Let, finally, $\widetilde{U}''_{k+1} \subset\subset U''_{k+1}$ be another open neighborhood of $M'_1 \cup \ldots \cup M'_{k+1}$ and $\psi \in C_0^\infty(U''_{k+1})$ a non–negative cut–off function with $\psi|\widetilde{U}''_{k+1} \equiv 1$. Then the function

$$g_{k+1} := \psi(\varphi(z'))H_{k+1}(z')$$

has the desired properties on $U'_{k+1} := \widetilde{U}''_{k+1} \cap \varphi^{-1}(\widetilde{U}''_{k+1})$. This finishes the inductive proof of Lemma 2.2. $\qquad\square$

D) By using Lemma 2.2 we, now, can at first easily finish the proof of Proposition 2.1 in the case $\mu_i \geq 4$ for all $i = 2, \ldots, n$. Namely, we get from the lemma for $k = L-1$ a nonpositive function $g \in C^3(\mathbb{C}^{n-1})$, weighted homogeneous of degree 1 with respect to $(\mu_2, \ldots, \mu_n)$, such that for any small enough $\epsilon > 0$ the function $P + \epsilon g$ is strictlyplurisubharmonic near $M'' \cap \Sigma$, where M'' is just the set of points, where P itself is not strictly plurisubharmonic. Therefore, for suitably chosen numbers $a, b \in (0, 1)$ the function

$$\tilde{P} := (1 - a)P + a(P + \epsilon g) - b\sigma$$

has the properties as desired in Proposition 2.1.

E) Finally, we still have to consider the case, where the rank p of the Levi form of P at the origin is ≥ 1. We may assume, that $\mu_2 = \cdots = \mu_{p+1} = 2$. We decompose P in the form

$$P(z') = P_2(z_2, \ldots, z_{p+1}) + 2\,\mathrm{Re}\sum_{j=2}^{p+1} z_j f_j(z_{p+2}, \ldots, z_n) + P''(z_{p+2}, \ldots, z_n)$$

with a polynomial P_2, weighted homogeneous of degree 1, complex polynomials $f_j, j = 2, \ldots, p+1$ of weighted degree of homogeneity $1/2$, and another homogeneous polynomial P'' of weighted degree 1 and depending only on $(z_{p+2}, \ldots, z_n)$. We also may assume, that no antiholomorphic terms occur in the f_j, because they can be swallowed up by a suitable biholomorphic change of coordinates which does not destroy the weighted homogeneity of any of the terms in the decomposition of P. Of course, we can apply Lemma 2.2 to the polynomials P_2 and P''. Therefore, the proof of Proposition 2.1 will be finished, if we show, that, in fact, $f_j = 0$ for each $j = 2, \ldots, p+1$. For this, we define for each such j and $v = (v_1, \ldots, v_3) \in \mathbb{C}^3$

$$g_j(v_2) := P(0, \ldots, v_2, 0, \ldots, 0)$$

where the variable v_2 is put at place number j. Furthermore, we put $d := \mu_{p+2} \cdots \mu_n$ and for a fixed $(z_{p+2}, \ldots, z_n) \in \mathbb{C}^{n-1}$

$$h_j(v_3) := f_j\left(v_3^{\frac{d}{\mu_{p+2}}} z_{p+2}, \ldots, v_3^{\frac{d}{\mu_n}} \right)$$

and

$$H_j(v_3) := P''\left(v_3^{\frac{d}{\mu_{p+2}}} z_{p+2}, \ldots, v_3^{\frac{d}{\mu_n}} \right)$$

Since P is plurisubharmonic, the domains

$$D_j := \left\{ v \in \mathbb{C}^3 \; : \; r_j(v) := \operatorname{Re} v_1 + g_j(v_2) + 2\operatorname{Re} v_2 h_j(v_3) + H_j(v_3) < 0 \right\}$$

are pseudoconvex. But on the other hand, the functions h_j are homogeneous of degree $\frac{d}{2}$ and H_j have degree of homogeneity d. A small calculation now shows, that, therefore, the mixed terms $v_2 h_j(v_3)$ would cause difficulties for the pseudoconvexity of D_j, unless they vanish. Hence $f_j \equiv 0$. This finishes the proof of the bumping lemma for weighted homogeneous domains. $\qquad\square$

As promised in part A) of the proof above, we will now show the following technical statement about special boundary systems in the sense of Catlin [5] (for all technical terms see this article).

Lemma 2.4 *Let $\Omega_0 = \{z : r_0(z =) \operatorname{Re} z_1 + P(z') < 0\}$ be as in Proposition 2.1. Then there is a boundary system for $\partial\Omega_0$ at 0 of rank $p = \operatorname{rank} \mathcal{L}_{\partial\Omega_0}(0)$ and codimension 0, such that its special functions $\tilde{r}_{p+2}, \ldots, \tilde{r}_n$ do not depend on z_1.*

Proof Let $\mathcal{M} = (1, 2, \ldots, m_{p+2}, \ldots, m_n)$ be the multitype of the boundary of Ω_0 at 0 and $\mathcal{B} = \{r_1, r_{p+2}, \ldots, r_n, L_{(2)}, \ldots, L_{(n)}\}$ be a boundary system. Then $\mathcal{M}$ is the commutator–multitype belonging to $\mathcal{B}$. Applying Proposition 3.6 of [5] with $\nu = n$ gives, that we have for all $k \in \{1, p+2, \ldots, n\}$

$$\frac{\partial^{|\alpha+\beta|} r_k}{\partial z^\alpha \partial \overline{z}^\beta}(0) = 0 \;\; \forall \alpha, \beta \in \mathbb{N}^n \; \text{with} \; \sum_{I=1}^n \frac{\alpha_i + \beta_i}{m_i} < \frac{1}{m_k}$$

Now we put $T_k := \{(\alpha, \beta) \in \mathbb{N}^n \times \mathbb{N}^n : \sum_{i=1}^n \frac{\alpha_i + \beta_i}{m_i} = \frac{1}{m_k}\}$ and

$$\tilde{r}_k(z) := \sum_{(\alpha, \beta) \in T_k} \frac{\partial^{|\alpha+\beta|} r_k}{\partial z^\alpha \partial \overline{z}^\beta}(0) z^\alpha \overline{z}^\beta$$

Then there is according to Proposition 4.4 of [5] a boundary system $\tilde{\mathcal{B}}$ at 0 with special functions just these $\tilde{r}_1, \tilde{r}_{p+2}, \ldots, \tilde{r}_n$. But, because $m_1 = 1$ and $m_k > 1$ for all $k \geq p+2$, the $\tilde{r}_k$ cannot depend on z_1 if $k \geq p+2$. $\qquad\square$

3 Proof of Theorem 1.9

We will need the following

Lemma 3.1 *Let $\Omega \subset \mathbb{C}^n$ be pseudoconvex with C^∞–smooth boundary. Furthermore, assume that $\partial\Omega$ is pseudoconvex extendable with finite order N at the point $z^0 \in \partial\Omega$. Then $\partial\Omega$ is of finite type at z^0 and one has the estimate*

$$\tau(\partial\Omega; z^0) \leq N$$

Remark *This statement is essentially contained in section 3 of [13]. For the convenience of the reader, we, nevertheless, give an explicit proof.*

Proof Let U be an open neighborhood of z^0, $r \in C^\infty(U)$ a defining function for Ω and $\rho : U \to \mathbb{R}$ a function of class C^3, which gives a pseudoconvex extension of $\partial\Omega$ at z^0. Suppose, furthermore, that $\gamma : \Delta \to \mathbb{C}^n$ is a holomorphic curve with $\gamma(0) = z^0$ touching $\partial\Omega$ at z^0 with order k, hence

$$\nu(r \circ \gamma, 0) = k\, \nu(\gamma - z^0, 0)$$

(Here, as usual, we mean by $\nu(\cdot, \cdot)$ orders of vanishing.)

We denote by ν^0 the exterior unit normal to $\partial\Omega$ at z^0. Since

$$|r(\gamma(\zeta))| < C_1 |\zeta|^{k\nu(\gamma,0)}$$

one has for all $0 < t << 1$

$$\sup_{|\zeta| \leq t} |r(\gamma(\zeta))| \leq C_1 t^{k\nu(\gamma,0)}$$

Hence, for a suitable constant $C_0 > 0$ independent of t, the curve

$$\Gamma(\zeta) := \gamma(\zeta) - C_0 t^{k\nu(\gamma,0)} \nu^0$$

lies inside Ω for $|\zeta| \leq t$. From the mean–value theorem for ρ we get immediately, with constants $C_1', C_2, C_2' > 0$ independent of t, the estimate

$$\begin{aligned}
\rho(\gamma(\zeta)) &= \rho(\Gamma(\zeta)) + \rho(\gamma(\zeta)) - \rho(\Gamma(\zeta)) \\
&< -C_2' |\Gamma(\zeta)|^N + C_1 t^{k\nu(\gamma,0)} \\
&\leq -C_2 |\zeta|^{N\nu(\gamma,0)} + C_1' t^{k\nu(\gamma,0)}
\end{aligned}$$

Namely, this follows from

$$|\Gamma(\zeta)| \geq |\gamma(\zeta)| - C_0 t^{k\nu(\gamma,0)} \geq C_3 |\zeta|^{\nu(\gamma,0)} - C_0 t^{k\nu(\gamma,0)}$$

with a suitable constant $C_3 > 0$. If we now would have $k > N$, our estimate would imply for all sufficiently small $t > 0$

$$\rho(\gamma(\zeta)) < 0 \ \text{ for } \ \frac{t}{2} < |\zeta| < t$$

and, hence,

$$\rho(\gamma(\zeta)) < 0 \ \forall \ \zeta \in \Delta(0,t) \setminus \{0\}$$

whereas $\gamma(0) \in \partial\Omega$. This is, however, a contradiction to the Hartogs disc theorem applied to the pseudoconvex region $\{\rho < 0\}$. Thus one has $k \leq N$. $\qquad\qquad\square$

We now come to the proof of Theorem 1.9, the main ingredient being the bumping in the weighted homogeneous case as obtained in Proposition 2.1. We will give the proof by going in a circle through the statements i)–iv). We start with

A) $iii) \implies iv)$: This follows directly from the definitions.

B) $iv) \implies i)$: We choose a realizing coordinate system $(z_1, \ldots, z_n)$ with respect to the admissible n-tupel of orders of extendability $(1, \mu_2, \ldots, \mu_n)$ for $\partial\Omega$ at z^0. Let $H_2 \subset \cdots \subset H_{n-1} \subset \mathbb{C}^n$ be a corresponding flag of complex linear subspaces through $z^0 = 0$. Then it follows from the Main Theorem of [5] for $2 \leq i \leq n$ together with Lemma 3.1 that

$$(3.23) \qquad m_i \leq \Delta_{n-i+1}(\partial\Omega, 0) \leq \tau(\partial(\Omega \cap H_i), 0) \leq N(\partial(\Omega \cap H_i), 0) = \mu_i$$

Hence one has $(1, \mu_2, \ldots, \mu_n) \geq \mathcal{M}(\partial\Omega, 0)$ with respect to the lexicographical order. Since $(1, \mu_2, \ldots, \mu_n)$ is an admissible weight, we must have equality everywhere in (3.23). In particular i) holds.

C) $i) \implies ii)$: We choose $(1, \mu_2, \ldots, \mu_n) = \mathcal{M}(\partial\Omega, 0)$. If $(z_1, \ldots, z_n)$ is a realizing coordinate system for $\mathcal{M}(\partial\Omega, 0)$, one can, obviously, find a local defining function r of the form (1.8) satisfying a)–c). We show, that d) also holds.

For this let $\gamma = (\gamma_1, \ldots, \gamma_n) : \Delta(0,1) \to \mathbb{C}^n$ be a non–trivial holomorphic curve through 0. We put $\widehat{\Delta}(\gamma, 0) := \frac{\nu(r_0 \circ \gamma, 0)}{\nu(\gamma, 0)}$. If $\nu(\gamma, 0) = \nu(\gamma_1, 0)$, then we get because of transversality $\nu(r_0 \circ \gamma, 0) = \nu(\gamma, 0)$, hence $\widehat{\Delta}(\gamma, 0) = 1$. Therefore, we may suppose, that there is an index $s \geq 2$, such that $\nu(\gamma, 0) = \nu(\gamma_s, 0)$. The numbers

$$k := \min_{2 \leq l \leq n} m_l \nu(\gamma_l, 0)$$

and the set

$$I := \{l \in \{2, \ldots, n\} : m_l \nu(\gamma_l, 0) = k\}$$

are well–defined and $I \neq \emptyset$. Suppose now, that $\widehat{\Delta}(\gamma, 0) > m_n$. Then we must have

Claim: $\nu(r_0 \circ \gamma, 0) > k$.

Namely, if $\nu(\gamma, 0) = \nu(\gamma_l, 0)$ for a certain $l \in I$, one has

$$\nu(r_0 \circ \gamma, 0) = \widehat{\Delta}(\gamma, 0)\nu(\gamma_l, 0) > m_n \nu(\gamma_l, 0) > m_l \nu(\gamma_l, 0) = k$$

If, on the other hand, $\nu(\gamma_l, 0) \neq \nu(\gamma, 0)$ for all $l \in I$, we choose a $j \in \{2, \ldots, n\} \setminus I$ with $\nu(\gamma, 0) = \nu(\gamma_j, 0)$ and obtain

$$\nu(r_0 \circ \gamma, 0) = \widehat{\Delta}(\gamma, 0)\nu(\gamma_j, 0) > m_j \nu(\gamma_j, 0) > k$$

This shows the Claim.

Since P does not contain pluriharmonic terms and $\operatorname{Re} z_1$ is harmonic, all homogeneous terms of degree $\leq k$ in $P(\gamma_2, \ldots, \gamma_n)$ and in $\operatorname{Re} \gamma_1$ must vanish. However, because of the definition of k, the homogeneous part $P^{(k)}$ of degree k in $P(\gamma_2, \ldots, \gamma_n)$ is

$$P^{(k)}(\zeta) = P(c_I(\zeta))$$

where $c_I : \mathbb{C} \to \mathbb{C}^{n-1}$ has components of the form $c_j(\zeta) := a_j \zeta^{\nu_j}$ for $j \in I$ and $c_j = 0$ for $j \notin I$. More precisely, $\nu_j = \nu(\gamma_j, 0)$ and a_j is the Taylor coefficient of degree ν_j of g_j at 0. We put $l_0 := \max I$ and distinguish between two cases:

1st case: $l_0 = n$ or $m_{l_0} = \cdots = m_n$. In this case, $\gamma_I(\zeta) := (0, c_I(\zeta))$ defines a nondegenerate holomorphic curve in $\mathbb{C}^n$ with $r(\gamma_I(\zeta)) = R_1(\gamma_I(\zeta))$. Because of property b) in ii), we get from this

$$|R_1(\gamma_I(\zeta))| \leq |\zeta|^{(1+\delta)k}$$

Since, moreover, for $j \in I$ the estimate $\nu_j = \frac{k}{m_j} = \frac{m_n}{m_j}\nu_n \geq \nu_n$ holds, we get $\nu(r \circ \gamma_I, 0) > k = m_n \nu_n = m_n \nu(\gamma_I, 0)$. Hence, $\tau(\partial\Omega, 0) > m_n$, in contradiction to the hypothesis.

2nd case: $l_0 \leq n-1$ and there is a $q \in \{l_0, \ldots, n-1\}$, such that $m_n \geq \cdots \geq m_{q+1} > m_q = \cdots = m_{l_0}$. Now we put

$$A := \left\{ (\gamma_I(\zeta), 0, \ldots, 0, z_{q+1}, \ldots, z_n) : (\zeta, z_{q+1}, \ldots, z_n) \in \mathbb{C}^{n-q+1} \right\}$$

Then A is an anlytic set of dimension $n - q + 1$ in 0. We show, that one has for suitable positive constants C, ϵ the estimate

$$(3.24) \qquad |r(z)| \leq C|z|^{(1+\epsilon)m_q} \quad \forall z \in A, \ z \text{ close to } 0$$

For this, consider $z := (\gamma_I(\zeta), 0, \ldots, z_{q+1}, \ldots, z_n) \in A$ with $(\zeta, z_{q+1}, \ldots, z_n) \in \mathbb{C}^{n-q+1}$ close enough to 0, and $z' := (c_I(\zeta), 0, \ldots, z_n)$. Then one has $r(z) = P(z') + R_1(z')$. Since $\nu_{l_0} \leq \nu_i$ for all $i \in I$, one has

$$|z| \approx |\zeta|^{\nu_{l_0}} + \sum_{l=q+1}^{n} |z_l|$$

With constants C, δ as in b) from ii) we can estimate

$$(3.25) \qquad |R_1(z')| \leq C\sigma^{1+\delta}(z')$$

$$\leq C' \left(|\zeta|^k + \sum_{l=q+1}^{n} |z_l|^{m_l} \right)^{1+\delta}$$

$$\leq C'' \left(|\zeta|^{\nu_{l_0}} + \sum_{l=q+1}^{n} |z_l| \right)^{(1+\delta)m_{l_0}}$$

$$\leq C''' |z|^{(1+\delta)m_q}$$

We still have to estimate $|P(z')|$. Because of the weighted homogeneity of P it is bounded from above by a sum of terms of the form

$$T = C \prod_{j \in I} |\zeta|^{\nu_j(\alpha_j + \beta_j)} \prod_{l=q+1}^{n} |z_l|^{\alpha_l + \beta_l}$$

where $C > 0$ is a constant depending on the coefficients of P and the a_j and where

$$\sum_{j \in I} \frac{\alpha_j + \beta_j}{m_j} + \sum_{l=q+1}^{n} \frac{\alpha_l + \beta_l}{m_l} = 1$$

Since $P(c_I(\zeta)) \equiv 0$, one has $B := \sum_{l=q+1}^{n} \frac{\alpha_l + \beta_l}{m_l} > 0$, hence also $B \geq \frac{1}{m_n}$. And since $m_l \geq m_{q+1} \forall l = q+1, \dots, n$, one obtains

$$\begin{aligned}
|T| &\leq C|\zeta|^{(1-B)k}|z|^{Bm_{q+1}} \\
&= C|\zeta|^{(1-B)m_{l_0}\nu_{l_0}}|z|^{Bm_{q+1}} \\
&\leq \tilde{C}|z|^{(1-B)m_{l_0}+Bm_{q+1}} \\
&= \tilde{C}|z|^{(1+\beta)m_q}
\end{aligned}$$

with $\beta := B(\frac{m_{q+1}}{m_q} - 1)$. Since $B \geq \frac{1}{m_n}$, (3.24) follows from this with $\epsilon := \frac{1}{m_n}(\frac{m_{q+1}}{m_q} - 1)$.

If, now, $H \subset \mathbb{C}^n$ is an arbitrary linear subspace of dimension q, it follows, that there is a non–degenerate holomorphic curve in $H \cap A$, which has with $\partial(\Omega \cap H)$ at 0 an order of contact $\geq (1 + \epsilon)m_q$. Therefore we get, that $\Delta_{n-q+1}(\partial\Omega, 0) > m_q$, a contradiction to the hypothesis of semiregularity.

D) $ii) \implies iii$): From Proposition 2.1 applied to the polynomial P, we obtain a C^3–smooth function $\tilde{P}$ on $\mathbb{C}^{n-1}$ with $\tilde{P} < P$, for which the function $\tilde{P} - 2\epsilon\sigma$ is plurisubharmonic for each sufficiently small $\epsilon > 0$. We fix $r_2 > 0$ and $\tilde{z}' \in \Delta^{n-1}(0, r_2)$ and put $d := \mu_2 \cdot \ldots \cdot \mu_n$. Then the domain

$$D_{\tilde{z}'} := \left\{ (v_1, v_2) : \left(v_1, v_2^{d/2}\tilde{z}_2, \dots, v_2^{d/2}\tilde{z}_{p+1}, v_2^{d/\mu_{p+2}}\tilde{z}_{p+2}, \dots, v_2^{d/\mu_n}\tilde{z}_n \right) \in \Omega \right\}$$

in $\mathbb{C}^2$ is pseudoconvex and has type d in 0, such that Lemma 1.5 of [13] can be applied in order to estimate the remainder term R_2 from (1.8) after the substitution made in the definition of the domain $D_{\tilde{z}'}$. We get with a suitable constant $C = C(r_0)$

$$\left| R_2\left(v_2^{d/2}\tilde{z}_2, \dots, v_2^{d/\mu_n}\tilde{z}_n \right) \right| \leq C|v_2|^{\frac{d}{2}+1}$$

for v_2 sufficiently close to 0. This implies, as can easily be seen,

$$(3.26) \qquad\qquad |R_2(\tilde{z}')| \leq C'\sigma(\tilde{z}')^{\frac{1}{2}+\frac{1}{d}}$$

From this, as it turns out, it follows, that for $A >> 1$ the function

$$\rho(z) := \operatorname{Re}\left(z_1 + Az_1^2 \right) + \tilde{P}(z') - \epsilon\sigma(z')$$

becomes an extending function for Ω at 0 with respect to the n–tuple $(1, \mu_2, \dots, \mu_n)$. The proof for this is very similar to the proof of Theorem 1 from [9]:

$$\begin{aligned}
\rho(z) &= \operatorname{Re} z_1 + A(\operatorname{Re} z_1)^2 - A(\operatorname{Im} z_1)^2 + \tilde{P}(z') - \epsilon\sigma(z') \\
&= r(z) - P(z') - R_1(z') - \operatorname{Im} z_1 R_2(z') - R_3(z') \\
&\quad + A(r(z) - P(z') - R_1(z') - \operatorname{Im} R_2(z') - R_3(z'))^2 \\
&\quad - A(\operatorname{Im} z_1)^2 + \tilde{P}(z') - \epsilon\sigma(z') \\
&\leq r(z) + |\operatorname{Im} z_1||R_2(z')| + C\sigma(z')^{1+\delta} + C(\operatorname{Im} z_1)^2 \\
&\quad + CA\left(r^2 + \sigma^2 + |z'|^2(\operatorname{Im} z_1)^2 + (\operatorname{Im} z_1)^4\right) \\
&\quad - \epsilon\sigma(z') - A(\operatorname{Im} z_1)^2
\end{aligned}$$

Here the constant $C > 0$ does not depend on A. If we now choose at first A large enough and then a small enough radius r_1 (depending on A), we, finally, get on $B(0, r_1)$ the estimate $\rho \leq r - \frac{\epsilon}{2}\sigma$. On the other hand, we trivially have the lower estimate

$$\rho \geq r - C|z'|^2$$

on $B(0, r_1)$. This shows, that iii) holds. The proof of Theorem 1.9 is now complete. $\qquad\square$

4 Proof of Theorems 1.6 and 1.8

The proofs of our Theorems 1.6 and 1.8 together with Corollary 1.7 are now simple.

Namely, with all the standard notation from above, we observe at first, that, according to iii) of Theorem 1.9, one has the equality $N(\partial\Omega, z^0) = m_n$. In addition, the definition of semiregularity gives $m_n = \tau(\partial\Omega, z^0)$. Hence, (1.3) holds.

Furthermore, as mentioned already in Remark a) after Theorem 1.9, the implication $i) \implies iii)$ of this theorem gives (1.4). The equality (1.5) of Corollary 1.7 then is a consequence of (1.1) and (1.4).

It remains to prove (1.6) of Theorem 1.8. For this, we remind the reader of the following characterization of the Bergman metric $B_G^2(z, X)$ of any bounded domain $G \subset\subset \mathbb{C}^n$ at a point $z \in G$ and applied to a vector $X \in \mathbb{C}^n$. Namely, we define

$$b_G^2(z, X) := \max\left\{ |(\partial f(z), X)|^2 : f \in L^2(G) \cap \mathcal{O}(G),\ f(z) = 0,\ \|f\|_{L^2(G)} = 1\right\}$$

Then one has the relation

$$(4.27) \qquad\qquad B_G^2(z, X) = \frac{b_G^2(z, X)}{K_G(z, z)}$$

As shown in (3.6) of [9], one has, in a suitable holomorphic coordinate system (z) with $z^0 = 0$ and the exterior unit normal vector e_1 to $\partial\Omega$ at 0 lying on the positive x_1–axis, for $a, t_0 > 0$ sufficiently small

$$\Delta_t := \Delta\left(-t, \frac{t}{2}\right) \times \prod_{i=2}^{n} \Delta\left(0, (at)^{\frac{1}{m_i}}\right) \subset \Omega$$

for all $0 < t < t_0$. If we denote by e_n the unit vector on the x_n–axis, we get from this and the inverse monotonicity of $b_G^2(z, X)$ as a function of G immediately

$$b_\Omega^2(-te_1, e_n) \leq b_{\Delta_t}^2(-te_1, e_n) = c(n, a)t^{-2 - \frac{2}{m_2} - \cdots - \frac{2}{m_{n-1}} - \frac{4}{m_n}}$$

This implies together with (1.5) the upper estimate $h_\Omega(0) \leq \frac{1}{m_n}$.

In order to prove the lower estimate for h_Ω we choose a suitable local exterior domain of comparison for Ω at 0. For this, we choose coordinates (z) and a defining function r for $\partial\Omega$ near $z^0 = 0$ as in ii) of Theorem 1.9. Then we put for positive numbers M and r_2 the function $r'(z) := (1 + Mx_1)r(z)$. As shown in (3.26), one has for a small $\delta > 0$ the estimate $|R_2(z')| \leq \sigma(z')^{\frac{1}{2}+\delta}$. From this we get with a suitable constant $C > 0$

$$
\begin{aligned}
r'(z) \; &= \; (1 + Mx_1)\,(x_1 + P(z') + R_1(z') + y_1 R_2(z') + R_3(z')) \\
&= \; x_1 + M(x_1^2 - y_1^2) + P(z') + R_1(z') \\
&\quad + (1 + Mx_1)(y_1 R_2(z') + R_3(z)) + Mx_1(P(z') + R_1(z')) + My_1^2 \\
&\geq \; x_1 + M(x_1^2 - y_1^2) + P(z') \\
&\quad - C\sigma(z')^{1+\delta} - (1 + Mr_2)\Big(|y_1|\sigma(z')^{\frac{1}{2}+\delta} + y_1^2\Big) \\
&\quad - Mr_2\Big(\sigma(z') + C\sigma(z')^{1+\delta}\Big) + My_1^2
\end{aligned}
$$

We choose $\epsilon > 0$ according to Proposition 2.1 and then make sure that M is sufficiently large and r_2 sufficiently small. Then we get on $B(0, r_2)$

$$r(z) \geq x_1 + M(x_1^2 - y_1^2) + P(z') - \frac{\epsilon}{2}\sigma(z')$$

and also

$$r'(z) \leq x_1 + M(x_1^2 - y_1^2) + C\sigma(z')$$

Hence, the mapping

$$F(z) := (z_1 + Mz_1^2, z')$$

maps the domain $\Omega \cap B(0, r_2)$ biholomorphically onto a domain G with

$$D_{int} := \{ z : \mathrm{Re}\, z_1 + C\sigma(z') < 0 \} \subset G \subset D_{ext} := \Big\{ z : \mathrm{Re}\, z_1 + \tilde{P}(z') - \frac{\epsilon}{2}\sigma(z') < 0 \Big\}$$

The methods of [14] apply to these comparison domains D_{int} and D_{ext}. Together with the localisation theorems from [8] and [21] we get for all z inside a cone $\Lambda \subset \Omega$ around the inner normal to $\partial\Omega$ at 0

$$
\begin{aligned}
B_\Omega^2(z, X) \; &\geq \; C_1 B_{\Omega \cap B(0, r_2)}^2(z, X) \\
&= \; C_1 B_G^2(F(z), F'(z)X) \\
&= \; \frac{C_1 b_G^2(F(z), F'(z)X)}{K_G(F(z), F(z))} \\
&\geq \; C_2 B_{D_{ext}}^2(F(z), F'(z)X) \frac{K_{D_{ext}}(F(z), F'(z)X)}{K_{D_{int}}(F(z), F(z))} \\
&\geq \; C_3 \left(\frac{|X_1|^2}{|z_1| + \sigma(z')} + \sum_{j=2}^{n} \frac{|X_j|^2}{(|z_1| + \sigma(z'))^{2/m_j}} \right)
\end{aligned}
$$

because for $z \in \Lambda$ the quotient $K_{D_{ext}}(F(z), F(z))/K_{D_{int}}(F(z), F(z))$ is bounded away from 0. Since one always has $|z_1| + \sigma(z') \leq C_4|r(z)|$ on the cone Λ, this implies the desired inequality $h_\Omega(0) \geq \frac{1}{m_n}$. $\qquad\qquad\qquad\qquad\qquad\qquad\qquad\qquad\qquad\qquad\qquad\qquad\square$

5 Construction of local peak functions

In this section, we will describe the main steps in the construction of a local Hölder continuous peak function on a C^∞–smooth pseudoconvex domain $\Omega \subset\subset \mathbb{C}^n$ near a point $z^0 \in \partial\Omega$, where $\partial\Omega$ is of semiregular type. This is, of course, the most important part of the proof of Theorem 1.10. Some technical details will be formulated as a lemma and proved later in section 7. The globalization will be postponed to the next section.

Let $\Omega \subset\subset \mathbb{C}^n$ be as above. According to Theorem 1.9 we can choose a local holomorphic coordinate system (z) centered at z^0, such that a defining function r of Ω is of the form

$$(5.28) \qquad r(z) = \operatorname{Re} z_1 + P(z') + R_1(z') + (\operatorname{Im} z_1)R_2(z') + R_3(z)$$

with properties a)—d) of part ii) of Theorem 1.10. We apply Proposition 2.1 to the polynomial P and obtain a weighted homogeneous C^3–function $\tilde{P}$ and an $\epsilon > 0$ as described there. Similar to part D) of section 3 one has for a sufficiently large constant $A >> 1$, that the function

$$\rho(z) := \operatorname{Re}\left(z_1 + Az_1^2\right) + \tilde{P}(z')$$

satisfies on a small ball $B = B(0, 2r_0)$ the estimate

$$(5.29) \qquad\qquad\qquad \rho(z) \leq r(z) - \epsilon\sigma(z')$$

In particular, the map

$$(5.30) \qquad\qquad\qquad F(z) := \left(z_1 + Az_1^2, z'\right)$$

throws $\Omega \cap B$ injectively and holomorphically into the domain

$$(5.31) \qquad\qquad G_0 := \left\{ w \in \mathbb{C}^n : \operatorname{Re} w_1 + \tilde{P}(w') + \frac{7}{8}\epsilon\,\sigma(w') < 0 \right\}$$

Putting

$$(5.32) \qquad\qquad\qquad \varphi(z) := \operatorname{Re} w_1 + \tilde{P}(w') + \frac{\epsilon}{2}\sigma(w')$$

gives

$$(5.33) \qquad\qquad F(\Omega \cap B) \subset G_0 \subset G_1 := \left\{ \varphi < -\frac{\epsilon}{4}\sigma \right\}$$

Since $\tilde{P}$ is weighted homogeneous with respect to $(m_2, \ldots, m_n)$, there is a constant $C_1 > 0$, such that one has for all positive t and all $w' \in \mathbb{C}^{n-1}$ the estimate

$$(5.34) \qquad\qquad\qquad \tilde{P}(w' + \xi') \leq \tilde{P}(w') + C_1 t$$

if $\xi' \in \mathbb{C}^{n-1}$ is chosen such that with the definition $m_0 := m_2 + \cdots + m_n$

$$|\xi_j| \leq \frac{t^{1/m_j}}{\left(1 + \sum_{l=2}^{n} \frac{|w_l|^2}{t^{2/m_l}}\right)^{m_0}}$$

for all $2 \leq j \leq n$. In this situation the method of the proof of Lemma 7 of [15] applies and it is not necessary to repeat this here. We get from it immediately

Lemma 5.1 *For each $t > 0$ there is an entire function g_t on $\mathbb{C}^{n-1}$ with $g_t(0) = 1$ which satisfies the following estimate with a constant C_2 independent of t:*

$$(5.35) \qquad |g_t(w')| \leq C_2 \left(1 + \sum_{l=2}^{n} \frac{|w_l|^2}{t^{2/m_l}}\right)^{m_0 n} \exp\left(\frac{1}{t}\varphi(0, w')\right)$$

We now put for $0 < t < 1$

$$(5.36) \qquad h_t(w) := \exp\left(\frac{w_1}{t}\right) g_t(w')$$

The properties of this important family of entire functions are put together in the following

Lemma 5.2 *There is a positive constant C_3 with*
    i)

$$\sup_{\overline{G}_1}|h_t| \leq C_3$$

    ii) For all $w \in G_1$ one has

$$\left|\frac{\partial h_t(w)}{\partial w_l}\right| \leq \frac{C_3}{t^{1/m_l}} \quad for \ all \ 1 \leq l \leq n$$

(with $m_1 := 1$)
    iii)

$$|h_t(w) - 1| \leq C_3 \sum_{j=1}^{n} \frac{|w_j|}{t^{1/m_j}} \quad for \ all \ w \in G_1$$

    iv) If one defines for $1 \leq a \leq 1/t$ the set W_a by

$$W_a := \left\{ w \in \mathbb{C}^n : a|\mathrm{Re}\, w_1| + \sum_{l=2}^{n} a^{1/m_j}|w_j| > 1 \right\}$$

one has the estimate

$$\sup_{W_a}|h_t(w)| \leq C_3 a t$$

We postpone the proof of this lemma to section 7 and show instead at first, how the family $\{h_t\}$ can be used for the construction of a local holomorphic peak function on the domain G_0 with respect to the point z^0 and Hölder continuous on $\overline{G}_0$.

Namely, we consider for each $k \in \mathbb{N}_0$ the function $H_k := h_{2^{-k}}$ and choose a number $c \in (\frac{3}{5}, 1)$ which will be further specified at a later point. With this we define the holomorphic function

$$(5.37) \qquad g := (1-c)\sum_{k=0}^{\infty} c^k H_k$$

for which we will show:

Lemma 5.3 *The function g of (5.37) is, for a suitable choice of the constant c, a local holomorphic peak function for the domain G_0 at 0. It is Hölder continuous on $\overline{G}_0$ near z^0.*

Proof 1) The holomorphicity of the function g on G_0 is clear from i) of Lemma 5.2 and the estimate (5.33). We now put for $k \in \mathbb{N}_0$

$$U_k := \left\{ w \in \mathbb{C}^n : 2^k |\mathrm{Re}\, w_1| + \sum_{L=2}^{n} 2^{\frac{k}{m_l}} |w_l| < 1 \right\}$$

Then we obviously have

$$S := \overline{G}_0 \backslash \{ (iy_1, 0) : y_1 \in \mathbb{R} \} \subset \bigcup_{m=0}^{\infty} U_m \backslash U_{m+1}$$

Let us, next, consider a point $w \in S$ and choose $m \in \mathbb{N}_0$ with respect to it such that $w = (w_1, w') \in U_m \backslash U_{m+1}$. We put $\widehat{w} := (\mathrm{Re}\, w_1, w')$. We get

$$(5.38) \qquad |g(w)| \leq (1-c) \sum_{k=0}^{\infty} c^k |H_k(w)|$$

$$= (1-c) \sum_{k=0}^{\infty} c^k |H_k(\widehat{w})|$$

$$\leq (1-c) \sum_{k=0}^{m} c^k + (1-c) \sum_{k=0}^{m} c^k |H_k(\widehat{w}) - 1|$$

$$+ (1-c) \sum_{k=m+1}^{\infty} c^k |H_k(\widehat{w})|$$

We will now go on by estimating the last two sums in this upper bound for $|g(w)|$. Namely, according to part iii) of Lemma 5.2 we have

$$|H_k(\widehat{w}) - 1| \leq C_3 \left(\sum_{l=2}^{n} 2^{k/m_l} |w_l| + 2^k |\mathrm{Re}\, w_1| \right)$$

Hence we get

$$(1-c)\sum_{k=0}^{m} c^k \left| H_k\left(\widehat{w}\right) - 1\right|$$

$$\leq (1-c)\left(\frac{(2c)^{m+1}-1}{2c-1}\left| \operatorname{Re} w_1\right| + C_3 \sum_{l=2}^{n} \frac{\left(2^{1/m_l}c\right)^{m+1}-1}{2^{1/m_l}c-1}\left| w_l\right|\right)$$

$$\leq 2c^{m+1}\frac{(1-c)\,C_3}{2^{1/m_n}c-1}\left(2^m \left| \operatorname{Re} w_1\right| + \sum_{l=2}^{n} 2^{m/m_l}\left| w_l\right|\right)$$

$$\leq \frac{4\,(1-c)\,C_3}{2^{1/m_n}c-1}c^{m+1}$$

Furthermore, part iv) of Lemma 5.2 implies for $a := 2^{m+1}$ immediately

$$\left| H_k(\widehat{w})\right| \leq C_3 2^{m+1-k}\ \forall k \geq m+1$$

Hence, we can estimate the last sum in (5.38) as follows:

$$\begin{aligned}
(1-c)\sum_{k=m+1}^{\infty} c^k \left| H_k\left(\widehat{w}\right)\right| &\leq (1-c)\,C_3 \sum_{k=m+1}^{\infty} 2^{m+1}\left(\frac{c}{2}\right)^k \\
&= (1-c)\,C_3 \frac{c^{m+1}}{2-c} \\
&\leq (1-c)\,C_3 c^{m+1}
\end{aligned}$$

Putting this together into (5.38) we get with a numerical constant $A > 0$

$$|g(w)| \leq 1 - c^{m+1} + A(1-c)c^{m+1}$$

Therefore, one has with a suitably chosen constant c

$$|g(w)| < 1$$

Because of continuity, this gives $|g| \leq 1$ everywhere on $\overline{G}_0$.

Furthermore, because of (5.37) and (5.36) together with Lemma 5.1 we get for all $y_1 \in (-1,1)$ the equality

$$\begin{aligned}
|g\left(iy_1,0\right)|^2 &= (1-c)^2 \left[\left(\sum_{k=0}^{\infty} c^k \cos\left(2^k y_1\right)\right)^2 + \left(\sum_{k=0}^{\infty} c^k \sin\left(2^k y_1\right)\right)^2\right] \\
&= (1-c)^2 \sum_{k,l=0}^{\infty} c^{k+l}\cos\left(\left(2^k - 2^l\right)y_1\right) \\
&\leq (1-c)^2 \sum_{k,l=0}^{\infty} c^{k+l} = 1
\end{aligned}$$

Obviously, equality holds in this estimate only if $\cos((2^k - 2^l)y_1) = 1$ for all $k, l \geq 0$. This is only the case for $y_1 = 0$. Therefore we get $|g(iy_1,0)| < 1$ for all $y_1 \in (-1,1)\setminus\{0\}$. Hence g is a local holomorphic peak function on G_0 with respect to 0.

2) It remains to show the Hölder continuity of the constructed peak function g on $\overline{G}_0$ near the point 0. Let for this B' be a small ball in $\mathbb{C}^n$ centered at 0 and choose $x, y \in G_0 \cap B'$. With the notation $\widehat{y} := (\operatorname{Re} y_1 + i \operatorname{Im} x_1, y_2, \ldots, y_n)$ one has

$$|x - \widehat{y}| \leq |\operatorname{Re}(x_1 - y_1)| + \sum_{l=2}^{n} |x_l - y_l| \leq 2|x - \widehat{y}|$$

and $|\widehat{y} - y| = |\operatorname{Im}(x_1 - y_1)|$. For $k \in \mathbb{N}_0$ and a number $b > 0$ which will be specified later we put

$$S_k := \left\{ \zeta \in \mathbb{C}^n : \frac{b}{2^{k+1}} < |\zeta| \leq \frac{b}{2^k} \right\}$$

We will now estimate at first $|g(x) - g(\widehat{y})|$ (and later $|g(\widehat{y}) - g(y)|$). Namely, if $x \neq \widehat{y}$, then there is a $m \in \mathbb{N}_0$ such that $x - \widehat{y} \in S_m$. We have

$$|g(x) - g(\widehat{y})| \leq (1 - c)\sum_{k=0}^{m} c^k |H_k(x) - H_k(\widehat{y})| + (1 - c) \sum_{k=m+1}^{\infty} c^k |H_k(x) - H_k(\widehat{y})|$$

In order to estimate this, we observe at first (see (5.32), that we can find a small positive number b (independent of m, x and y), such that

$$\varphi(\tau x + (1 - \tau)\widehat{y}) \leq C_3 2^{-m} \leq C_3 2^{-k} \ \forall \tau \in [0, 1] \ and \ \forall k \leq m$$

Hence, the mean value theorem together with part iii) of Lemma 5.2 gives

$$|H_k(x) - H_k(\widehat{y})| \leq A_1 2^k |x - \widehat{y}|$$

This implies

$$\sum_{k=0}^{m} c^k |H_k(x) - H_k(\widehat{y})| \leq A_2 (2c)^{m+1} |x - \widehat{y}|$$

On the other hand, one has, since $|H_k| \leq C_3$ for all k, the estimate

$$\sum_{k=m+1}^{\infty} c^k |H_k(x) - H_k(\widehat{y})| \leq 2C_3 \frac{c^{m+1}}{1 - c} \leq 2 \frac{C_3}{b(1 - c)} (2c)^{m+1} |x - \widehat{y}|$$

This gives with constants A_3 and A_4

$$\begin{aligned}
|g(x) - g(\widehat{y})| &\leq A_3 (2c)^{m+1} |x - \widehat{y}| \\
&= A_3 (2c)^{m+1} |x - \widehat{y}|^{1-\eta} |x - \widehat{y}|^{\eta} \\
&\leq 2cA_3 b^{1-\eta} (2^\eta c)^m |x - \widehat{y}|^{\eta} \\
&\leq A_4 |x - y|^{\eta}
\end{aligned}$$

if only η is chosen, such that $2^\eta c \leq 1$.

Now we come to the estimate of $|g(\widehat{y}) - g(y)|$. One has because of (5.36) and Lemma 5.1

$$\begin{aligned} |H_k(\widehat{y}) - H_k(y)| &= 2|H_k(y)|\left|\sin\left(2^{k-1}\operatorname{Im}(x_1 - y_1)\right)\right| \\ &\leq 2C_3\left|\sin\left(2^{k-1}\operatorname{Im}(x_1 - y_1)\right)\right| \end{aligned}$$

Hence

$$|g(\widehat{y}) - g(y)| \leq 2(1-c)C_3\sum_{k=0}^{\infty} c^k\left|\sin\left(2^{k-1}\operatorname{Im}(x_1 - y_1)\right)\right|$$

It follows as above, that also this term is bounded from above by $A_5|x-y|^{\eta}$ if only $2^{\eta}c \leq 1$. This finishes the proof of Lemma 5.3. $\qquad\qquad\square$

Coming back to the map F of (5.30) and its relation to the domain G_0 (see (5.31)), we get from Lemma 5.3 immediately:

Corollary 5.4 *Let g be the function defined by (5.37) with c chosen such that Lemma 5.3 holds, then the function*

$$(5.39) \qquad\qquad g_{z^0}(z) := g\left(z_1 + Az_1^2, z'\right)$$

is a local holomorphic peak function on Ω near z^0, which is Hölder continuous on $\overline{\Omega}$ near z^0.

6   Globalization

The local peak functions g_{z^0} of Corollary 5.4 can easily be globalized on Ω by the usual $\overline{\partial}$–technique, in order to prove Theorem . We just indicate the major steps. Namely, the global version h_{z^0} of g_{z^0} can be writte in the form

$$(6.40) \qquad\qquad h_{z^0} = \frac{\dfrac{\chi}{g_{z^0}-1} - u + 1}{\dfrac{\chi}{g_{z^0}-1} - u - 1}$$

where χ is a smooth cut–off function supported in B (Here B is a ball as it was chosen for (5.29) to hold) and $\equiv 1$ on a concentric ball $B' \subset\subset B$. The function u in (6.40) is chosen in the following way:

Since $F^{-1}(G_0) \supset (\overline{\Omega} \cap \overline{B}) \setminus \{z^0\}$ and g_{z^0} is holomorphic on $F^{-1}(G_0)$ and since, furthermore, we assume in Theorem , that $\overline{\Omega}$ has a Stein neighborhood basis, we can find a pseudoconvex domain $D \supset \overline{\Omega}$ (with smooth boundary) such that

$$\alpha := \frac{\overline{\partial}\chi}{g_{z^0} - 1}$$

defines a C^{∞}–smooth, $\overline{\partial}$–closed $(0,1)$–form on D. We may assume, that $\overline{B} \subset D$. As function u we now take a C^{∞}–solution of the equation $\overline{\partial}u = \alpha$ on D. Obviously, we, therefore, also may assume, that $\operatorname{Re} u > 0$ on $\overline{\Omega}$. It is easy to see, that then the function h_{z^0} from (6.40) is a local holomorphic peak function for Ω near z^0.

It remains to show the Hölder continuity of h_{z^0} on $\overline{\Omega}$ near z^0. For this, we observe at first, that the absolute value of the function

$$\varphi(w) := \chi(w) - (u(w) + 1)(g_{z^0} - 1)$$

is, on B, bounded from below by a positive constant b. Let now $x, y \in \Omega$. If they lie in B, we have

$$h_{z^0}(x) - h_{z^0}(y) \;=\; \frac{-2}{\varphi(x)\varphi(y)}(\chi(x) - \chi(y))(g_{z^0}(y) - 1) + \chi(y)(g_{z^0}(x) - g_{z^0}(y))$$
$$+ (g_{z^0}(x) - 1)(g_{z^0}(y) - 1)(u(x) - u(y))$$

This implies immediately:

$$|h_{z^0}(x) - h_{z^0}(y)| \leq C|x - y|^\eta$$

with the same Hölder exponent η as for g_{z^0}.

If $x \notin B$ or $y \notin B$, we even have

$$|h_{z^0}(x) - h_{z^0}(y)| \leq C|x - y|$$

This finishes the proof of Theorem 1.10. $\qquad\qquad\square$

7 Proof of Lemma 5.2

A decisive technical tool for showing Theorem 1.10 was Lemma 5.2. We, now, come to the proof of this lemma. For this, we will use all the previous notations, in particular those introduced in section 5.

For part i) of Lemma 5.2 we observe, that for all $v' \in \mathbb{C}^{n-1}$ with $\max_{2 \leq j \leq n} |v_j| \geq 1$

$$\sigma(v') \geq \frac{1}{2n}\sum_{j=2}^{n} |v_j|^2$$

since $m_l \geq 2 \;\forall\; 2 \leq l \leq n$. Hence, one has for any $w \in G_1$ with $\max_{2 \leq j \leq n} \frac{|w_j|}{t^{1/m_j}} \geq 1$ the estimate

$$
\begin{aligned}
\frac{1}{t}\varphi(w) \;&\leq\; \frac{1}{2t}\varphi(w) - \frac{\epsilon}{8t}\sigma(w') \\
&=\; \frac{1}{2t}\varphi(w) - \frac{\epsilon}{8}\sigma\left(\frac{w_2}{t^{1/m_2}}, \ldots, \frac{w_n}{t^{1/m_n}}\right) \\
&\leq\; \frac{1}{2t}\varphi(w) - \frac{\epsilon}{16n}\sum_{j=2}^{n} \frac{|w_j|^2}{t^{2/m_j}}
\end{aligned}
$$

Now we get from (5.35)

$$(7.41) \quad |h_t(w)| \leq C_2 \left(1 + \sum_{l=2}^{n} \frac{|w_l|^2}{t^{2/m_l}} \right)^{m_0 n} \exp \left(\frac{1}{2t}\varphi(w) - \frac{\epsilon}{16n} \sum_{j=2}^{n} \frac{|w_j|^2}{t^{2/m_j}} \right)$$

Hence, if we put

$$C_2' := \sup_{x \geq 0} (1 + x)^{m_0 n} \exp \left(-\epsilon \frac{x}{16n} \right)$$

we get

$$|h_t(w)| \leq C_2 C_2' \exp \left(\frac{\varphi(w)}{2t} \right)$$

If, on the other hand, $\max_{2 \leq j \leq n} \frac{|w_j|}{t^{1/m_j}} \leq 1$, one has

$$|h_t(w)| \leq C_2 (n+1)^{m_0 n} \exp \frac{\varphi(w)}{2t}$$

according to the estimate (7.41). This proves i).

Now we come to the proof of part ii) of Lemma 5.2. We write

$$R_j := \frac{t^{1/m_j}}{\left(1 + \sum_{l=2}^{n} \frac{|w_l|^2}{t^{2/m_l}} \right)^{m_0}}$$

for $2 \leq j \leq n$ and $R_1 := t$. Furthermore, we denote for $w \in G_1$ by $\Delta(w)$ the polycylinder

$$\Delta(w) := \Delta(w_1, R_1) \times \Delta(w_2, R_2) \times \cdots \times \Delta(w_n, R_n)$$

Inequality (5.34) easily gives together with (5.32)

$$\varphi(v) \leq \varphi(w) + C_4 t$$

on $\Delta(w)$ with a constant C_4 independent of t. This, however, gives together with (7.41) the estimate

$$\max_{\Delta(w)} |h_t| \leq C_5 e^{\varphi(w)/2t}$$

with a constant C_5 which is independent of w and t. Part ii) follows directly, if one uses the Cauchy estimates on $\Delta(w)$.

In order to show part iii) we put for $\tau \in (0, 1)$

$$S_\tau(w) := \left(\tau w_1, \tau^{1/m_2} w_2, \ldots, \tau^{1/m_n} w_n \right)$$

Then one obviously has

$$\begin{aligned} h_t(w) - 1 &= \int_0^1 \left(\frac{d}{d\tau} h_t(S_\tau(w)) \right) d\tau \\ &= \sum_{j=1}^{n} \frac{w_j}{m_j} \int_0^1 \frac{\partial h_t}{\partial w_j} (S_\tau(w)) \tau^{\frac{1}{m_j} - 1} d\tau \end{aligned}$$

Since $S_\tau(w) \in G_1$ for $\tau > 0$ and $w \in G_1$, part ii) can be applied to $\frac{\partial h_t}{\partial w_j}(S_\tau(w))$. This gives iii).

Let, finally, for the proof of iv) $\gamma \in (0, 1)$ be a number which will be further specified at a later stage. If $w \in W_a$ is an arbitrary point, there are two cases:

1st case: One has $\sum_{j=2}^n a^{1/m_j} |w_j| \leq \gamma$.

Observe that always $\tilde{P}(w') + \frac{\epsilon}{2}\sigma(w') \leq C_6\sigma(w') \leq C_7\frac{\gamma^2}{a}$ with constants C_6, C_7 independent of γ and a. Since $|\operatorname{Re} w_1| > \frac{1-\gamma}{a}$, we get under the assumption that $\operatorname{Re} w_1 \geq 0$ the inequalities

$$\frac{1-\gamma}{a} < \operatorname{Re} w_1 = \varphi(w) - \tilde{P}(w') - \frac{\epsilon}{2}\sigma \leq C_7\frac{\gamma^2}{a}$$

This is a contradiction if the constant γ is chosen small enough. Hence we must have $\operatorname{Re} w_1 < -\frac{1-\gamma}{a}$. After possibly further decreasing γ this implies

$$\varphi(w) \leq \operatorname{Re} w_1 + C_7\frac{\gamma^2}{a} < -\frac{1}{2}\frac{1-\gamma}{a}$$

In a similar way as in the proof of part i) one deduces

$$|h_t(w)| \leq C_2 C_2'' \exp\left(-\frac{1-\gamma}{4a\,t}\right) \leq C_3\,a\,t$$

with a universal constant C_2''.

2nd case: One has $\sum_{j=2}^n a^{1/m_j} |w_j| > \gamma$.

Because of the estimate

$$\sum_{j=2}^n \frac{|w_j|^2}{t^{2/m_j}} \geq \frac{\gamma^2}{n}\frac{1}{(a\,t)^{2/m_n}}$$

we get in this case in total analogy to the proof of part i) the estimate

$$\begin{aligned}
|h_t(w)| &\leq C_2 C_2'' \exp\left(-C_8(\gamma)\sum_{j=2}^n \frac{|w_j|^2}{t^{2/m_j}}\right) \\
&\leq C_2 C_2''' \exp\left(-C_9(\gamma)(a\,t)^{-2/m_n}\right)
\end{aligned}$$

where the number C_2''' is given by

$$C_2''' := \sup_{x>0}(1 + x)^{m_0 n} \exp\left(-C_8(\gamma)x\right)$$

with positive constants $C_8(\gamma)$ and $C_9(\gamma)$. But, in addition, one has

$$\exp\left(-C_9(\gamma)(a\,t)^{-2/m_n}\right) \leq C_3'a\,t$$

with a universal constant $C_3' > 0$. Alltogether we obtain the claim of iv). $\square$

8 Appendix

In Theorem 1.2 of this article we stated the inequality

$$(8.42) \qquad\qquad 2e(\partial\Omega, 0) \leq g(\partial\Omega, 0)$$

It comes from Theorem 2 of [9], where its proof was indicated based on Lemma 3.1 of [9] and the methods of [12]. As was kindly pointed out to us by T. Ohsawa, this proof of inequality (8.42) needs an additional explanation in order to become completely clear. The point is the following (we use the definitions and basic notations of [9]):

The growth exponent $g(\partial\Omega, 0)$ is defined by admitting nontangential approach within Ω to the boundary point 0 (see Definition 1 of [9]). On the other hand, the extendability exponent $e(\partial\Omega, 0)$ is, in suitable coordinate systems near 0, approximated by distinguished flags $H_1 \subset \cdots \subset H_n$ of linear subspaces of $\mathbb{C}^n$ (see Definition 2 of [9]). The proof for (8.42) as given in [9] makes it completely clear, that the necessary lower estimate for the Bergman kernel $K_\Omega(z)$ holds for nontangential approach to 0 within $\Omega \cap H_1$. However, it does not show, how this estimate follows for all $z \in \Omega$ lying inside an arbitrary open acute cone Λ in Ω around the inner normal on $\partial\Omega$ at 0. We use this opportunity to explain this point in detail by proving the following

Proposition 8.1 *Let $\Omega = \{z \in \mathbb{C}^n : r(z) < 0\} \subset\subset \mathbb{C}^n$ be a pseudoconvex domain with C^∞–smooth boundary and a C^∞ defining function r. Let $z_0 \in \partial\Omega$. We choose a holomorphic coordinate system z centered at z_0, such that $\partial r(0) = dz_1$ and put $H_k := \{z \in \mathbb{C}^n : z_{k+1} = \cdots = z_n = 0\}$ for $1 \leq k \leq n$. For $0 < A < 1$ we denote $\Lambda_A := \{z \in \mathbb{C}^n : |\operatorname{Re} z_1| > A|z|\}$. Furthermore, we suppose, that there is a constant $R > 0$ and for all $k = 2, \ldots, n$ an extending function ρ_k on $H_k \cap B(0, R)$ for $\Omega \cap H_k$ of order μ_k with $\mu_2 \leq \cdots \leq \mu_n$. Then there are constants $C_0 > 0$ and $0 < R_0 < 1$, such that one has for all $z_* \in \Omega \cap \Lambda_A \cap \partial B(0, R_0)$ and all $0 < t < t_0$ (here $0 < t_0 << 1$ is chosen suitably small) the estimate*

$$(8.43) \qquad\qquad K_\Omega(tz_*) \geq \frac{C_0}{t^{2\left(1+\sum_{k=2}^n \frac{1}{\mu_k}\right)} \left(\log \frac{1}{t}\right)^{n-1}}$$

for the Bergman kernel K_Ω of Ω.

Remark *Due to recent work of T. Ohsawa (see [22]), it is possible to avoid the logarithmic term in (8.43), since Lemma 8.4 below can be replaced by an extension theorem without logarithmic terms. This, however, does not change the estimate (8.42).*

For the convenience of the reader we recall the following

Definition 8.2 *Let $D \subset\subset \mathbb{C}^n$ be a pseudoconvex domain with C^∞–smooth boundary and $0 \in \partial D$. If U is a neighborhood of 0, then a function $\rho \in C^3(U)$ is called an extending function of order μ, if it has the following properties:*

i) $\rho(0) = 0$;

ii) The set $\{\rho = 0\}$ is a smooth hypersurface in U, which is pseudoconvex from the side $\{\rho < 0\}$;

iii) for suitable constants $c_1, c_2 > 0$ the estimate

$$-c_1\left(\operatorname{dist}(z, \partial D) + |z|^2\right) \leq \rho(z) \leq -c_2|z|^\mu$$

is satisfied on $D \cap U$.

The rest of this Appendix will be devoted to the proof of Proposition 8.1, which will be reduced to the application of a certain lemma (Lemma 8.4) about the extension of holomorphic L^2–functions with weights.

Proof We put $m := \mu_n + 1$ and define the mapping $\varphi : \mathbb{C}^n \longrightarrow \mathbb{C}^n$ by

$$\varphi(w) := (-z_{*1}w_1, w' + (w_1^m - (-t)^m + t)z_*')$$

Here $t > 0$ is arbitrary. Then φ is biholomorphic and

$$\varphi^{-1}(v) = \left(-\frac{v_1}{z_{*1}}, v' - \left(\left(-\frac{v_1}{z_{*1}}\right)^m - (-t)^m + t\right)z_*'\right)$$

In particular, one has $\varphi^{-1}(tz_*) = -te_1$ for $t > 0$, where e_1 is the first unit vector in $\mathbb{C}^n$. We calculate

$$\frac{\partial(r \circ \varphi)}{\partial w_1}(v) = -z_{*1}\frac{\partial r}{\partial z_1} \circ \varphi(v) + mv_1^{m-1}\sum_{k=2}^{n}\frac{\partial r}{\partial z_k} \circ \varphi(v)z_{*k}$$

If $|v| < R' << R$, we have $\operatorname{Re}\frac{\partial r}{\partial z_1}(\varphi(v)) \geq \frac{1}{2}$, $|\frac{\partial r}{\partial z_k}(\varphi(v))| \leq 2$ for all $k \leq 2$, and $|\operatorname{Im}\frac{\partial r}{\partial z_1} \circ \varphi(v)| < A^2/6$. Hence

$$
\begin{aligned}
\operatorname{Re}\frac{\partial(r \circ \varphi)}{\partial w_1}(v) &\geq \frac{1}{2}\operatorname{Re}z_{*1} - mR'^{m-1}2nR_0 - A^2R_0 \\
&\geq \left(\frac{1}{2}A - 2nmR'^{m-1} - \frac{1}{6}A^2\right)R_0 \\
&\geq \frac{1}{4}AR_0
\end{aligned}
$$

(8.44)

if R' has been chosen suitably. Notice, that $\operatorname{Re}z_{*1} < 0$. Hence, for $\eta > 0$, which will be specified later, we have

$$\operatorname{Re}\frac{\partial(r \circ \varphi)}{\partial w_1}(w - \eta te_1) \geq \frac{1}{4}AR_0$$

if $|w| < R'/2$ and $t < R'/2\eta$. Therefore, one has for these w and t_1

$$
\begin{aligned}
r \circ \varphi(w - t\eta e_1) &= r \circ \varphi(w) - 2\eta t\operatorname{Re}\int_0^1 \frac{\partial}{\partial z_1}(r \circ \varphi)(w - st\eta e_1)\,ds \\
&< r \circ \varphi(w) - \frac{1}{4}\eta AR_0 t
\end{aligned}
$$

We now put $\Omega^* := \varphi^{-1}(\Omega)$ and $F(w) := w - \eta t e_1$. Furthermore, we write $\Omega_\alpha^* := \{\tilde{w} : r(\varphi(\tilde{w})) < -\alpha\}$ for any $0 < \alpha < 1$. Because of $\Omega^* = \{r \circ \varphi < 0\}$ we then have for $\alpha := \frac{1}{4}\eta A R_0 t$ the inclusion

$$F(\Omega^* \cap B(0, R'/2)) \subset \Omega_\alpha^* \cap B(0, R')$$

The localization theorem for the Bergman kernel gives

$$K_{\Omega^*}(-te_1) \geq cK_{\Omega^* \cap B(0,R'/2)}(-te_1)$$

if $t < R'/4$. Hence we have

(8.45)
$$\begin{aligned}
K_\Omega(-tz_*) = K_{\Omega^*}(-te_1)|z_{*1}|^{-2} \\
\geq cK_{\Omega^* \cap B(0,R'/2)}(-te_1) \\
= cK_{F(\Omega^* \cap B(0,R'/2))}(F(-te_1)) \\
\geq cK_{\Omega_\alpha^* \cap B(0,R')}(F(-te_1)) \\
= cK_{\Omega_\alpha^* \cap B(0,R')}(-(1+\eta)te_1)
\end{aligned}$$

for $t < R'/2\eta$.

In order to continue, we need the following lemma, which is an easy consequence of the mean value inequality

Lemma 8.3 *The extending function ρ_k satisfies after possibly shrinking R on all of $B(0, R) \cap H_k$ the inequality*

(8.46)
$$\frac{1}{C_1}\left(r(w) - |w|^2\right) \leq \rho_k(w) \leq C_1(r(w) - |w|^{\mu_k})$$

with a constant C_1 not depending on k.

We now define on $H_k \cap B(0, R)$ the function

$$\rho_k^*(w) := \rho_k(-z_{*1}w_1, w'') \quad (w'' = (w_2, \ldots, w_k))$$

If R is chosen as in Lemma 8.3, $R < R_0$, and if we fix $C_2 > 1$ such that $|r(v) - r(\tilde{v})| \leq C_2|v - \tilde{v}|$ for all $v, \tilde{v} \in B(0, R)$, we get together with (8.46)

$$\begin{aligned}
\rho_k^*(w) &= \rho_k(-z_{*1}w_1, w'') \\
&< C_1\left(r(-z_{*1}w_1, w'', 0) - |(-z_{*1}w_1, w'')|^{\mu_k}\right) \\
&= C_1 r \circ \varphi(w) + C_1(r(-z_{*1}w_1, w'', 0) - r \circ \varphi(w)) - C_1|(-z_{*1}w_1, w'')|^{\mu_k} \\
&< C_1 r \circ \varphi(w) + C_1 C_2(|w_1|^m + 2t)|z_*'| - \frac{C_1}{2}|z_{*1}|^{\mu_k}|w_1|^{\mu_k} - \frac{C_1}{2}|w''|^{\mu_k} \\
&< C_1 r \circ \varphi(w) + C_1\left[C_2 R_0|w_1|^{m-\mu_k} - \frac{1}{2}(AR_0)^{\mu_k}\right]|w_1|^{\mu_k} - \frac{C_1}{2}|w''|^{\mu_k} \\
&\quad + 2C_1 C_2 R_0 t
\end{aligned}$$

After shrinking R even further, this becomes

(8.47) $$\rho_k^*(w) < C_1 r \circ \varphi(w) - C_3|(w_1, w'')|^{\mu_k} + 2C_1 C_2 R_0 t$$

with a constant C_3 independent of t. If now $\alpha := \eta A R_0 t/4n$ and $w \in \Omega_\alpha^*$ we get from (8.47)

(8.48) $$\rho_k^*(w) < (-C_1\alpha + 2C_1 C_2 R_0 t) - C_3|(w_1, w'')|^{\mu_k}$$

At this point, we decide to fix $\eta := 8nC_2/A$. This turns (8.48) into the decisive estimate

(8.49) $$\rho_k^*(w) < -C_3|(w_1, w'')|^{\mu_k}$$

Hence, as in the sketch of the proof of Theorem 2 in [9] we are again in a situation, where we can apply the following extension lemma for holomorphic L^2–functions (see Lemma 3.1 from [9])

Lemma 8.4 *Let $k \in \{2, \dots, n\}$. Then there is a constant $K > 0$, such that any holomorphic function $f_{(k-1)}$ on $\Omega_{\tilde\alpha}^* \cap H_{k-1} \cap B(0, R)$ $(\tilde\alpha := (k-2)\eta R_0 A t/4n)$ with*

$$I_{k-1}\big(f_{(k-1)}\big) := \int_{\Omega_{\tilde\alpha}^* \cap H_{k-1} \cap B(0,R)} \big|f_{(k-1)}\big|^2 \prod_{l=k}^n |\rho_l^*|^{\frac{2}{\mu_l}} \, d\lambda_{2k-2} < \infty$$

admits a holomorphic extension $f_{(k)}$ to the set $\Omega_\alpha^* \cap H_k \cap B(0, R)$ $(\alpha := (k-1)\eta A R_0 t/4n)$ for which

$$I_k\big(f_{(k)}\big) \le K I_{k-1}\big(f_{(k-1)}\big) \log \frac{1}{t}$$

(Notice, that for the proof of this lemma no lower estimate for ρ_k^* is needed.)
By iteration of Lemma 8.4 we get

Lemma 8.5 *There is a constant $L > 0$, such that each holomorphic function f on $\Omega^* \cap H_1 \cap B(0, R)$ with*

$$I(f) := \int_{\Omega^* \cap H_1 \cap B(0,R)} |f|^2 \prod_{l=2}^n |\rho_l^*|^{\frac{2}{\mu_l}} \, d\lambda_2 < \infty$$

admits a holomorphic extension $\widehat{f}$ to the domain $\Omega_{\tilde\alpha}^* \cap B(0, R) \supset \Omega_\alpha^* \cap B(0, R')$ (here the parameters are $\tilde\alpha := (n-1)\eta A R_0 t/4n$ and $\alpha := \eta A R_0 t/4$), such that

(8.50) $$\int_{\Omega_\alpha^* \cap B(0,R')} \big|\widehat{f}\big|^2 \, d\lambda_{2n} \le LI(f)\left(\log \frac{1}{t}\right)^{n-1}$$

In order to finish the proof of Proposition 8.1, we only need to find the right holomorphic function f on $\mathcal{O}(\Omega^* \cap H_1 \cap B(0, R))$ to start with when we apply Lemma 8.5. For this we estimate at first ρ_k^* from below. From (8.46) we get on $H_k \cap B(0, R)$:

$$
\begin{aligned}
\rho_k^*(w_1, w'') &= \rho_k(-z_{*1}w_1, w'') \\
&> \frac{1}{C_1}\left(r(-z_{*1}w_1, w'', 0) - (1 + R_0^2)|w|^2\right) \\
&= \frac{1}{C_1} r \circ \varphi(w_1, w'', 0) + \frac{1}{C_1}(r(-z_{*1}w_1, w'', 0) - r \circ \varphi(w_1, w'', 0)) \\
&\quad - \frac{1 + R_0^2}{C_1}|w|^2 \\
&> \frac{1}{C_1} r \circ \varphi(w_1, w'', 0) - \frac{C_2}{C_1}(|w_1|^m + 2t)R_0 - \frac{1 + R_0^2}{C_1}|w|^2 \\
&> \frac{1}{C_1} r \circ \varphi(w) - C_4|w|^2 - C_4 t
\end{aligned}
$$

with a universal constant C_4.

Let now h be a holomorphic function on $\Omega^* \cap H_1 \cap B(0, R)$ with

   i) $\operatorname{Re} h(w_1) \leq \gamma(r \circ \varphi(w_1, 0) - |w_1|^2)$,

   ii) $\gamma t' \leq |h(-t')| \leq \frac{1}{\gamma} t'$ for $0 < t' < R'/\eta$,

   iii) $|\partial h/\partial w_1| \geq \gamma$, where γ only depends on R_0 and A, but not on t.

(Such a family of functions obviously exists.)

From the above estimate for $\rho_k^*(w_1, w'')$ we get for all $w = (w_1, 0) \in \Omega^* \cap H_1 \cap B(0, R)$ because of i)

$$
\begin{aligned}
|\rho_k^*(w_1, 0)| &= -\rho_k^*(w_1, 0) \leq -\frac{1}{C_1} r \circ \varphi(w_1, 0) + C_4|w_1|^2 + C_4 t \\
&\leq \left(\frac{1}{C_1} + C_4\right)\frac{1}{\gamma}|\operatorname{Re} h(w_1)| + C_4 t
\end{aligned}
$$

if $\rho_k^*(w_1, 0) < 0$.

If on the other hand $\rho_k^*(w_1, 0) \geq 0$, we get from (8.47)

$$
|\rho_k^*(w_1, 0)| = \rho_k^*(w_1, 0) \leq 2C_1 C_2 R_0 t
$$

This means, that one has in both cases on $\Omega^* \cap H_1 \cap B(0, R)$ the estimate

$$
|\rho_k^*(w_1, 0)| \leq C_5|h(w_1) - C_6 t|
$$

with $C_5 := (\frac{1}{C_1} + C_4)\frac{1}{\gamma}$ and a suitable constant C_6 which only depends on the constants C_1, C_2, R_0, C_4, C_5 (notice, that $\operatorname{Re} h < 0$).

Finally, we define

$$
f(w_1) := \left(\frac{\sqrt{t}}{t - h(w_1)}\right)^2 \frac{1}{(C_6 t - h(w_1))^{e'}}
$$

with $e' := \sum_{k=2}^n 1/\mu_k$. This gives because of iii)

$$
|f(w)|^2 \prod_{k=2}^n |\rho_k^*(w_1, 0)|^{2/\mu_k} \leq \frac{C_5^{2e'}}{\gamma^2} \frac{t^2}{|t - h(w_1)|^4}\left|\frac{\partial h}{\partial w_1}\right|^2
$$

Hence

$$I(f) = \int_{\Omega^* \cap H_1 \cap B(0,R)} |f(w_1)|^2 \prod_{k=2}^{n} |\rho_k^*(w_1,0)|^{2/\mu_k} \, d\lambda_2 \leq \frac{C_5^{2e'}}{\gamma^2} t^2 \int_{\{\operatorname{Re}\zeta < 0\}} \frac{d\lambda_2}{|t - \zeta|^4}$$

But

$$\int_{\{\operatorname{Re}\zeta < 0\}} \frac{d\lambda_2}{|t - \zeta|^4} = \int_{-\infty}^{0} \int_{-\infty}^{\infty} \frac{dy}{\left((t-x)^2 + y^2\right)^2} \, dx = I_0 \int_{-\infty}^{0} \frac{dx}{(t-x)^3} = \frac{1}{2} \frac{I_0}{t^2}$$

with

$$I_0 = \int_{\mathbb{R}} \frac{dy}{(1 + y^2)^2}$$

This shows, that $I(f) \leq C_7$ independent of t.

Now we associate with f a function $\widehat{f}$ as in Lemma 8.5. We get

$$\left\| \widehat{f} \right\|_{L^2(\Omega_\alpha^* \cap B(0,R'))}^2 \leq C_8 \left(\log \frac{1}{t} \right)^{n-1}$$

with $\alpha := \eta A R_0 t / 4$ and a constant C_8 independent of t. Furthermore, we have because of
ii)

$$\left| \widehat{f}(-(1 + \eta) t e_1) \right|^2 \geq \frac{C_9}{t^{2+2e'}}$$

Together with (8.45) we, alltogether, get

$$K_\Omega(t z_*) \;\geq\; c K_{\Omega_\alpha^* \cap B(0,R')}(-(1 + \eta) t e_1) \geq \frac{\left| \widehat{f}(-(1 + \eta) t e_1) \right|^2}{\left\| \widehat{f} \right\|_{L^2(\Omega_\alpha^* \cap B(0,R'))}}$$

$$\geq \;\frac{C_9}{C_8 t^{2+2\sum_{k=2}^{n} \frac{1}{\mu_k}} \left(\log \frac{1}{t} \right)^{n-1}}$$

This finishes the proof of Proposition 8.1. $\qquad\qquad\qquad\qquad\qquad\qquad\qquad \Box$

Bibliography

[1] d'Angelo, J. P.: Real hypersurfaces, orders of contact and applications. Ann. of Math. **115** (1982), 625—637.

[2] Boas, H., Straube, E.: On equality of line type and variety type of real hypersurfaces in $\mathbb{C}^n$. J. Geom. Anal. **2** (1992), 95—98.

[3] Catlin, D.: Necessary conditions for the subellipticity and hypoellipticity for the $\bar\partial$–Neumann problem on pseudoconvex domains. In: Recent Developments in Several Complex Variables. Annals of Mathematics Studies **100** (1981), 93—100.

[4] Catlin, D.: Global regularity for the $\bar\partial$–Neumann problem. Proc. Symp. Pure Math. **41** (1984), 39—49.

[5] Catlin, D.: Boundary invariants of pseudoconvex domains. Ann. of Math. **120** (1984), 529—586.

[6] Catlin, D.: Estimates of invariant metrics on pseudoconvex domains of dimension two. Math. Z. **200** (1989), 429—466.

[7] Diederich, K., Fornæss, J. E.: Pseudoconvex domains: Existence of Stein neighborhoods. Duke Math. J. **44** (1977), 641—662.

[8] Diederich, K., Fornæss, J. E., Herbort, G.: Boundary behavior of the Bergman metric. Proc. Symp. Pure Math. **41** (1984), 59—67.

[9] Diederich, K.,Herbort, G.: Geometric and analytic boundary invariants on pseudoconvex domains. Comparison results. J. Geom. Analysis **3** (1993), 237—267.

[10] Diederich, K.: Geometric and analytic invariants for pseudoconvex hypersurfaces. A series of lectures. Preprint 1993.

[11] Diederich, K., Herbort, G.: Pseudoconvex domains of semiregular type II. In preparation.

[12] Diederich, K., Herbort, G., Ohsawa, T.: The Bergman kernel on uniformly extendable pseudoconvex domains. Math. Ann. **273** (1986), 471—478.

[13] Fornæss, J. E., Sibony, N.: Construction of plurisubharmonic functions on weakly pseudoconvex domains. Duke Math. J. **58** (1989), 633—655.

[14] Herbort, G.: Über das Randverhalten der Bergmanschen Kernfunktion und Metrik für eine spezielle Klasse schwach pseudokonvexer Gebiete des $\mathbb{C}^n$. Math. Z. **184** (1983), 193—202.

[15] Herbort, G.: Invariant metrics and peak functions on pseudoconvex domains of homogeneous finite diagonal type. Math. Z. **209** (1992), 223—243.

[16] McNeal, J.: Lower Bounds on the Bergman metric near a point of finite type. Ann. Math. **136** (1992), 339—360.

[17] McNeal, J.: Convex domains of finite type. J. Func. Anal. **108** (1992), 361—373.

[18] McNeal, J.: Estimates on te Bergman kernel on convex domains. To appear in Adv. Math. 1993.

[19] McNeal, J., Stein, E. M.: Mapping properties of the Bergman projection on convex domains of finite type. Duke Math. J. **73** (1994), 177–199.

[20] Noell, A.: Peak functions on weakly pseudoconvex domains in $\mathbb{C}^n$. Several Complex Variables: Proceedings of the Mittag–Leffler–Institute 1987—1988. Mathematical Notes No. 38, Princeton University Press 1993.

[21] Ohsawa, T.: Boundary behavior of the Bergman kernel function. Publ. R.I.M.S., Kyoto Univ. **16** (1984), 897—902.

[22] Ohsawa, T.: On the extension of L^2–holomorphic functions III: negligible weights. Preprint 1993 (to appear in Math. Z. 1994).

[23] Yu, J. Y.: Multitype of convex domains. Indiana Univ. Math. J. **41** (1992), 837—849.

Addendum:

On March 15, 1994 the authors were kindly informed by E. Straube of the existence of a preprint of J.Y. Yu entitled "Peak functions on pseudoconvex domains" and containing similar results as this article.

E. Straube also pointed out to us, that he and H. Boas knew the results of J. McNeal from [16], when they proved their version of Theorem 1.4.

Surfaces de Riemann de bord donne dans $\mathbb{C}P^n$

Pierre Dolbeault and Gennadi Henkin

Introduction

Soit X une variété analytique complexe, de dimension complexe n. Soit γ une courbe réelle fermée orientée, ou plus généralement une 1-chaîne fermée de classe C^k, alors $b\gamma = 0$. S'il existe une 1-chaîne holomorphe S de $X \setminus spt\gamma$, ayant une extension simple à X que l'on note encore S telle que $bS = \gamma$, on dit que γ est le bord de S. La 1-chaîne γ étant donnée, on cherche une condition nécessaire et suffisante pour que γ soit le bord d'une 1-chaîne holomorphe S (problème du bord).

Pour $X = \mathbb{C}^n$, ou plus généralement une variété kählérienne et si γ est une courbe, le problème du bord est un cas spécial du problème de Plateau puisqu'un ensemble analytique complexe S avec $bS = \gamma$ minimise localement l'aire dans l'espace de toutes les chaînes rectifiables de même bord γ [8]. Dans le cas $X = \mathbb{C}^n$, c'est aussi la recherche de l'enveloppe polynômiale de γ (ou enveloppe d'holomorphie de γ par rapport à $\mathcal{O}(\mathbb{C}^n)$).

Une condition nécessaire à la solution du problème du bord est la suivante : pour toute 1-forme différentielle holomorphe φ de X, $(d''\varphi = 0)$, on a
$\int_\gamma \varphi = <\gamma, \varphi> = 0$, puisque

$$< \gamma, \varphi > \ = \ < bS, \varphi > \ = \ - < dS, \varphi > \ = \ < S, d\varphi > \ = \ < S, d''\varphi > \ = \ 0 \ ;$$

on dit alors que γ satisfait à la condition du moment.

On remarque que cette condition est vide dans $\mathbb{C}P^n$ car, alors, $\varphi = 0$. Dans $\mathbb{C}^n \cong \mathbb{C}P^n \setminus \mathbb{C}P^{n-1}$, John Wermer a montré ([16], [17], 1958) que cette condition est suffisante lorsque γ est l'image homéomorphe d'un cercle C de $\mathbb{C}$ par une application holomorphe d'un voisinage de C dans $\mathbb{C}$. Le problème a été repris par E. Bishop et H. Royden [2], [12]. L'hypothèse d'analyticité réelle de γ a été affaiblie en différentiabilité C^1 par Bishop (1963) [3] et G. Stolzenberg (1966) [15]. Le résultat de H. Alexander (1971) sur les enveloppes d'arcs rectifiables [1] permet de prolonger le théorème de Wermer pour un arc fermé rectifiable. Une nouvelle démonstration dans le cas où γ est une courbe C^1 sauf sur un fermé de mesure de Hausdorff 1-dimensionnelle nulle est due à Harvey et Lawson (1975) [9], [8].

Dans $\mathbb{C}P^2$, l'exemple suivant, essentiellement dû à B. Lawson a été introduit en 1984, puis rédigé mais non publié : soit $f : \mathbb{C} \to \mathbb{C}P^2$,
$t \mapsto (t - 1, t(t - 1), te^t)$. Pour $\rho \gg 0$, toute droite projective rencontre
$S = f(B(0, \rho))$ qui a pour bord $\gamma = f(bB(0, \rho))$. De sorte que, bien que γ soit contenue dans $\mathbb{C}^2 \approx \mathbb{C}P^2 \setminus Q$, où Q est une droite projective telle que $\gamma \cap Q = \emptyset$, γ ne satisfait pas à la condition du moment dans $\mathbb{C}^2$. Cet exemple montre que le problème dans $\mathbb{C}P^n$ a un sens. La solution du problème du bord dans $\mathbb{C}P^2$ a été annoncée dans [6].

Cet article est organisé somme suit :

La section 1 contient l'énoncé principal qui donne une condition nécessaire et suffisante pour qu'une 1-chaîne fermée γ de classe C^2 de $\mathbb{C}P^n (n \geq 2)$ soit le bord d'une 1-chaîne holomorphe S de $\mathbb{C}P^n \setminus spt\gamma$. Cet énoncé est démontré complètement dans le cas $n = 2$ et est annoncé pour $n \geq 3$ avec une démonstration de la nécessité dans la section 2 pour $n \geq 2$ et un plan de démonstration de la suffisance dans la section 6 pour $n \geq 3$.

Les sections 3 et 4 établissent la suffisance de la condition dans $\mathbb{C}P^2$ en construisant des fonctions méromorphes (section 3) dont les diviseurs définissent la 1-chaîne holomorphe cherchée (section 4).

La section 5 montre l'indépendance de la condition de la section 1 par rapport aux changements de coordonnées permis et aux projections permises.

La section 7 donne une interprétation de S comme enveloppe d'holomorphie d'une courbe analytique réelle γ dans un ouvert de Stein convenable de $\mathbb{C}P^n$.

1 Définitions, notations et résultat principal

Soit $\mathbb{C}P^n$ l'espace projectif complexe, de dimension $n \geq 2$, muni de la métrique de Fubini-Study par rapport à laquelle les notions de masse d'un courant et de mesure de Haussdorff k-dimensionnelle $\mathcal{H}^k$ seront définies.

1.1 Définitions et notations

On appelle 1-*chaîne fermée* ou 1-*cycle* une combinaison linéaire γ, à coefficients entiers, de courants d'intégration sur des courbes réelles, fermées, orientées, de classe C^2 de l'espace projectif $\mathbb{C}P^n$ possédant la propriété suivante : le support de γ, $spt\gamma$ contient un fermé τ tel que $\mathcal{H}^1(\tau) = 0$ et que $spt\gamma \setminus \tau$ soit une sous-variété C^2 de dimension 1 ; alors $d\gamma = b\gamma = 0$. On appelle 1-chaîne holomorphe d'un ouvert U de $\mathbb{C}P^n$ une combinaison linéaire, à coefficients entiers, de courants d'intégration sur des sous-ensembles analytiques complexes irréductibles, de dimension complexe 1, i.e. des courbes analytiques complexes irréductibles de U ou des surfaces de Riemann immergées dans D.

S'il existe une 1-chaîne holomorphe S de $\mathbb{C}P^n \setminus spt\gamma$ ayant une extension simple à $\mathbb{C}P^n$, encore notée S, telle que $bS = \gamma$, on dira que γ est le bord de S. On cherche à résoudre le problème du bord suivant : γ étant donnée, trouver S telle que $bS = \gamma$.

Soient $(w_0, w_1, ..., w_n)$ des coordonnées homogènes dans $\mathbb{C}P^n$,
$Q = \{w_0 = 0\}$ l'hyperplan à l'infini, tels que $spt\gamma \cap Q = \emptyset$. On considère les coordonnées affines $(z_k = \dfrac{w_k}{w_0}; k = 1, ..., n)$ dans $\mathbb{C}P^n \setminus Q \approx \mathbb{C}^n$. On pose :

$$\nu = (w_1, ..., w_{n-1}) \ ; \ \zeta = (z_1, ..., z_{n-1}) \ ; \ \xi \in \mathbb{C} \ ; \ \eta = (\eta_1, ..., \eta_{n-1}) \in \mathbb{C}^{n-1} ;$$

$$\eta.\nu = \sum_{\ell=1}^{n-1} \eta_\ell w_\ell \ ; \ \eta.\zeta = \sum_{\ell=1}^{n-1} \eta_\ell z_\ell$$

On considère la forme linéaire $\tilde{g} = w_n - \xi w_0 - \eta.\nu$ et $g = \dfrac{\tilde{g}}{w_0} = z_n - \xi - \eta.\zeta$.

On désigne par $D(\xi,\eta)$ l'hyperplan projectif d'équation $\tilde{g} = 0$ et d'équation non homogène $g = 0$ dans $\mathbb{C}P^n \setminus Q$. Quand (ξ,η) varie dans $\mathbb{C}^n$, $D(\xi,\eta)$ décrit un ouvert dense de $\mathcal{G} = (\mathbb{C}P^n)'$.

1.2. Théorème principal.— *Dans les notations ci-dessus, les deux conditions suivantes sont équivalentes :*

(i) *γ est le bord d'une 1-chaîne holomorphe, de masse finie, de $\mathbb{C}P^n \setminus spt\gamma$;*

(ii) *il existe un point $(\xi^*,\eta^*) \in \mathbb{C} \times \mathbb{C}^{n-1}$ au voisinage duquel la fonction vectorielle*

$$G(\xi,\eta) = \frac{1}{2\pi i} \int_\gamma \zeta \frac{dg}{g}$$

est égale à

$$\sum_{j=1}^{N^+} f_j^+(\xi,\eta) - \sum_{j=1}^{N^-} f_j^-(\xi,\eta)$$

où les fonctions $f_j = f_j^\pm$, de composantes scalaires $f_{jk}(k=1,...,n-1)$, sont holomorphes en (ξ,η) et satisfont au système d'équations aux dérivées partielles de l'onde de choc

$$f_{jk} \frac{\partial f_j}{\partial \xi} = \frac{\partial f_j}{\partial \eta_k}.$$

1.2.1. Remarques.

1. La condition (ii) est invariante par changement de coordonnées permis (section 5).

2. La démonstration de la nécessité de (ii) est complète. La suffisance de la condition est démontrée dans le cas $n = 2$, annoncée dans le cas $n \geq 3$ avec un schéma de démonstration.

1.2.2. Proposition.— *γ étant donnée, deux solutions du problème du bord diffèrent d'une 1-chaîne algébrique projective.*

Démonstration.— Soient S_1, S_2 deux 1-chaînes holomorphes distinctes telles que $bS_1 = \gamma = bS_2$. Alors $S_1 - S_2$ est un courant rectifiable, de type $(n-1, n-1)$, d-fermé puisque $d(S_1 - S_2) = -b(S_1 - S_2) = 0$; de plus $\mathcal{H}^3(spt(S_1 - S_2)) = 0$. D'après le théorème de structure de Harvey-Shiffman, [10], [8], $S_1 - S_2$ est une 1-chaîne holomorphe de $\mathbb{C}P^n$, i.e. $S_1 - S_2 = \sum n_\ell[W_\ell]$ où, pour tout ℓ, $n_\ell \in \mathbb{Z}$ et $[W_\ell]$ est le courant d'intégration sur l'ensemble analytique complexe irréductible W_ℓ de $\mathbb{C}P^n$. D'après le théorème de Chow [7], W_ℓ est une sous-variété algébrique projective de dimension 1. $\qquad\square$

1.3. Corollaire.— *Si $N^+ = 0 = N^-$, la condition (ii) équivaut à la condition du moment dans*

$\mathbb{C}^n \cong \mathbb{C}\,P^n \setminus D(\xi^*, \eta^*)$ *et le théorème se réduit à celui de Wermer* ([17], *Theorems* 1.1 *and* 1.2).

Démonstration.

$$G(\xi, \eta) = \frac{1}{2\pi i} \int_\gamma \zeta \frac{dg}{g} = \sum_{j=1}^{N^+} f_j^+(\xi, \eta) - \sum_{j=1}^{N^-} f_j^-(\xi, \eta)$$

$$D(\xi^*, \eta^*) = \{w_n - \xi^* w_0 - \eta^* . \nu = 0\}.$$

Prenons les coordonnées homogènes de manière que $D(\xi^*, \eta^*)$ soit l'hyperplan à l'infini, alors $\xi^* = \infty$ et ξ au voisinage de ξ^* signifie $|\xi| > A \gg 0$.

Soit G_k la k-ième composante de G, $k = 1, ..., n-1$.

Alors $\quad -(2\pi i)G_k(\xi, \eta) = \displaystyle\int_\gamma z_k \frac{dg}{\xi - (z_n - \eta.\zeta)} \; ; \; dg = dz_n - \sum_{\ell=1}^{n-1} \eta_\ell dz_\ell.$

On choisit A pour que $spt\gamma \subset B(0, \frac{A}{2})$

$$\frac{1}{\xi - (z_n - \eta.\zeta)} = \frac{1}{\xi} \frac{1}{1 - \xi^{-1}(z_n - \eta.\zeta)} = \sum_{p=0}^{+\infty} \frac{1}{\xi^{p+1}} (z_n - \eta.\zeta)^p.$$

Pour $|\xi| > A$, la série des $(1/\xi)$ ci-dessus converge normalement.

La condition **(ii)** $G(\xi, \eta) = 0$ pour $|\xi| > A$ et $|\eta - \eta^*| < \alpha$ entraîne

$$\int_\gamma z_k (z_n - \sum_{\ell=1}^{n-1} \eta_\ell z_\ell)^p dz_m = 0 \;\; \text{avec} \;\; k = 1, ..., n-1; m = 1, ..., n$$

d'où, η étant arbitraire assez voisin de η^*,

$$\int_\gamma z_k z_n^a \zeta^b dz_m = 0.$$

C'est-à-dire l'intégrale, sur γ, de tout monôme en $(z_1, ..., z_n)$ ayant z_k en facteur ($k = 1, ..., n-1$) par rapport à dz_m est nulle.

Mais $\displaystyle\int_\gamma z_n^q dz_n = \frac{1}{q+1} \int_\gamma d(z_n^{q+1}) = 0$ et pour $q \geq 1$,

$$\int_\gamma z_n^q dz_k = \int_\gamma z_n^q \frac{dz_k}{dz_n} dz_n = \int_\gamma \frac{\partial}{\partial z_n}(z_k z_n^q) dz_n - q \int_\gamma z_k z_n^{q-1} dz_n \; ;$$

dans le dernier membre, le premier terme est nul d'après la formule de Stokes, le second est nul d'après ce qui précède.

Pour $q = 0$, $\displaystyle\int_\gamma dz_k = 0$ d'après la formule de Stokes. $\square$

1.4. Corollaire.— *Si γ est définie par une courbe orientable, il existe une orientation pour laquelle le courant d'intégration γ est le bord du courant d'intégration défini par une surface de Riemann immergée dans $\mathbb{C}P^n$.*

1.5. Remarque.— On peut établir le théorème dans le cas où les composantes de la chaîne γ sont des courbes C^1 en dehors de fermés $\mathcal{H}^1$-négligeables.

2   Condition nécessaire

2.1. Préliminaires.— Supposons d'abord que γ soit la chaîne définie par une courbe orientée et qu'il existe une 1-chaîne holomorphe S définie par une courbe analytique complexe, notée aussi S, de $\mathbb{C}P^n \setminus \gamma$ telle que $\mathcal{H}^2(S) < +\infty$. On suppose que S a une extension simple à $\mathbb{C}P^n$ (notée aussi S) telle que $\gamma = bS$.

Pour un ouvert dense de $\mathcal{G}$, les points d'intersection de $D(\xi, \eta)$ et de la courbe analytique complexe S sont en nombre fini, car ils forment un ensemble discret compact, les ensembles $(\bar{S})_{\mathbb{C}P^n}$ et $D(\xi, \eta)$ étant compacts dans $\mathbb{C}P^n$, et ils sont contenus dans $\mathbb{C}P^n \setminus Q \cong \mathbb{C}^n$; on les numérote de façon arbitraire.

Soit $f_j^+(\xi, \eta) \in \mathbb{C}^{n-1}$, $j = 1, ..., N^+$ telle que $(f_j^+(\xi, \eta) \, ; \, \xi + \eta.f_j^+(\xi))$ soit le j-ième point d'intersection de $D(\xi, \eta)$ et de S.

Pour Q appartenant à un ouvert dense de $\mathcal{G}$, il existe un nombre fini de points d'intersection $q_s, s = 1, ..., N^-$ de Q et de S.

2.2. Lemme.— *Dans les notations ci-dessus, pour $D(\xi, \eta)$ et Q dans un ouvert dense convenable de $\mathcal{G}$, les fonctions $f_j^+(\xi, \eta)$ et les fonctions*

$$f_s^-(\xi, \eta) = (1 - \eta.\nu)^{-1}[\xi\nu + \frac{d\nu}{dw_0} + (\eta.\frac{d\nu}{dw_0})\nu - (\eta.\nu)\frac{d\nu}{dw_0}](q_s)$$

sont localement holomorphes et satisfont à la relation

$$(1) \qquad G(\xi, \eta) = (1/2\pi i)\int_\gamma \zeta\frac{dg}{g} = \sum_{j=1}^{N^+} f_j^+(\xi, \eta) - \sum_{s=1}^{N^-} f_s^-(\xi, \eta).$$

Démonstration.

$$\int_\gamma \zeta\frac{dg}{g} = \int_\gamma \frac{\nu}{w_0}\frac{w_0}{\tilde{g}}d(\frac{\tilde{g}}{w_0}) = \int_\gamma \frac{\nu}{\tilde{g}}(\frac{d\tilde{g}}{w_0} - \frac{\tilde{g}}{w_0^2}dw_0) = \int_\gamma \frac{\nu}{w_0}\frac{d\tilde{g}}{\tilde{g}} - \nu\frac{dw_0}{w_0^2}.$$

La forme différentielle $\omega_1 = \dfrac{\nu}{w_0}\dfrac{d\tilde{g}}{\tilde{g}} - \nu\dfrac{dw_0}{w_0^2}$ restreinte à S est méromorphe ; à distance finie, i.e. pour $w_0 \neq 0$, elle a pour pôles simples les points $p_j = (f_j^+(\xi, \eta) \, ; \, \xi + \eta.f_j^+(\xi))$ et, à l'infini, i.e. pour $w_0 = 0$, les pôles doubles q_s. On considère, sur S, les disques centrés aux points ci-dessus, traces sur S des boules de $\mathbb{C}P^n$, centrées en ces points, de rayon ε assez petit ; soit B la réunion de ces disques de S. Alors l'extension simple à $\mathbb{C}P^n$ du courant d'intégration sur $S \setminus B$ a pour bord $bS - bB$.

Si B_j est le disque centré en p_j, de rayon ε, on a : $\lim\limits_{\varepsilon \to 0} \int_{bB_j} \omega_1 = 2\pi i f_j^+(\varepsilon, \eta)$.

Au voisinage de q_s, on a $\nu = \nu(q_s) + \dfrac{d\nu}{dw_0}(q_s)w_0 + \cdots$; alors B_s' étant le disque de rayon ε centré en q_s, on a :

$$\int_{bB_s'} \omega_1$$

$$= \int_{bB_s'} \frac{1}{w_0}[\nu(q_s) + \frac{d\nu}{dw_0}(q_s)w_0 + \cdots]\frac{d\tilde{g}}{\tilde{g}} - \int_{bB_s'} [\nu(q_s) + \frac{d\nu}{dw_0}(q_s)w_0 + \cdots]\frac{dw_0}{w_0^2}$$

$$= \int_{bB'_s} (\nu \frac{d\tilde{g}}{dw_0} \frac{1}{\tilde{g}})(q_s) \frac{dw_0}{w_0} - \nu(q_s) \int_{bB'_s} \frac{dw_0}{w_0^2} - \frac{d\nu}{dw_0}(q_s) \int_{bB'_s} \frac{dw_0}{w_0} + 0(\varepsilon)$$

$$= 2\pi i [\nu \frac{d\tilde{g}}{dw_0} \frac{1}{\tilde{g}} - \frac{d\nu}{dw_0}](q_s) + 0(\varepsilon)$$

Mais $\tilde{g} = w_n - \xi w_0 - \eta.\nu$; $\dfrac{d\tilde{g}}{dw_0} = -\xi - \eta.\dfrac{d\nu}{dw_0}$ et $w_0 = 0$ en q_s ; on choisit $w_n = 1$,
alors

$$f_s^-(\xi,\eta) = -\lim_{\varepsilon \to 0} \frac{1}{2\pi i} \int_{bB'_j} \omega_1 = \nu(q_s) \frac{\xi + \eta.\dfrac{d\nu}{dw_0}}{1 - \eta.\nu}(q_s) + \frac{d\nu}{dw_0}(q_s)$$

$$= \frac{1}{1 - \eta.\nu}[\xi\nu + \frac{d\nu}{dw_0} + (\eta.\frac{d\nu}{dw_0})\nu - (\eta.\nu)\frac{d\nu}{dw_0}](q_s).$$

La forme ω_1 est d-fermée sur $S \setminus B$, alors la valeur du courant $b(S \setminus B)$ sur ω_1 est nulle ;
de plus $bS = \gamma$ et $bB = \sum_j bB_j + \sum_k bB'_k$, donc

$$\int_\gamma \omega_1 = \sum_j f_j^+ - \sum_k f_k^-.$$

□

2.2.1. Lemme.— *Dans les notations du lemme 2.2., on a*

$$(1/2\pi i) \int_\gamma \frac{dg}{g} = N^+ - N^-.$$

Démonstration.

$$\int_\gamma \frac{dg}{g} = \int_\gamma \frac{w_0}{\tilde{g}}.d(\frac{\tilde{g}}{w_0}) = \int_\gamma \frac{w_0}{\tilde{g}}(\frac{d\tilde{g}}{w_0} - \frac{\tilde{g}}{w_0^2}dw_0) = \int_\gamma \frac{d\tilde{g}}{\tilde{g}} - \int_\gamma \frac{dw_0}{w_0} ;$$

la première intégrale est égale au nombre N^+ des points d'intersection de $D(\xi,\eta)$ et de S,
la seconde au nombre N^- des points d'intersection de la droite de l'infini $w_0 = 0$ et de S.

□

2.3. Lemme de Darboux.— *Les fonctions vectorielles f_j^+ et f_j^- satisfont au système
d'équations aux dérivées partielles*

$$(2) \qquad\qquad f_{jk} \frac{\partial f_j}{\partial \xi} = \frac{\partial f_j}{\partial \eta_k} ; \; k = 1, ..., n-1,$$

où f_{jk} est la k-ième composante scalaire du vecteur f_j.

Remarques.

1) Dans le cas $n = 2$, (2) est une équation scalaire ; le lemme pour les fonctions f_j^+ se
trouve implicitement dans ([5], chapitre X, p. 154).

2) Dans le cas $n = 2$, **(2)** est l'équation de l'onde de choc dans laquelle le temps est η_1 et ξ la coordonnée d'espace.

Démonstration.

1) Pour f_j^+ ; on omet l'indice supérieur $+$. Soit $H(\zeta; z_n)$ une fonction holomor- phe sur un ouvert U de $\mathbb{C}^n$ à valeurs dans $\mathbb{C}^{n-1}$ telle que $S = H^{-1}(0)$. Les points d'intersection $(f_j(\xi, \eta) \; ; \; \xi + \eta.f_j(\xi, \eta))$ satisfont à
$\mathcal{H}(\xi, \eta) = H(f_j \; ; \; \xi + \eta.f_j) \equiv 0$, donc $d\mathcal{H} \equiv 0$, i.e. les n applications linéaires

$$\frac{\partial \mathcal{H}}{\partial \xi}, \; \frac{\partial \mathcal{H}}{\partial \eta_k} \; : \; \mathbb{C} \to \mathbb{C}^{n-1}, \; k = 1, ..., n-1$$

satisfont au système d'équations linéaires

$$\frac{\partial \mathcal{H}}{\partial \xi} = D_\zeta H \circ \frac{\partial f_j}{\partial \xi} + \frac{\partial H}{\partial z_n}(1 + \eta.\frac{\partial f_j}{\partial \xi}) = 0$$

$$\frac{\partial \mathcal{H}}{\partial \eta_k} = D_\zeta H \circ \frac{\partial f_j}{\partial \eta_k} + \frac{\partial H}{\partial z_k}(f_{jk} + \eta.\frac{\partial f_j}{\partial \eta_k}) = 0 \; ; \; k = 1, ..., n-1.$$

Comme le système ci-dessus a une solution non nulle en $D_\zeta H(\zeta; \xi + \eta.\zeta)$ et $\frac{\partial H}{\partial z_n}(\zeta; \xi + \eta.\zeta)$, il existe $\lambda_k \in \mathbb{C}^*$ tel que, pour $k = 1, ..., n-1$,

(i) $\dfrac{\partial f_j}{\partial \xi} = \lambda_k \dfrac{\partial f_j}{\partial \eta_k}$

(ii) $1 + \eta.\dfrac{\partial f_j}{\partial \xi} = \lambda_k(f_{jk} + \eta.\dfrac{\partial f_j}{\partial \eta_k})$.

Donc **(i)**, **(ii)** non trivial équivaut à

(iii) $(f_{jk} + \eta.\dfrac{\partial f_j}{\partial \eta_k})\dfrac{\partial f_j}{\partial \xi} = (1 + \eta.\dfrac{\partial f_j}{\partial \xi})\dfrac{\partial f_j}{\partial \eta_k}$.

Mais, d'après **(i)**, on a :

$$(\eta.\frac{\partial f_j}{\partial \eta_k})\frac{\partial f_j}{\partial \xi} = (\eta.\frac{\partial f_j}{\partial \eta_k})\lambda_k\frac{\partial f_j}{\partial \eta_k} = (\eta.\lambda_k\frac{\partial f_j}{\partial \eta_k})\frac{\partial f_j}{\partial \eta_k} = (\eta.\frac{\partial f_j}{\partial \xi})\frac{\partial f_j}{\partial \eta_k}.$$

Alors **(iii)** équivaut à

$$(2) \qquad\qquad f_{jk}\frac{\partial f_j}{\partial \xi} = \frac{\partial f_j}{\partial \eta_k} \; ; \; k = 1, ..., n-1.$$

2) Pour f_s^- ; on omet l'indice supérieur $-$.

On a : $f_s(\xi) = \dfrac{1}{1 - \eta.\nu}[\xi\nu + \dfrac{d\nu}{dw_0} + (\eta.\dfrac{d\nu}{dw_0})\nu - (\eta.\nu)\dfrac{d\nu}{dw_0}](q_s)$

Posons $(1 - \eta.\nu)^{-1} = [\;]$. Alors

$\dfrac{\partial f_s}{\partial \xi} = [\;]\nu$

$\dfrac{\partial f_s}{\partial \eta_k} = w_k[\;]^2(\xi\nu + \dfrac{d\nu}{dw_0} + (\eta.\dfrac{d\nu}{dw_0})\nu - (\eta.\nu)\dfrac{d\nu}{dw_0}) + [\;](\dfrac{dw_k}{dw_0}\nu - w_k\dfrac{d\nu}{dw_0})$

$$= [\;]^2 \nu(\xi w_k + w_k(\eta.\frac{d\nu}{dw_0})) + (1 - \eta.\nu)\frac{dw_k}{dw_0})$$

$$+ w_k[\;]^2(\frac{d\nu}{dw_0} - (\eta.\nu)\frac{d\nu}{dw_0} - (1 - \eta.\nu)\frac{d\nu}{dw_0})$$

$$= f_{sk}.\frac{\partial f_s}{\partial \xi}.$$

$\square$

2.4. Proposition.— *Dans les notations du théorème* **1.2**, **(i)** *entraîne* **(ii)**.

Démonstration.— Si γ est défini par une courbe connexe orientée fermée, cela résulte des lemmes **2.2** et **2.3**. Le cas général où γ est une 1-chaîne, $S = \sum n_\ell[W_\ell]$ et $bS = \gamma$ se traite de la même façon : pour tout ℓ, on considère les intersections de $D(\xi, \eta)$ et de Q avec W_ℓ avec la multiplicité n_ℓ. $\square$

3   Condition suffisante dans $\mathbb{C}P^2$: construction d'une fonction méromorphe

Dans ce cas les fonctions G et f_j sont scalaires, η est scalaire et
$g = z_2 - \xi - \eta z_1$. La démonstration est une extension de celle de Harvey-Lawson dans le cas hypersurface [8].

3.1. Définitions et notations

3.1.1. Pour tout point $a \in \mathbb{C}P^2$, on considère la projection $\pi^a : \mathbb{C}P^2 \setminus \{a\} \to \Delta$ de centre a sur une droite projective Δ quelconque ne passant pas par a. Le choix des coordonnées homogènes dans $\mathbb{C}P^2$ est soumis à la seule condition que la droite à l'infini $Q = \{w_0 = 0\}$ ne rencontre pas $spt\gamma$, donc elle peut être prise dans un ouvert dense de $\mathcal{G}$. On prend le centre de projection a sur Q. Fixer a de coordonnées $(0, a_1, a_2)$ revient à fixer η tel que $a_2 - \eta a_1 = 0$. La projection de tout point $(z_1, z_2) \in \mathbb{C}P^2 \setminus Q \cong \mathbb{C}^2$ sur Δ est définie par $\xi = z_2 - \eta z_1$.

3.1.2. Toute 1-chaîne γ fermée s'écrit $\gamma = \sum_{p=1}^{P} \mu_p[V_p]$ où V_p est une courbe C^2 fermée orientée et où $[V_p]$ désigne le courant d'intégration sur V_p, μ_p est un entier. On dira que V_p est une composante de γ.

3.1.3. Lemme.— *Dans les notations de* **3.1.2.**, *pour tout* $p = 1, ..., P$, *pour presque tout couple*
$(a, Q; a \in Q)$, π^a *définit une immersion de* V_p *dans* Δ *et une immersion de* $V = spt\gamma \setminus \tau$
dans Δ, *i.e.* $\pi^a(V_p) = \Gamma_p^a$ *et* $\pi^a V$ *sont, localement des courbes de classe* C^2 *ou la réunion de deux courbes* C^2.

Démonstration.— Il suffit de considérer une courbe γ de classe C^2 contenue dans $\mathbb{C}P^2 \setminus Q \cong \mathbb{C}^2$. L'ensemble des droites réelles de $\mathbb{C}^2$ tangentes à γ constitue la courbe γ^* duale de γ, de classe C^1 dans $(\mathbb{R}^4)'$. Les droites projectives de $\mathbb{C}P^2$ passant par a ont pour traces dans $\mathbb{C}^2$ les droites affines complexes de direction donnée ; elles constituent dans $(\mathbb{C}^2)'$ un hyperplan complexe H, de dimension réelle 2, i.e. un sous-espace affine de dimension réelle 2 de $(\mathbb{R}^4)'$. Pour presque tout point a, on a $H \cap \gamma^* = \emptyset$.

Les bisécantes de γ dans $\mathbb{C}^2$ constituent une sous-variété C^2 de dimension 2 dans $(\mathbb{R}^4)'$ qui, pour presque tout a, rencontre H en un ensemble discret de points à distance finie, donc un ensemble fini. $\qquad\square$

3.1.4. Dans la suite, on supposera la conclusion du lemme **3.1.3.** valide et on notera $\Gamma^a = \pi^a(spt\gamma)$; Q étant choisi, $a \in Q$ est déterminé par η, alors on posera $\pi^a = \pi_\eta$ et $\Gamma^a = \Gamma_\eta$, $\Gamma_p^a = \Gamma_{p\eta}$.

3.1.5. Mise en place géométrique. On fait un changement de coordonnées dans $\mathbb{C}P^2$ pour que $(\xi^*, \eta^*) = (0, 0)$ (5.2.2) et on suppose que π_0 est une immersion pour chaque V_p, alors $\mathbb{C}P^1 \setminus \Gamma_{p0}$ a un nombre fini de composantes connexes D_{pq} et une petite perturbation de η laisse fixe ce nombre. On pose $\mathbb{C}P^1 \setminus \Gamma_{p0} = \cup_{q=0}^{Q_p} D_{pq}$ et on suppose que $\xi = 0$ appartient à une composante connexe D_0 de $\mathbb{C}P^1 \setminus \Gamma_0$, contenue dans $\cap_p D_{p0}$.

3.2. Proposition.— *La condition* **(ii)** *entraîne que, pour $|\eta|$ et $|\xi|$ assez petits et pour tout $m \in \mathbb{N}$, on a*

$$G_m(\xi, \eta) = (1/2\pi i) \int_\gamma z_1^m g^{-1} dg = C_m(\xi, \eta) + P_m(\xi, \eta)$$

où $C_m(\xi, \eta) = \sum_{j=1}^{N^+} (f_j^+(\xi, \eta))^m - \sum_{s=1}^{N^-} (f_s^-(\xi, \eta))^m$ *et où* $P_m(\xi, \eta)$ *est un polynôme en ξ de degré* $\leq m - 2$, *nul si* $m = 0, 1$.

La démonstration repose sur le

3.2.1. Lemme.— *Pour tout $m \in \mathbb{N}$, la fonction $G_m(\xi, \eta)$ satisfait à*

$$\partial G_m/\partial \eta = (m/m + 1)(\partial G_{m+1}/\partial \xi).$$

Démonstration.— On a $G_m(\xi, \eta) = (1/2\pi i) \int_\gamma z_1^m (z_2 - \xi - \eta z_1)^{-1} d(z_2 - \eta z_1)$

$(\partial G_m/\partial \eta) = (1/2\pi i) \int_\gamma z_1^{m+1}(z_2 - \xi - \eta z_1)^{-2} d(z_2 - \eta z_1)$
$$-(1/2\pi i) \int_\gamma z_1^m (z_2 - \xi - \eta z_1)^{-1} dz_1$$

$d(z_1^{m+1}(z_2 - \xi - \eta z_1)^{-1}) = -z_1^{m+1}(z_2 - \xi - \eta z_1)^{-2} d(z_2 - \eta z_1)$
$$+(m+1)z_1^m(z_2 - \xi - \eta z_1)^{-1} dz_1$$

$\int_\gamma d(z_1^{m+1}(z_2 - \xi - \eta z_1)^{-1}) = <b\gamma, z_1^{m+1}(z_2 - \xi - \eta z_1)^{-1}> = 0$ car $b\gamma = 0$.

Alors
$\int_\gamma z_1^{m+1}(z_2 - \xi - \eta z_1)^{-2} d(z_2 - \eta z_1) = (m+1) \int_\gamma z_1^m(z_1 - \xi - \eta z_1)^{-1} dz_1$, donc
$(\partial G_m/\partial \eta) = (1/2\pi i)(m/m + 1) \int_\gamma z_1^{m+1}(z_2 - \xi - \eta z_1)^{-2} d(z_2 - \eta z_1)$
$= (m/m + 1)(\partial G_{m+1}/\partial \xi).$ $\qquad\square$

3.2.2. Lemme.— *La condition* **(ii)** *entraîne que, pour $|\eta|$ et $|\xi|$ assez petits et pour tout $m \in \mathbb{N}$, on a :*

$$\partial f_j^m/\partial \eta = (m/m + 1)(\partial f_j^m/\partial \xi) \quad avec \quad f_j = f_j^\pm.$$

Démonstration.— On a $f_j(\partial f_j/\partial\xi) = \partial f_j/\partial\eta$.
Pour $m \geq 1$, $(\partial f_j^m/\partial\eta) = m f_j^{m-1}(\partial f_j/\partial\eta) = m f_j^m(\partial f_j/\partial\xi)$
$= (m/m+1)(\partial f_j^{m+1}/\partial\xi)$.
Pour $m = 0$, $P_0(\xi,\eta)$ est une constante entière, la solution S est définie localement par la
condition **(ii)** (voir 3.3.1. et 4.3.1.), alors d'après 2.2.1., $P_0 = 0$. □

3.2.3. Démonstration de 3.2.

3.2.2 entraîne, dans les mêmes hypothèses $\partial C_m/\partial\eta = (m/m+1)(\partial C_{m+1}/\partial\xi)$. Supposons, pour $m \in \mathbb{N}^*$, la relation

$$(3) \qquad\qquad \partial^{m-1}G_m/\partial\xi^{m-1} = \partial^{m-1}C_m/\partial\xi^{m-1}.$$

D'après les lemmes 3.2.1 et 3.2.2,
$\partial^m G_{m+1}/\partial\xi^m = \partial^{m-1}/\partial\xi^{m-1}(\partial G_{m+1}/\partial\xi)$
$= \partial^{m-1}/\partial\xi^{m-1}(m+1/m)(\partial G_m/\partial\eta)$
$= (m+1/m)(\partial/\partial\eta)(\partial^{m-1}/\partial\xi^{m-1}G_m) = (m+1/m)(\partial/\partial\eta)(\partial^{m-1}C_m/\partial\xi^{m-1})$
$= (m+1/m)(\partial^{m-1}/\partial\xi^{m-1})(\partial/\partial\eta)C_m$
$= (\partial^m/\partial\xi^m)C_{m+1}$, ce qui établit **(3)** pour $m + 1$.
En outre **(3)** est satisfaite pour $m = 1$ d'après la condition **(ii)**.
Pour $m = 0$, on a : $G_0(\xi,\eta) = (1/2\pi i)\displaystyle\int_\gamma g^{-1}dg \in \mathbb{Z}$ et $C_0(\xi,\eta) \in \mathbb{Z}$. □

3.3. Construction d'une fonction méromorphe $R(\xi, w)$

Dans cette section, on suppose $\eta = 0$ et on pose $f_j^\pm(\xi,0) = f_j^\pm(\xi)$.

3.3.1. Construction locale.— D'après la condition **(ii)**, il existe $\varepsilon > 0$ tel que les fonctions
$f_j^\pm(\xi)$ satisfassent

$$G(\xi,0) = \sum_{j=1}^{N^+} f_j^+(\xi) - \sum_{j=1}^{N^-} f_j^-(\xi)$$

pour $|\xi| < \varepsilon$. Alors la fonction

$$(4) \qquad\qquad R(\xi,w) = \Pi_{j=1}^{N^+}(w - f_j^+(\xi))\,\Pi_{s=1}^{N^-}(w - f_s^-(\xi))^{-1}$$

est rationnelle en w sur $B(0,\varepsilon) \times \mathbb{C}$.

Soit $\rho > 0$ tel que $|f_j^\pm(\xi)| < \rho$ pour tout j et $|\xi| < \varepsilon$. Alors dans
$B' = \mathbb{C} \setminus \overline{B(0,\rho)} \subset \mathbb{C}$ (coordonnée w), on a $|f_j^\pm/w| < 1$. Pour des déterminations
continues convenables de log,

$$\log R(\xi,w) = \sum_j \log w(1 - (f_j^+(\xi)/w)) - \sum_k \log w(1 - (f_k^-(\xi)/w))$$

$$= C_0 \log w - \sum_j \sum_{m=1}^{+\infty}(1/m)(f_j^+/w)^m + \sum_k \sum_{m=1}^{\infty}(1/m)(f_k^-/w)^m.$$

C_0 est une constante entière. A cause de la convergence normale de la série en w dans B', on peut modifier l'ordre des termes et

$$R(\xi, w) = w^{C_0} exp - \sum_{m=1}^{+\infty} (1/m)C_m(\xi)w^{-m}.$$

3.3.2. Extension des coefficients $C_m(\xi)$.

$G_m(\xi) = G_m(\xi, 0)$ est définie sur $\mathbb{C} \setminus \Gamma_0$; $P_m(\xi)$ est un polynôme en ξ défini à partir du point
$0 \in D_0 \subset \cap_p D_{p0}$ mais qui a un sens pour tout $\xi \in \mathbb{C}$, donc $G_m(\xi) - P_m(\xi)$ qui coïncide avec $C_m(\xi) = C_m(\xi, 0)$ pour $|\xi| < \varepsilon$ existe globalement sur $\mathbb{C} \setminus \Gamma_0$, d'où la première assertion du lemme suivant :

3.3.3. Lemme.

— *Pour $\eta = 0$, chaque coefficient $C_m(\xi)$, défini pour $|\xi| < \varepsilon$ a une extension holomorphe, notée encore $C_m(\xi)$, à $\mathbb{C} \setminus \Gamma_0$, telle que*
$C_m = G_m - P_m$. De plus, il existe une constante réelle β telle que, pour tout $\alpha > 1$ et pour tout $|\xi| \le \alpha$, on ait $|C_m(\xi)| \le \beta^m \alpha^m$.

Démonstration de la dernière assertion.

On a $\gamma = \sum_{p=1}^P \mu_p \gamma_p$ avec $\gamma_p = [V_p]$, courant d'intégration sur la variété V_p. Soit
$G_{mp}(\xi) = \int_{V_p} z_1^m (z_2 - \xi)^{-1} dz_2$, alors $G_m(\xi) = \sum_{p=1}^P \mu_p G_{mp}(\xi)$.
On va estimer $|G_{mp}(\xi)|$. D'après le lemme 3.1.3., il existe un recouvrement de $\Gamma_{0p} = \pi_0 V_p$ par des ouverts connexes $U_{\nu p}$ de $\mathbb{C}$ assez petits possédant les propriétés suivantes :

(a) $U_{\nu p} \cap \Gamma_{0p}$ est un arc connexe L_p ou la réunion de deux arcs connexes L_p', L_p'' se coupant transversalement en un seul point.

(b) $U_{\nu p} \setminus L_p$ a deux composantes connexes et il en est de même de $U_\nu \setminus L_p'$ et $U_\nu \setminus L_p''$. Alors au-dessus d'un tel arc L_p, L_p' ou L_p'', V_p est le graphe d'une fonction $z_1(z_2)$.

On considère une partition C^∞ de l'unité $(\chi_{\nu p})$ subordonnée au recouvrement $(U_{\nu p})$. Alors

$$G_{mp}(\xi) = \int_{V_p} z_1^m (z_2 - \xi)^{-1} dz_2 = \int_{\Gamma_{0p}} \sum_\nu \chi_{\nu p}(z_1(z_2), z_2) z_1^m(z_2)(z_2 - \xi)^{-1} dz_2.$$

On a $|G_{mp}(\xi)| \le C \sum_\nu \|\chi_{\nu p}(z_1(z_2), z_2) z_1^m(z_2)\|_{C^1(\Gamma_{0p})}$ où C est une constante universelle car $\chi_{\nu p}$ est à support compact dans $\mathbb{C}$ ([8], theorem 3.18), d'où l'existence d'une constante $\beta_{(p)}$ telle que, pour tout $\xi \in \mathbb{C}$, on ait

$$|G_{mp}(\xi)| < \beta_{(p)}^m$$

$G_m(\xi)$ étant la combinaison linéaire des $G_{mp}(\xi)$ à coefficients μ_p, il existe une constante β_0 telle que pour tout $\xi \in \mathbb{C}$, on ait

(a)
$$|G_m(\xi)| \le \beta_0^m.$$

Pour $|\xi| \le \dfrac{\varepsilon}{2}$, on a $|C_m(\xi) = \sum_{j=1}^{N^+}(f_j^+)^m - \sum_{s=1}^{N^-}(f_s^-)^m|$; soit
$\beta_1 = (N^+ + N^-) \sup_{|\xi| \le \frac{\varepsilon}{2}, j} |f_j^\pm(\xi)|$, alors $|C_m(\xi)| \le \beta_1^m.$

Pour $|\xi| \leq \dfrac{\varepsilon}{2}$, la relation $P_m(\xi) = G_m(\xi) - C_m(\xi)$ entraîne l'existence d'une constante β_2 telle que

$|P_m(\xi)| \leq \beta_2^m$. Le polynôme P_m en ξ étant de degré $\leq m - 2$, en considérant les valeurs de P_m pour $(m - 1)$ points ξ quelconques tels que $|\xi| \leq \dfrac{\varepsilon}{2}$, on obtient une majoration des modules des coefficients de P_m par $\dfrac{\beta_3^m}{m - 1}$ pour une constante β_3 convenable.

Alors, pour une constante $\alpha > 1$ quelconque, pour $|\xi| < \alpha$, on a

(b) $$|P_m(\xi)| \leq \beta_3^m \alpha^m.$$

Compte-tenu de **(a)** et de **(b)**, on obtient une constante β telle que, pour tout $|\xi| < \alpha$, on ait

$$|C_m(\xi)| \leq \beta^m \alpha^m.$$

□

3.3.4. Notations.— Soient $Sing\ \Gamma_{0p}$ l'ensemble (fini) des points de Γ_{0p} au voisinage desquels Γ_{0p} n'est pas une courbe lisse et $Sing\ \Gamma_0 = \cup_{p=1}^{P} Sing\ \Gamma_{0p}$. Soit D_δ une composante connexe de $\mathbb{C} \setminus spt\Gamma_0$. Soit $Sing\ \pi_0\ \mathcal{V}$ l'ensemble des points au voisinage desquels $\pi_0\mathcal{V}$ n'est pas une courbe lisse ; c'est un ensemble de mesure 1-dimensionnelle nulle. Alors $Sing\ \Gamma_0$ est contenu dans $\pi_0\tau\ \cup\ Sing\ \pi_0\mathcal{V}$.

3.3.5. Lemme.— *Dans les notations de* **3.3.4**, $C_m(\xi)$ *a une extension continue à* $\bar{D}_\delta \setminus Sing\Gamma_0$.

Démonstration.— $C_m(\xi)$ se comporte comme $G_m(\xi) = \sum_{p=1}^{P} \mu_p G_{mp}(\xi)$. Alors la conclusion résulte d'une assertion du théorème de Plemelj-Sokhotski ([8], 3.18) appliqué à $G_{mp}(\xi)$. □

3.3.6. Proposition.— *Sur* $(B(0, \alpha) \setminus \Gamma_0) \times (\mathbb{C} \setminus B(0, \beta\alpha))$, *la fonction*

(5) $$R(\xi, w) = w^{C_0} exp - \sum_{m=1}^{+\infty} m^{-1} C_m(\xi) w^{-m}$$

est une fonction holomorphe et, pour tout δ *(cf. 3.3.4), elle a une extension continue* R^δ *à* $(B(0, \alpha)\ \cap\ (\bar{D}_\delta \setminus Sing\Gamma_0)) \times (\mathbb{C} \setminus B(0, \beta\alpha))$.

Démonstration.— C_m est holomorphe sur $B(0, \alpha) \setminus \Gamma_0$ d'après sa définition et est continue sur chaque $\bar{D}_\delta \setminus Sing\ \Gamma_0$ (lemme 3.3.5). D'après le lemme 3.3.3, la convergence de la série **(5)** en w est normale sur tout compact, donc sa somme $R(\xi, w)$ est holomorphe dans son domaine de définition ; en outre, pour tout δ, elle a une extension continue à $(\bar{D}_\delta \setminus Sing\ \Gamma_0) \times (\mathbb{C} \setminus B(0, \beta\alpha))$. □

Rappelons la variante suivante du théorème de E.E. Levi ([8], Coroll. 2.9) :

3.3.7. Lemme.— *Soient* Ω *un ouvert connexe de* $\mathbb{C}$ *(coordonnée* ξ*),* $\alpha > 0$, $B' = (\mathbb{C} \setminus B(0, \alpha)) \subset \mathbb{C}$ *(coordonnée* w*),* $F \in \mathcal{O}(\Omega \times B')\ \cap\ C(\bar{\Omega} \times B')$, E *une partie de* Ω. *Si pour tout* $\xi \in E$, *la fonction holomorphe* $F_\xi(w) = F(\xi, w)$ *a une extension rationnelle à* $\mathbb{C}$ *et si* E *est un ouvert non vide, alors* F *a une extension méromorphe à* $\Omega \times \mathbb{C}$, *rationnelle en* w, *à coefficients continus sur* $\bar{\Omega}$.

3.3.8. Proposition.— *Dans les notations de* **3.3.2** *et de* **3.3.6**, $R^0(\xi, w)$ *est rationnelle en* w *dans* $D_0 \times \mathbb{C}$, *à coefficients holomorphes dans* D_0 *et continus dans* $\bar{D}_0 \setminus Sing\Gamma_0$.

Démonstration.— $\xi = 0$ appartient à D_0 (3.1.5) ; on applique le lemme 3.3.7 à $F = R^0$ défini dans 3.3.6 avec $E = B(0, \varepsilon)$ (3.3.1) et compte tenu de **(4)**. $\qquad\square$

3.3.9. Lemme.— *On pose* $\alpha' = \beta\alpha$. *Alors*
(i) *Pour tout* $\alpha > 1$, *la série* $R(\xi, w)$ *est égale à* $\sum_{m=-\infty}^{C_0} A_m(\xi)w^m$ *et converge uniformément sur tout compact de* $(B(0, \alpha) \setminus \Gamma_0) \times (\mathbb{C} \setminus B(0, \alpha'))$.
(ii) *Tout coefficient* A_m *est une fonction polynômiale d'un nombre fini de* $\mathbb{C}_m$.

Démonstration.— Résulte de 3.3.6 par réarrangement des termes. $\qquad\square$

3.3.10. Lemme.— *La fonction* R *possède les propriétés suivantes :*
(i) *La restriction de* A_m *à* D_δ *se prolonge continûment à* $\bar{D}_\delta \setminus Sing\Gamma_0$ *en une fonction* $A_{\delta m}$.
(ii) *La restriction de* $R(\xi, w)$ *à* $(B(0, \alpha) \cap D_\delta) \times (\mathbb{C} \setminus B(0, \alpha'))$ *se prolonge continûment à*
$(B(0, \alpha) \cap \bar{D}_\delta \setminus Sing\Gamma_0) \times (\mathbb{C} \setminus B(0, \alpha'))$ *en une fonction* $R^\delta(\xi, w)$.
(iii) *Pour tout* $\xi \in \bar{D}_\delta \setminus Sing\Gamma_0$, $R^\delta(\xi, w)$ *est holomorphe en* w *sur* $\mathbb{C} \setminus B(0, \alpha')$ *et a pour développement de Laurent* $R^\delta(\xi, w) = \sum_{m=-\infty}^{C_0} A_{\delta m}(\xi)w^m$.
(iv) *Soient* $Reg\Gamma_0 = \Gamma_0 \setminus (\pi_0\tau \cup Sing\pi_0\mathcal{V})$ *et* Γ_* *une composante connexe de* $Reg\Gamma_0 \cap \bar{D}_\delta \cap \bar{D}_{\delta'}$ *contenue dans* $\pi_0(V_p)$, *alors* $\pi_0^{-1}(\Gamma_0) \cap spt\gamma$ *est le graphe d'une fonction* $h(\xi) \in C^2(\Gamma_*)$ *et*
$R^\delta(\xi, w) = (w - h(\xi))^{\pm\mu_p} R^{\delta'}(\xi, w)$ *sur* $(B(0, \alpha) \cap \Gamma_*) \times (\mathbb{C} \setminus B(0, \alpha'))$ *où l'exposant* $+\mu_p$ *figure si* Γ_* *est contenu dans le bord de* D_δ *et l'exposant* $-\mu_p$ *si* Γ_* *est dans le bord de* D'_δ.
(v) *Pour tout point* $\xi \in \bar{D}_\delta \setminus Sing\,\Gamma_0$, $R^\delta(\xi, w)$ *est rationnelle en* w *et ses coefficients sont holomorphes en* ξ *sur* D_δ *et continus sur* $\bar{D}_\delta \setminus Sing\,\Gamma_0$.

Démonstration.

(i) résulte de l'expression de A_m en fonction des C_ℓ (3.3.9 (ii)) et du lemme 3.3.5.
(ii) est la seconde assertion de la Proposition 3.3.6.
(iii) résulte, comme la première assertion de la Proposition 3.3.6, de l'estimation des coefficients C_m (lemme 3.3.3).
(iv) résulte du fait que $\mathbb{C} \setminus (\pi_0\tau \cup Sing\pi_0\mathcal{V})$ est connexe puisque
$\mathcal{H}^1(\pi_0\tau \cup Sing\pi_0\mathcal{V}) = 0$, de la formule du saut dans le théorème de Plemelj-Sokhotski, compte tenu de la multiplicité μ_p de V_p.
(v) est vérifiée dans D_0 (Proposition 3.3.8) et ensuite, en passant d'un D_δ à un D'_δ adjacent i.e. en traversant le bord d'un D_{pq} composante connexe de $\mathbb{C}P^1 \setminus \Gamma_{p,0}$, à l'aide de (iv) et du théorème de Levi généralisé ([8], corollaire 3.20). $\qquad\square$

4   Condition suffisante dans $\mathbb{C}P^2$: construction de la 1-chaîne holomorphe S

4.1. Définition de la fonction R_η

La fonction R est définie sur $\mathbb{C}^2 \setminus \pi_0^{-1}(\Gamma_0) \cong (\mathbb{C} \setminus \Gamma_0) \times \mathbb{C}$, comme on le voit en faisant croître α indéfiniment dans le lemme 3.3.10.

Faisons varier le centre a de la projection π^a (voir 3.1.1.) dans la droite de l'infini Q, ce qui revient à faire varier η ; choisissons η de sorte que la conclusion du lemme 3.1.3. soit valide.

On a établi le lemme 3.3.10. pour $\eta = 0$ fixé. D'après la condition (ii) du théorème 1.2., il existe $\alpha > 0$ tel que, pour tout $|\eta| < \alpha$ satisfaisant à la condition ci-dessus, la construction de la section 3.3. soit valide ; sur $H_\eta = \mathbb{C}^2 \setminus \pi_\eta^{-1}(\Gamma_\eta) \cong (\mathbb{C} \setminus \Gamma_\eta) \times \mathbb{C}$, on note R_η la fonction définie comme R sur $\mathbb{C}^2 \setminus \pi_0^{-1}(\Gamma_0)$. On a

$$\log R_\eta(\xi, w) = C_0(\xi, \eta) \log w - \sum_{m=1}^{+\infty} (1/m) C_m(\xi, \eta) w^{-m}.$$

4.2. Propriétés des coefficients $C_m(\xi, \eta)$

4.2.1. Comme dans le cas $\eta = 0$, C_0 est un entier localement constant (3.3.1). **4.2.2. Lemme.—** *Pour $|\eta| < \alpha$ et $\xi \in \mathbb{C} \setminus \Gamma_\eta$, on a*

$$(6) \qquad\qquad (\partial C_m / \partial \eta) = (m/m + 1)(\partial C_{m+1} / \partial \xi).$$

Démonstration.— Comme dans 3.3.3., on a

$$(7) \qquad\qquad C_m(\xi, \eta) = G_m(\xi, \eta) - P_m(\xi, \eta).$$

D'après 3.2.3., au voisinage de $(0, 0)$, C_m satisfait à la condition (6). D'après 3.2.1., G_m satisfait à
$(\partial G_m / \partial \eta) = (m/m + 1)(\partial G_{m+1} / \partial \xi)$ globalement pour $|\eta| < \alpha$ et pour tout $\xi \in \mathbb{C} \setminus \Gamma_\eta$.
Alors d'après (7), $P_m(\xi, \eta)$ satisfait, au voisinage de $(0, 0)$ à

$$(8) \qquad\qquad (\partial P_m / \partial \eta) = (m/m + 1)(\partial P_{m+1} / \partial \xi),$$

P_m, étant un polynôme en ξ, satisfait à (8) globalement en ξ. Alors (6) résulte de (7).   □

4.3. Définition de S dans H_η

4.3.1. La projection π_η applique tout point $(z_1, z_2) \in \mathbb{C}^2$ en ξ sur la droite projective Δ (voir 3.1.1) tel que $z_2 - \eta z_1 - \xi = 0$. La fonction R_η satisfait à

$$(9) \qquad\qquad \log R_\eta(\xi, z_1) = C_0 \log z_1 - \sum_{m=1}^{+\infty} (1/m) C_m(\xi, \eta) z_1^{-m}.$$

Mais $\xi = z_2 - \eta z_1$, alors le second membre de (9) est

$$\varphi_\eta(z_1, z_2) = C_0 \log z_1 - \sum_{m=1}^{\infty} (1/m) C_m(z_2 - \eta z_1, \eta) z_1^{-m}.$$

On note $\mathcal{R}_\eta(z_1, z_2)$ la fonction déduite de $R_\eta(\xi, z_1)$ par remplacement de ξ par $z_2 - \eta z_1$, alors $\varphi_\eta(z_1, z_2) = \log \mathcal{R}_\eta(z_1, z_2)$.

D'après le théorème de Poincaré-Lelong [8], $S_\eta = (i/\pi) d' d'' \log |\mathcal{R}_\eta|$ est une 1-chaîne holomorphe dans $H_\eta = \mathbb{C}^2 \setminus \pi_\eta^{-1}(\Gamma_\eta)$.

4.3.2. Lemme.— *Dans les hypothèses de* 4.1.*, on a $S_\eta = S_0$ dans $H_\eta \cap H_0$.*

Démonstration.— Dans H_η, pour $|z_1|$ assez grand, on a :

$(\partial \varphi_\eta / \partial \eta)(z_1, z_2)$

$$= - \sum_{m=1}^{+\infty} (1/m)[(\partial C_m/\partial \xi)(-z_1) + (\partial C_m/\partial \eta)](z_2 - \eta z_1, \eta) z_1^{-m}$$

$$= (\partial C_1/\partial \xi) + \sum_{m=1}^{+\infty} ((1/m+1)(\partial C_{m+1}/\partial \xi) - (1/m)(\partial C_m/\partial \eta)) z_1^{-m}$$

$$= (\partial C_1/\partial \xi)(z_2 - \eta z_1, \eta),$$

d'après (**6**).

Alors $\mathcal{R}_\eta^{-1}(\partial \mathcal{R}_\eta/\partial \eta)(z_1, z_2) = (\partial \varphi_\eta/\partial \eta)(z_1, z_2) = (\partial C_1/\partial \xi)(z_2 - \eta z_1, \eta)$ dans H_η, par unicité.

Soit $\mathcal{A} = \mathcal{A}(z_1, z_2, \eta)$ une fonction holomorphe pour $|\eta| < \alpha$ et $(z_1, z_2) \in G_\alpha = \cap_{|\eta|<\alpha} H_\eta$; on pose $\mathcal{R}_\eta' = e^{\mathcal{A}} \mathcal{R}_\eta$; on a

$(\partial \mathcal{R}_\eta'/\partial \eta) = e^{\mathcal{A}} \mathcal{R}_\eta[(\partial \mathcal{A}/\partial \eta) + (1/\mathcal{R}_\eta)(\partial \mathcal{R}_\eta/\partial \eta)] =$
$e^{\mathcal{A}} \mathcal{R}_\eta[(\partial \mathcal{A}/\partial \eta) + (\partial C_1/\partial \xi)].$

Choisissons $\mathcal{A}$ telle que

$$(\mathbf{10}) \qquad (\partial \mathcal{A}/\partial \eta)(z_1, z_2, \eta) = -(\partial C_1/\partial \xi)(z_2 - \eta z_1, \eta).$$

(**10**) est une équation différentielle en η, pour $|\eta| < \alpha$, dont le second membre est holomorphe en (z_1, z_2) dans G_α. Alors toute solution de (**10**)

$$\mathcal{A}(z_1, z_2, \eta) = - \int (\partial C_1/\partial \xi)(z_2 - \eta z_1, \eta) d\eta + \lambda(z_1, z_2),$$

où λ est une fonction holomorphe dans G_α, est une fonction holomorphe pour $|\eta| < \alpha$ et $(z_1, z_2) \in G_\alpha$; $e^{\mathcal{A}}$ est holomorphe sans zéro et $\mathcal{R}_\eta' = e^{\mathcal{A}} \mathcal{R}_\eta$ est indépendant de η. En outre, $(i/\pi) d' d'' \log |\mathcal{R}_\eta'| = (i/\pi) d' d'' \log |\mathcal{R}_\eta| = S_\eta$, donc pour tout $|\eta| < \alpha$, on a $S_\eta = S_0$ dans G_α.

Si α est assez petit et $|\eta| < \alpha$, alors S_η contient un nombre fini, uniformément borné, de composantes irréductibles, en vertu du lemme 3.1.3.

Si α est assez petit, toute composante irréductible de S_η et de S_0 rencontre G_α.

Alors S_η et S_0 coïncident, non seulement dans G_α, mais aussi dans $H_\eta \cap H_0$.					$\square$

4.4. Définition de S dans $\mathbb{C}P^2 \setminus spt\gamma$.

4.4.1. Proposition.—*Dans les hypothèses de* **4.1***, il existe une 1-chaîne holo- morphe S à support dans $\mathbb{C}P^2 \setminus spt\gamma$, prolongeant S_η pour tout η assez petit.*

4.4.2. Lemme.— $S' = \cup_{|\eta|<\alpha} S_\alpha$ *est une 1-chaîne holomorphe qui prolonge S_0 de* $\mathbb{C}^2 \setminus \pi_0^{-1}(\Gamma_0)$ *à* $\mathbb{C}^2 \setminus spt\gamma$.

Démonstration.— D'après le lemme 4.3.2. qui est valide pour $|\eta'| < \alpha$ au lieu de 0, on a $S_\eta = S_{\eta'}$ dans $H_\eta \cap H_{\eta'}$.
D'autre part on a

$$\tag{11} \cup_{|\eta|<\alpha} H_\eta = \mathbb{C}^2 \setminus spt\gamma,$$

en effet soit $z \in \mathbb{C}^2 \setminus spt\gamma$, la projection de centre z de $\mathbb{C}P^2$ sur Q envoie la courbe $spt\gamma$ sur une courbe L_z de Q ; il suffit de choisir le centre de projection a sur Q en dehors de L_z avec η satisfaisant aux hypothèses de **4.1.**, pour que z appartienne à $H_\eta = \mathbb{C}^2 \setminus \pi_\eta^{-1}(\Gamma_\eta)$, ce qui établit la relation **(11)** puis le lemme. $\qquad\square$

4.4.3. Lemme.— *S' se prolonge en une 1-chaîne holomorphe unique S dans $\mathbb{C}P^2 \setminus spt\gamma$.*

Démonstration.— Nous allons montrer que le support de S' est de volume localement fini au voisinage de Q. Le volume est évalué à l'aide de la métrique de Fubini-Study de $\mathbb{C}P^2$. Le support de S est un revêtement fini à nombre fixe de feuillets au-dessus de chaque composante connexe D_δ de $\mathbb{C} \setminus \Gamma_\eta$ pour la projection π_η. Alors, comme conséquence de l'inégalité de Wirtinger, $sptS$ est de volume 2-dimensionnel fini au voisinage de Q, donc d'après le théorème de Bishop-Stoll [4], [14], a une extension holomorphe à $\mathbb{C}P^2 \setminus spt\gamma$ qui coïncide avec $\overline{(sptS')}_{\mathbb{C}P^2}$, donc S', définie à partir des composantes irréductibles de $sptS'$ avec des multiplicités finies, s'étend en une chaîne holomorphe bien déterminée S à $\mathbb{C}P^2 \setminus spt\gamma$. $\qquad\square$

4.4.4. La proposition 4.4.1. résulte des lemmes 4.4.2. et 4.4.3.

4.5. Lemme.— *Dans les notations de 4.4., S a une extension simple S à $\mathbb{C}P^2$ telle que $dS = \gamma$.*

Démonstration.— $sptS$ est de volume localement fini au voisinage de $spt\gamma$, puisque c'est un revêtement fini à nombre fixe de feuillets au-dessus de chaque composante connexe de $\mathbb{C}\setminus\Gamma_\eta$ pour la projection π_η (cf. la démonstration de 4.4.3.). Alors S a une extension simple S à $\mathbb{C}P^2$.

Comme S est fermée dans $\mathbb{C}P^2 \setminus spt\gamma$, on a $sptbS \subset spt\gamma$.

S, donc S, étant localement plats [8], bS est localement plat, de support contenu dans $spt\gamma$; d'après le théorème du support on a $bS = k\gamma$ où k est une fonction L^1_{loc} ; de plus, $bbS = 0$ entraîne que k est constant et S, donc aussi bS sont rectifiables, alors $k \in \mathbb{Z}$.

S est définie à partir de S_0, elle-même définie par la fonction $R_0(\xi, z_1)$, on a :

$$\tag{12} \frac{dR_0}{R_0} = C_0 \frac{dz_1}{z_1} - \sum_{m=1}^{+\infty} C_m(\xi) z_1^{-m-1} dz_1.$$

Au voisinage d'un point lisse ξ_0 de $\Gamma_0 = spt\pi_0(\gamma)$, d'après le lemme 3.3.10. (iv), on a :

$$\tag{13} R^\delta(\xi, z_1) = (z_1 - h(\xi))^{\mu_p} R^{\delta'}(\xi, z_1)$$

pour un choix convenable de δ et de δ'.

Dans les notations de la section 1, pour $\eta = 0$, le saut de

$$(1/2\pi i) \int_\gamma g^{-1} dg = (1/2\pi i) \int_\gamma \frac{dz_2}{z_2 - \xi}$$

en ξ_0, est μ_p d'après la formule de l'indice. D'après **(12)**, on a :

$$C_0 = C_0(\xi) = (1/2\pi i) \int_{|z_1 - h(\xi)| = \varepsilon} \frac{dR_0}{R_0}.$$

Le saut de C_0 est μ_p d'après **(13)**.

D'après la démonstration du lemme 2.2. et de la proposition 2.4., et d'après la proposition 3.2. pour $m = 0$, $(1/2\pi i) \int_{bS} \dfrac{dz_2}{z_2 - \xi}$ est égale à la somme de P_0 et du nombre de points d'intersection C_0^δ de $D(\xi, 0)$ et de S, comptés avec leurs multiplicités, pour $\xi \in D_\delta$; donc les sauts des intégrales de $\dfrac{dz_2}{z_2 - \xi}$ en ξ_0 par rapport à $bS = k\gamma$ et par rapport à γ sont égaux, il en résulte que $k = 1$. $\qquad\qquad\square$

5 Invariance de la condition (ii) par changement de coordonnées et par projection

5.1. Changements de coordonnées permis

Un changement de coordonnées projectives de $\mathbb{C}P^n$ est dit permis s'il satisfait aux conditions suivantes :

(α) la droite de l'infini $w_0 = 0$ est conservée ;

(β) le changement de coordonnées restreint à

$\mathbb{C}^n : (z_1, ..., z_n) \mapsto (u_1, ..., u_n)$ induit un changement de coordonnées dans $\mathbb{C}^{n-1} : (z_1, ..., z_{n-1}) \mapsto (u_1, ..., u_{n-1})$. Alors

$$(14) \qquad \text{pour} \quad p = 1, ..., n-1, \quad \text{on a} \quad u_p = a_p^1 z_1 + \cdots + a_p^{n-1} z_{n-1} + b_p,$$

$$\text{pour} \quad p = n, \ u_n = a_n^1 z_1 + \cdots + a_n^{n-1} z_{n-1} + a_n^n z_n + b_n, \quad \text{avec} \quad a_n^n \neq 0.$$

Le changement de coordonnées **(14)** est obtenu par une suite de changements de coordonnées de la forme

$u_p = a z_p$ où $a \in \mathbb{C}$,

$u_p = z_p + b$ où $b \in \mathbb{C}$,

$u_p = z_p + z_q, q \neq p, q \neq n$, pour $p \in [1, ..., n]$,

les autres coordonnées étant inchangées.

5.2. Invariance par changement de coordonnées permis

5.2.1. Proposition.— *Dans les notations de* 1.2, *la condition* **(ii)** *est invariante par un changement de coordonnées permis.*

Démonstration.— Dans les notations de 1.1 et 1.2, on a
$g = z_n - \xi - \sum_{\ell=1}^{n-1} \eta_\ell z_\ell$; $\eta = (\eta_1, ..., \eta_{n-1})$; G_k étant la k-ième composante de G, $f_{jk}^{\pm}$
la k-ième composante de $f_j^{\pm}$, on a, d'après la condition **(ii)** :

$$G_k(\xi, \eta) = (1/2\pi i) \int_\gamma z_k \frac{dg}{g} \; ; \; k = 1, ..., n-1.$$

Au voisinage du point $(\xi^*, \eta^*) \in \mathbb{C} \times \mathbb{C}^{n-1}$, $G_k(\xi, \eta)$ est égale à

$$\sum_{j=1}^{N^+} f_{jk}^+(\xi, \eta) - \sum_{j=1}^{N^-} f_{jk}^-(\xi, \eta)$$

où les $f_{jk}^{\pm}$ sont holomorphes et satisfont au système

$$f_{jk} \frac{\partial f_{j\ell}}{\partial \xi} = \frac{\partial f_{j\ell}}{\partial \eta_k} \quad \ell = 1, ..., n-1.$$

On considère d'abord les cas où $p \neq n$. Alors, on peut prendre $p = 1, q = 2$.
(a) $u_1 = az_1$: en posant $\eta_1' = \eta_1/a$, on a

$$aG_1(\xi, \eta_1, ..., \eta_{n-1}) = (1/2\pi i) \int_\gamma u_1 \frac{d(z_n - \xi - \eta_1' u_1 - \sum_{\ell=2}^{n-1} \eta_\ell z_\ell)}{z_n - \xi - \eta_1' u_1 - \sum_{\ell=2}^{n-1} \eta_\ell z_\ell}$$

$$= \sum_{j=1}^{N^+} a f_{j1}^+ - \sum_{j=1}^{N^-} a f_{j1}^-.$$

$$G_k(\xi, \eta_1, ..., \eta_{n-1}) = (1/2\pi i) \int_\gamma z_k \frac{d(z_n - \xi - \eta_1' u_1 - \sum_{\ell=2}^{n-1} \eta_\ell z_\ell)}{z_n - \xi - \eta_1' u_1 - \sum_{\ell=2}^{n-1} \eta_\ell z_\ell}$$

$$= \sum_{j=1}^{N^+} f_{jk}^+ - \sum_{j=1}^{N^-} f_{jk}^-,$$

avec $k = 2, ..., n-1$.
On pose :

$$\varphi_{j1}^{\pm} = a f_{j1}^{\pm} \; ; \; \varphi_{jk}^{\pm} = f_{jk}^{\pm}, \; k = 2, ..., n-1 \; ; \; g' = z_n - \xi - \eta_1' u_1 - \sum_{\ell=2}^{n-1} \eta_\ell z_\ell \; ;$$

alors :

$$G_1'(\xi, \eta_1', \eta_2, ..., \eta_{n-1}) = (1/2\pi i) \int_\gamma u_1 \frac{dg'}{g'} = \sum_j \varphi_{j1}^+ - \sum_j \varphi_{j1}^- \; ;$$

$$G_k'(\xi, \eta_1', \eta_2, ..., \eta_{n-1}) = (1/2\pi i) \int_\gamma u_k \frac{dg'}{g'} = \sum_j \varphi_{jk}^+ - \sum_j \varphi_{jk}^-, \; k = 2, ..., n-1.$$

Les conditions

$$f_{j1}\frac{\partial f_{j1}}{\partial \xi} = \frac{\partial f_{j1}}{\partial \eta_1} \; ; \; f_{j1}\frac{\partial f_{j\ell}}{\partial \xi} = \frac{\partial f_{j\ell}}{\partial \eta_1}$$

$$f_{jk}\frac{\partial f_{j1}}{\partial \xi} = \frac{\partial f_{j1}}{\partial \eta_k} \; ; \; f_{jk}\frac{\partial f_{j\ell}}{\partial \xi} = \frac{\partial f_{j\ell}}{\partial \eta_k}$$

équivalent, respectivement à :

$$\varphi_{j1}\frac{\partial \varphi_{j1}}{\partial \xi} = \frac{\partial \varphi_{j1}}{\partial \eta_1'} \; ; \varphi_{j1}\frac{\partial \varphi_{j\ell}}{\partial \xi} = \frac{\partial \varphi_{j\ell}}{\partial \eta_1}$$

$$\varphi_{jk}\frac{\partial \varphi_{j1}}{\partial \xi} = \frac{\partial \varphi_{j1}}{\partial \eta_k} \; ; \varphi_{jk}\frac{\partial \varphi_{j\ell}}{\partial \xi} = \frac{\partial f_{j\ell}}{\partial \eta_k} \; ; \; k,\ell = 2,...,n-1$$

car $\dfrac{\partial \varphi_{j1}}{\partial \eta_1'} = a\dfrac{\partial f_{j1}}{\partial \eta_1}\cdot\dfrac{\partial \eta_1}{\partial \eta_1'} = a^2\dfrac{\partial f_{j1}}{\partial \eta_1}$ et $\varphi_{j1} = af_{j1}$.

On a la condition **(ii)** avec $\eta'^*_1 = \dfrac{\eta_1^*}{a}$.

(b) $u_1 = z_1 + b$

$$G_1(\xi,\eta) = (1/2\pi i)\int_\gamma (u_1 - b)\frac{dg'}{g'} \quad \text{avec} \quad g' = z_n - (\xi - \eta_1 b) - \eta_1 u_1 - \sum_{\ell=2}^{n-1} \eta_\ell z_\ell$$

$$= (1/2\pi i)\int_\gamma u_1\frac{dg'}{g'} - (1/2\pi i)\int_\gamma b\frac{dg'}{g'}.$$

On pose $\xi' = \xi - \eta_1 b$.

Mais $(1/2\pi i)\displaystyle\int_\gamma \frac{dg'}{g'} = N^+ - N^-$, d'après le lemme 2.2.1.

Alors $G_1'(\xi',\eta) = (1/2\pi i)\displaystyle\int_\gamma u_1\frac{dg'}{g'} = \sum_{j=1}^{N^+}\varphi_{j1}^+ - \sum_{j=1}^{N^-}\varphi_{j1}^-$ où $\varphi_{j1}^\pm = f_{j1}^\pm - b$; les relations

relatives à $G_k(\xi,\eta)$ étant inchangées, pour $k = 2,...,n-1$; on omet $\pm$.

De plus : $\dfrac{\partial \varphi_{j1}}{\partial \xi'} = \dfrac{\partial f_{j1}}{\partial \xi}$; $\dfrac{\partial \varphi_{j1}}{\partial \eta_1} = \dfrac{\partial f_{j1}}{\partial \eta_1} - \dfrac{\partial f_{j1}}{\partial \xi}b$, d'où

$$\varphi_{j1}\frac{\partial \varphi_{j1}}{\partial \xi'} = (f_{j1} - b)\frac{\partial f_{j1}}{\partial \xi} = \frac{\partial f_{j1}}{\partial \eta_1} - b\frac{\partial f_{j1}}{\partial \xi} = \frac{\partial \varphi_{j1}}{\partial \eta_1} \qquad ;$$

les autres relations différentielles se vérifient de la même façon, de sorte que la condition **(ii)** est satisfaite avec $\xi'^* = \xi^* - \eta_1^* b$.

(c) $u_1 = z_1 + z_2$

$$G_1(\xi,\eta) + G_2(\xi,\eta) = \int_\gamma u_1\frac{d(z_n - \xi - \eta_1 u_1 - \eta_2' z_2 - \cdots - \eta_{n-1} z_{n-1})}{(z_n - \xi - \eta_1 u_1 - \eta_2' z_2 - \cdots - \eta_{n-1} z_{n-1})}$$

$$= \sum_{j=1}^{N^+}(f_{j1}^+ + f_{j2}^+) - \sum_{j=1}^{N^-}(f_{j1}^- + f_{j2}^-)$$

$$= G_1'(\xi,\eta_1,\eta_2',...,\eta_{n-1}) \text{ avec } \eta_2' = \eta_2 - \eta_1$$

$$(f_{j1} + f_{j2})(\xi,\eta_1,\eta_2,...,\eta_{n-1}) = \varphi_{j1}(\xi,\eta_1,\eta_2',...,\eta_{n-1})$$

$$= (f_{j1} + f_{j2})(\xi,\eta_1,\eta_1 + \eta_2',...,\eta_{n-1})$$

$$\frac{\partial \varphi_{j1}}{\partial \eta_1} = \frac{\partial f_{j1}}{\partial \eta_1} + \frac{\partial f_{j2}}{\partial \eta_1} + \frac{\partial f_{j1}}{\partial \eta_2} + \frac{\partial f_{j2}}{\partial \eta_2} \; ; \; \frac{\partial \varphi_{j1}}{\partial \eta_2'} = \frac{\partial f_{j1}}{\partial \eta_2} + \frac{\partial f_{j2}}{\partial \eta_2} \; ;$$

$$\frac{\partial \varphi_{j1}}{\partial \xi} = \frac{\partial f_{j1}}{\partial \xi} + \frac{\partial f_{j2}}{\partial \xi} \; ;$$

$$\varphi_{j1}\frac{\partial \varphi_{j1}}{\partial \xi} = (f_{j1} + f_{j2})(\frac{\partial f_{j1}}{\partial \xi} + \frac{\partial f_{j2}}{\partial \xi}) = \frac{\partial f_{j1}}{\partial \eta_1} + \frac{\partial f_{j1}}{\partial \eta_2} + \frac{\partial f_{j2}}{\partial \eta_1} + \frac{\partial f_{j2}}{\partial \eta_2} = \frac{\partial \varphi_{j1}}{\partial \eta_1}$$

$$\varphi_{j2}\frac{\partial \varphi_{j1}}{\partial \xi} = f_{j2}(\frac{\partial f_{j1}}{\partial \xi} + \frac{\partial f_{j2}}{\partial \xi}) = \frac{\partial f_{j1}}{\partial \eta_2} + \frac{\partial f_{j2}}{\partial \eta_2} = \frac{\partial \varphi_{j1}}{\partial \eta_2'} \qquad\qquad ;$$

les autres relations différentielles se vérifient de la même façon, de sorte que la condition **(ii)** est satisfaite avec $\eta_2'^* = \eta_2^* - \eta_1^*$.

(a') $u_n = az_n, a \neq 0$; posons : $\xi' = a\xi$; $\eta_\ell' = a\eta_\ell$, $\ell = 1, ..., n-1$;
$g' = u_n - \xi' - \sum_{\ell=1}^{n-1} \eta_\ell' z_\ell$;
$\eta' = (\eta_1', ..., \eta_{n-1}')$; $\varphi_{jk}^{\pm}(\xi', \eta') = f_{jk}^{\pm}(a\xi, a\eta)$; on a :

$$G_k'(\xi', \eta') = (1/2\pi i)\int_\gamma z_k \frac{dg'}{g'} = G_k(a\xi, a\eta) = \sum_{j=1}^{N^+} \varphi_{jk}^+(\xi', \eta') - \sum_{j=1}^{N^-} \varphi_{jk}^-(\xi', \eta').$$

Alors $\dfrac{\partial \varphi_{jk}^{\pm}}{\partial \xi'} = a\dfrac{\partial f_{jk}^{\pm}}{\partial \xi}$; $\dfrac{\partial f_{jk}^{\pm}}{\partial \eta_\ell'} = a\dfrac{\partial f_{jk}^{\pm}}{\partial \eta_\ell}$, donc les $\varphi_{jk}^{\pm}$ satisfont, par rapport à ξ', η', aux mêmes relations différentielles que les $f_{jk}^{\pm}$ par rapport à ξ, η, i.e. la condition **(ii)** est satisfaite dans les nouvelles coordonnées ; en outre, $\xi'^* = a\xi^*$ et $\eta'^* = a\eta^*$.

(b') $u_n = z_n + b$

$$G_k(\xi, \eta) = (1/2\pi i)\int_\gamma z_k \frac{d(u_n - \eta_1 z_1 - \cdots - \eta_{n-1}z_{n-1})}{(u_n - (\xi + b) - \eta_1 z_1 - \cdots - \eta_{n-1}z_{n-1})}$$

$$= \sum_{j=1}^{N^+} \varphi_{jk}^+(\xi', \eta) - \sum_{j=1}^{N^-} \varphi_{jk}^-(\xi', \eta)$$

avec $\xi' = \xi + b$ et $\varphi_{jk}^{\pm}(\xi', \eta) = f_j^{\pm}(\xi', \eta)$. Les relations différentielles par rapport à $\xi' = \xi + b$ et η subsistent pour $\varphi_{jk}^{\pm}$ au voisinage de $(\xi'^* = \xi^* + b, \eta^*)$.

(c') $u_n = z_n + z_1$

$$G_k(\xi, \eta) = (1/2\pi i)\int_\gamma z_k \frac{d(u_n - (\eta_1 - 1)z_1, ..., \eta_{n-1}z_{n-1})}{u_n - \xi - (\eta_1 - 1)z_1, ..., \eta_{n-1}z_{n-1}}$$

$$= \sum_{j=1}^{N^+} \varphi_{jk}^+(\xi, \eta') - \sum_{j=1}^{N^-} \varphi_{jk}^-(\xi, \eta')$$

avec $\eta_1' = \eta_1 - 1$; $\eta_\ell' = \eta_\ell, \ell = 2, ..., n-1$ et $\varphi_{jk}^{\pm}(\xi, \eta') = f_{jk}^{\pm}(\xi, \eta')$;

$$\frac{\partial \varphi_{jk}^{\pm}}{\partial \eta_1'}(\xi, \eta') = \frac{\partial f_{jk}^{\pm}}{\partial \eta_1}(\xi, \eta),$$

d'où les relations différentielles satisfaites par les $\varphi_{jk}^{\pm}$ au voisinage de ξ^*,
$\eta'^*_1 = \eta^*_1 - 1$, $\eta'^*_\ell = \eta^*_\ell$, $\ell = 2, ..., n-1$. $\square$

5.2.2. Corollaire.— *Il existe un changement de coordonnées permis tel que, dans les nouvelles coordonnées, la condition* (ii) *soit valide au voisinage de* $\xi^* = 0$, $\eta^* = 0$.

Démonstration.— Cela résulte des évaluations de ξ'^*, η'^* faites dans la démonstration de 5.2.1. $\square$

5.3. Invariance par projection dans un sous-espace $P \cong \mathbb{C}P^2$ de $\mathbb{C}P^n$

5.3.1. Définition d'une projection permise π_A. On considère une projection de $\mathbb{C}P^n$ sur un sous-espace projectif P, de dimension complexe 2. Chaque projetante, étant un sous-espace projectif qui coupe P en un point, est de dimension $(n-2)$; elle est déterminée par un sous-espace projectif fixe A, disjoint de P et par un point de $\mathbb{C}P^n \setminus A$, donc A est de dimension $(n-3)$. On désigne la projection par π_A et A est appelé le centre de π_A. On suppose, en outre, que A est contenu dans l'hyperplan de l'infini Q de $\mathbb{C}P^n$; π_A restreinte à $\mathbb{C}P^n \setminus Q \cong \mathbb{C}^n$ est une projection π de $\mathbb{C}^n$ sur $\mathbb{C}^2 \cong P \setminus P \cap Q$. S'il existe un système de coordonnées (v_1, v_2) de $\mathbb{C}^2$ tel que π restreinte à $\mathbb{C}^{n-1}$ (de coordonnées $(z_1, ..., z_{n-1})$) est une projection sur $\mathbb{C}$ (de coordonnée v_1), on dit que π_A est une projection permise. Alors π est définie par

$$v_1 = a_1 z_1 + \cdots + a_{n-1} z_{n-1} + a_0$$

$$v_2 = z_n + b_1 z_1 + \cdots + b_{n-1} z_{n-1} + b_0$$

où les coefficients a_ℓ, b_ℓ sont des constantes complexes, pour $\ell = 0, 1, ..., n-1$ et où le rang de la matrice des coefficients de $z_1, ..., z_n$ est 2.

5.3.2. Proposition.— *Dans les notations de 1.2. et de 5.3.1., pour toute projection permise* π_A:
$\mathbb{C}P^n \to P \cong \mathbb{C}P^2$, *satisfaisant à la condition* $(\star)$ *ci-dessous, la 1-chaîne* $\pi(\gamma)$ *satisfait à la condition* (ii) *dans* P.

Démonstration.— Dans P, considérons l'intégrale

$$G(\xi', \eta') = (1/2\pi i) \int_{\pi(\gamma)} v_1 \frac{d(v_2 - \eta' v_1)}{v_2 - \xi' - \eta' v_1}$$

$$= (1/2\pi i) \int_\gamma (a_1 z_1 + \cdots + a_{n-1} z_{n-1} + a_0) \frac{d(z_n - \eta_1 z_1 - \cdots - \eta_{n-1} z_{n-1})}{z_n - \xi - \eta_1 z_1 - \cdots - \eta_{n-1} z_{n-1}}$$

avec $\xi = (\xi' + \eta' a_0 - b_0)$; $\eta_\ell = (\eta' a_\ell - b_\ell)$, $\ell = 1, ..., n-1$, alors

$$G(\xi', \eta') = (1/2\pi i) \int_\gamma (a_1 z_1 + \cdots + a_{n-1} z_{n-1} + a_0) \frac{d(z_n - \eta_1 z_1 - \cdots - \eta_{n-1} z_{n-1})}{z_n - \xi - \eta_1 z_1 - \cdots - \eta_{n-1} z_{n-1}}$$

$$= \sum_{j=1}^{N^-} \varphi_j^+(\xi', \eta') - \sum_{j=1}^{N^-} \varphi_j^-(\xi', \eta')$$

avec $\varphi_j^{\pm}(\xi', \eta') = (a_1 f_{j1}^{\pm} + \cdots + a_{n-1} f_{jn-1}^{\pm} - a_0)(\xi, \eta_1, ..., \eta_{n-1})$.

D'après la démonstration de la proposition 5.2.1., les fonctions $\varphi_j^{\pm}(\xi', \eta')$ satisfont aux relations différentielles de la condition (ii) au voisinage d'un point convenable (ξ'^*, η'^*) de $\mathbb{C}^2$. Les coefficients a_ℓ, b_ℓ doivent satisfaire à la condition :

($\star$) $\eta_\ell = \eta' a_\ell - b_\ell$ est dans un voisinage donné de η_ℓ^* pour $\ell = 1, ..., n-1$ et $\xi = (\xi' + \eta' a_0 - b_0)$ dans un voisinage donné de ξ^*.

6 Schéma de démonstration de la suffisance de la condition (ii) dans $\mathbb{C}P^n$

6.1. On suppose que la 1-chaîne γ satisfait à la condition **(ii)** dans $\mathbb{C}P^n$. Alors dans les hypothèses de la proposition 5.3.2., pour toute projection permise π_A de $\mathbb{C}P^n$ sur un sous-espace projectif P de dimension 2, la condition **(ii)** est satisfaite par $\pi_A(\gamma)$ dans P. D'après le théorème 1.2 pour $n = 2$, il existe une 1-chaîne holomorphe S' de $P \setminus spt\pi_A(\gamma)$ telle que $bS' = \pi(\gamma)$. Pour construire une 1-chaîne holomorphe de $\mathbb{C}P^n \setminus spt\gamma$ telle que $\gamma = bS$, on va utiliser une méthode de projection analogue à celle qui a été introduite par Harvey, Lawson et Shiffman ([8], [9], [10]).

6.2. Définitions et notations.— Soient $a = (a_1, ..., a_{n-1})$; $b = (b_1, ..., b_{n-1})$. Pour (a, b) fixe, on considère la projection permise

$$\pi_b^a : \quad \begin{matrix} \mathbb{C}^n & \longrightarrow & \mathbb{C}^2 = P \setminus P \cap Q \\ z & \longmapsto & (v_1, v_2) \end{matrix} \quad \text{avec} \quad \begin{cases} v_1 & = a_1 z_1 + \cdots + a_{n-1} z_{n-1} \\ v_2 & = b_1 z_1 + \cdots + b_{n-1} z_{n-1} + z_n \end{cases}$$

Dans les notations de 5.3.2., prenons $\xi' = \xi$, $\eta' = 0$, alors la condition ($\star$) de la proposition 5.3.2. est satisfaite avec $b_\ell = \eta_\ell$ pour η_ℓ au voisinage de η_ℓ^* ; $\ell = 1, ..., n-1$; elle subsiste pour η' assez petit.

Dans P, on applique la méthode décrite dans les sections 3 et 4. La condition **(ii)** est valide au voisinage de $(\xi, 0)$. Pour cela, on considère le centre de la projection $\pi_0 : \mathbb{C}^2 \to \mathbb{C}$, dans $P \cap Q$ déterminé $\eta' = 0$, on a alors, dans P la forme linéaire $g' = v_2 - \xi$.

On considère les coefficients $C_m(\xi)$, $m \in \mathbb{N}$, définis d'abord localement par les fonctions scalaires $\varphi_j^\pm(\xi) = a.f_j^\pm = \sum_{\ell=1}^{n-1} a_\ell\, f_{j\ell}(\xi, \eta)$. Les coefficients $C_m(\xi)$ ont une extension à tout $\xi \in \mathbb{C} \setminus \pi_0 \circ \pi_b^a(spt\gamma)$.

On remarque que la projection π_0 est :

$$\begin{matrix} \mathbb{C}^2 & \longrightarrow & \mathbb{C} \\ (v_1, v_2) & \longmapsto & v_2 \end{matrix} \quad \text{de sorte que} \quad \pi_0 \circ \pi_b^a : \quad \begin{matrix} \mathbb{C}^n & \longrightarrow & \mathbb{C} \\ z & \longmapsto & v_2 \end{matrix}$$

est indépendante de a et sera notée π_b.

On en déduit la fonction

$$\varphi^{ab}(\xi, v_1) = C_0 \log v_1 - \sum_{m=1}^{+\infty} (1/m) C_m(\xi) v_1^{-m}$$

telle que $exp\varphi^{ab}(\xi, v_1)$ prolonge la fonction

$$(15) \qquad \prod_{j=1}^{N^+} (v_1 - \varphi_j^+(\xi)) \prod_{j=1}^{N^-} (v_1 - \varphi_j^-(\xi))^{-1}$$

au complémentaire de $\pi_0^{-1}(\pi_b(spt\gamma))$ pour ξ assez petit et v_1 assez grand.

Considérons la fonction de z,

$$\varphi^{ab}[z] = C_0 \log(a_1 z_1 + \cdots + a_{n-1} z_{n-1})$$
$$- \sum_{m=1}^{+\infty} (1/m) C_m (b_1 z_1 + \cdots + b_{n-1} z_{n-1} + z_n)(a_1 z_1 + \cdots + a_{n-1} z_{n-1})^{-m}$$

La fonction $\mathcal{R}^{ab}[z] = exp\varphi^{ab}[z]$ a une extension méromorphe à

$$K^{ab} = \mathbb{C}^n \setminus \pi_b^{-1}(\pi_b(spt\gamma)).$$

Désignons par s la 1-chaîne holomorphe, définie par les fonctions vectorielles $f_j^{\pm}(\xi, \eta)$ au voisinage de (ξ^*, η^*), avec $\eta = b$, i.e. par les graphes

$$(z_1, ..., z_{n-1}) = f_j^+(\xi, b) \; ; \; z_n = \xi + \sum_{\ell=1}^{n-1} b_\ell f_{j\ell}^+(\xi, b) \text{ avec la multiplicité } + 1$$

$$(z_1, ..., z_{n-1}) = f_j^-(\xi, b) \; ; \; z_n = \xi + \sum_{\ell=1}^{n-1} b_\ell f_{j\ell}^-(\xi, b) \text{ avec la multiplicité } - 1$$

s est le diviseur de la fonction méromorphe (15).

On en déduit que s est contenue dans la $(n-1)$-chaîne holomorphe

$$S^{ab} = (i/\pi)d'd'' \log |\mathcal{R}^{ab}[z]|$$

dans le sens suivant : le support de s est contenu dans le support de S^{ab} ; les multiplicités des composantes irréductibles respectives des supports des deux chaînes sont compatibles.

6.3. Les étapes de la démonstration sont les suivantes :

6.3.1. L'intersection d'ensembles analytiques complexes de dimension $(n-1)$ a un sens ; on peut en donner un à l'intersection d'un nombre fini de $(n-1)$-chaînes holomorphes, en tenant compte des multiplicités. $S_b = \cap_a S^{ab}$ est une 1-chaîne holomorphe de $K^b = \cap_a K^{ab}$ qui prolonge s ; a décrivant un ensemble fini de points de $\mathbb{C}^{n-1}$.

6.3.2. Dans $K^{b_1} \cap K^{b_2}$, on a $S_{b_1} = S_{b_2}$. Alors la réunion $S = \cup_b S_b$ a un sens et est une 1-chaîne de $\mathbb{C}^n \setminus spt\gamma$.

6.3.3. S se prolonge à $\mathbb{C}P^n \setminus spt\gamma$ à travers l'hyperplan à l'infini Q, grâce au théorème de Bishop-Stoll ([4], [14]).

6.3.4. S a une extension simple $\mathcal{S}$ à $\mathbb{C}P^n$ telle que $b\mathcal{S} = \gamma$ (cf. 4.5).

7   Application : enveloppe d'holomorphe de γ

7.1. Enveloppe d'holomorphie dans un ouvert de Stein

Soit X une variété analytique complexe ; γ une courbe C^2. Supposons :
1) Il existe une solution du problème du bord dans X, i.e. une courbe analytique complexe S de $X \setminus \gamma$ dont le courant d'intégration a une extension simple notée encore S telle que $\gamma = bS$;
2) l'existence d'un domaine Ω de X contenant l'adhérence $\bar{S}$ de S dans X. Alors, on sait que γ satisfait à la condition du moment pour les fonctions de $\mathcal{O}(\Omega)$.

Réciproquement, si γ est contenue dans un ouvert de Stein Ω de X et satisfait à la condition du moment pour les fonctions de $\mathcal{O}(\Omega)$, d'après le théorème de Remmert [11] il existe un plongement $j : \Omega \to \mathbb{C}^N$ pour N assez grand ; d'après le théorème de Wermer, il existe une courbe analytique complexe $\sum$ dans $\mathbb{C}^N$ telle que $b\sum = j(\gamma)$, en outre $\sum \subset j(\Omega)$, alors pour $S = j^{-1}(\sum)$, on a $bS = \gamma$; d'après Wermer ([17], theorems 1.1., 1.2.), $\bar{\Sigma}$ est l'enveloppe d'holomorphie de $j(\gamma)$ dans $\mathbb{C}^N$; mais $\mathcal{O}(\mathbb{C}^N) \cong (\mathcal{O}(\Omega))$, donc $\bar{S}$ est l'enveloppe d'holomorphie de γ dans Ω.

7.2. Soit γ une courbe fermée C^ω (analytique réelle) de $\mathbb{C}P^n$ telle qu'il existe une courbe analytique complexe S de $\mathbb{C}P^n \setminus \gamma$ de bord γ sans composante connexe compacte. Alors S a une extension analytique complexe S' contenant $\bar{S}$. Soit γ_ε une courbe C^ω, petite déformation de γ dans S', contenue dans $S' \setminus S$, alors il existe une courbe analytique complexe S'' telle que $bS'' = \gamma_\varepsilon$ et que $\bar{S} \subset S'' \subset S'$, S'' est un ensemble analytique complexe fermé de la variété $\mathbb{C}P^n \setminus \gamma_\varepsilon$; d'après le théorème de Siu [13], il existe un domaine de Stein Ω de $\mathbb{C}P^n \setminus \gamma_\varepsilon$ tel que $\Omega \supset S''$. Alors, d'après 7.1., $\bar{S}$ est l'enveloppe d'holomorphie de γ dans Ω, d'où

7.2.2. Proposition.— *Si γ est une courbe analytique réelle de $\mathbb{C}P^n$ satisfaisant à la condition (ii) du théorème 1.2., il existe une solution S du problème du bord $bS = \gamma$ dans $\mathbb{C}P^n$ ayant pour adhérence $\bar{S}$ l'enveloppe d'holomorphie de γ dans un ouvert de Stein Ω contenu dans $\mathbb{C}P^n$.*

Bibliography

[1]   Alexander, H.: Polynomial approximation and hulls in sets of finite linear measure in $\mathbb{C}^n$. Amer. Jl. of Math., **93** (1971), 65–74.

[2]   Bishop, E.: Analyticity in certain Banach algebras. Trans. A.M.S., **102** (1962), 507–544.

[3]   Bishop, E.: Holomorphic completions, analytic continuations and the interpolation of seminorms. Ann. of Math. **78** (1963), 468–500.

[4]   Bishop, E.: Conditions for the analyticity of certain sets. Michigan Math. J. **11**, (1964), 289–304.

[5]   Darboux, Théorie des surfaces,I, 2e éd.. Gauthier-Villars, Paris, (1914).

[6]   Dolbeault, P. et Henkin, G.: Surface de Riemann de bord donné dans $\mathbb{C}P^2$. Comptes rendus Ac. Sci. Paris, **316** (1993), 27–32.

[7]   Griffiths, P. and Harris, J.: Principles of algebraic geometry. John Wiley and Sons, New York, 1978.

[8]   Harvey, R.: Holomorphic chains and their boundaries. Proc. Symp. Pure Math. **30**, Part 1 (1977), 309–382.

[9]   Harvey, R. and Lawson, B.: On boundaries of complex analytic varieties I. Ann. of Math., **102** (1975), 233–290.

[10] Harvey, R. and Shiffman, B.: A characterization of holomorphic chains. Ann. of Math. (2), **99** (1974), 553–587.

[11] Remmert, R.: Sur les espaces holomorphiquement séparables et holomorphiquement convexes. Comptes rendus Ac. Sci. Paris, **243** (1956), p. 118–121.

[12] Royden, H.: Algebras of bounded analytic functions on Riemann surfaces. Acta Math., **114** (1965), 113–142.

[13] Siu, Y.T.: Every Stein subvariety admits a Stein neigh-bourhood. Inv. Math. **38** (1976/77), 89–100.

[14] Stoll, W.: Über die Fortsetzbarkeit analytischer Mengen endlichen Oberflächeninhaltes. Arch. Math. **9** (1958), 167–175.

[15] Stolzenberg, G.: Uniform approximation on smooth curves. Acta Math., **115** (1966), 185–198.

[16] Wermer, J.: Function rings and Riemann surfaces. Ann. of Math. **67** (1958), 45–71.

[17] Wermer, J.: The hull of a curve in $\mathbb{C}^n$. Ann. of Math., **68** (1958), 550–561.

Subvarieties of homogeneous and almost homogeneous manifolds

Alan Huckleberry

Dedicated to Professor P. Dolbeault

The general context of this paper is that of compact subvarieties of a complex manifold X which is equipped with the action of a Lie group G of holomorphic transformations. Here we restrict our attention to the case where X is homogeneous or at least almost homogeneous. For example, in Chapter 2 we consider the *analytic hypersurfaces* $\mathcal{H}(X)$, i. e., the 1-codimensional complex analytic subsets, in a homogeneous space $X = G/H$. Although the results are primarily formulated in the language of complex geometry, we are mainly interested in determining the complex analytic objects which define the hypersurfaces, e.g., holomorphic functions coming from associated Stein manifolds, rational functions from related projective varieties, Θ-functions on abelian groups, Fourier-Jacobi series,.... Difficulties arise, for example, because the ambient manifolds are in general non-compact, non-Kählerian, and the relevant cohomology groups are infinite-dimensional.

One of the main tools for the study of subvarieties of a homogeneous space is the use of the differential forms defined by smoothing the associated currents. We present a unified treatment of the smoothing procedure in Chapter 1. In particular, we give details of unpublished results of W. Richthofer ([51]) on the positivity of the smooth (p,p)-form associated to a positive (p,p)-current.

In Chapter 3 we consider hypersurfaces in a compact almost homogeneous manifold. Other than presenting a treatment of the basic results, we prove a characterization of almost homogeneous Moishezon 3-folds. This result (Bochum Dissertation of M. Martin ([41])) is typical of one of the goals of our approach of the subject in that it is a reduction to the algebraic case where combinatorial tools are available.

Finally, in Chapter 4 we present recent results of the author and J. Winkelmann ([32]) on compact subvarieties of a quotient $X = G/\Gamma$ of a complex Lie group by a discrete subgroup. Roughly speaking, unless X is a torus, such a subvariety Z tends to have high-codimension. For example, if G is simple and Γ is cocompact, then $codim(Z) \geq \sqrt{dim(X)}$. One of the main consequences of our considerations is a proof of a non-abelian version of the *Bloch conjecture*: Let X be a compact parallelizable complex manifold and $f : \mathbb{C} \longrightarrow X$ a holomorphic map. Then the complex analytic Zariski closure $\overline{f(\mathbb{C})}$ of the image is an orbit of a connected complex Lie subgroup of $Aut_{\mathcal{O}}(X)$.

1   Smoothing currents on homogeneous spaces

In complex geometry, and in particular for the applications in this paper, *currents* arise by integration over subvarieties. We say that a current T is smoothed by a differential form ω if

$$T(\varphi) = \int \omega \wedge \varphi$$

for every test form φ. In general it is difficult to derive properties of the differential form from those of the current. For example, integration over a complex subvariety always has a certain degree of positivity, whereas the associated differential form may even be negative.

Our goal here is to present the smoothing procedure on homogeneous spaces. The main result, due to W. Richthofer ([51])[1], is that the positivity of a (p,p)-current is inherited by the associated differential form (*Theorem 1.3.1*).[2] Since the proofs in this subject require a certain degree of complicated notation and technical work, we thought it prudent to give Richthofer's complete unified treatment.

1.1   Smoothing currents on a Lie group

Let G be a connected real Lie group with Lie algebra $\mathfrak{g}$. Regard the elements of $\mathfrak{g}$ (resp., $\mathfrak{g}^*$) as left-invariant vector fields (resp., 1-forms) on G. Let $((\mu_1, \dots , \mu_n))$ be a basis of $\mathfrak{g}^*$ and define $\mu_G := \mu_1 \wedge \dots \wedge \mu_n$. Take dg to be the measure associated to μ_G.

For $g \in G$ let $r_g : G \longrightarrow G$ be right-translation and define the *modular function* Δ_G on G by

$$r_g^* \mu_G = \Delta_G(g^{-1})\mu_G.$$

If T is a current on G, then $T_y(\varphi(\cdot, y))$ means that T is applied to whatever is inside the brackets as a function of y. With χ (resp., φ) we will always denote a smooth function (resp., form) with compact support on the manifold under consideration. If χ is such a function and T is a current on G, then

$$\chi * T := \int_G \chi(g) l_{g^{-1}}^* T \, dg$$

is called a *smoothing of T*, and if

$$\chi * T(\varphi) = \int T_\chi \wedge \varphi$$

for all test forms φ, then we say that the differential form T_χ represents the smoothing.

a. Distributions on G

Let T be a distribution (considered as 0-current) on G and define the function

$$h(g) := T_y[\chi(gy^{-1})\Delta_G(y)\mu_G(y)].$$

---

[1] We wish to thank W. Richthofer for allowing us to use a significant part of his heretofore unpublished paper [51].

    [2] The case of $(1, 1)$-currents and discrete isotropy was handled in [46] (see also [9]). Recently F. Berteloot gave another proof in the $(1, 1)$-case ([6]).

Lemma 1.1.1 *The function h is smooth.*

Proof Let $\xi \in \mathfrak{g}$ and $\xi_t(g)$ be an orbit of the one parameter group associated to ξ, i.e.,

$$(\xi h)(g) = \lim_{t \longrightarrow 0} \frac{1}{t}[h(\xi_t(g)) - h(g)]$$

$$= \lim_{t \longrightarrow 0} T_y \frac{1}{t}[\chi(\xi_t(g)y^{-1})\Delta_G(y) - \chi(gy^{-1})\Delta_G(y)]\mu_G(y) \ .$$

Consider the argument of T_y in the formula above as a function $\varphi_t(y)$. Then

$$\lim_{t \longrightarrow 0} \varphi_t(y) = \Delta_G(y)\xi\chi(gy^{-1})\mu_G(y)$$

in $C^\infty(G)$. Therefore we obtain

$$\lim_{t \longrightarrow 0} T_y(\varphi_t(y)) = T_y(\xi\chi(gy^{-1})\Delta_G(y)\mu_G(y)) \ .$$

Hence, by iteration we can conclude that h is C^∞ (see also [15], 17.10.1). $\qquad\square$

Proposition 1.1.2 *The function h represents a smoothing of T, i.e., for $f \in C_c^\infty(G)$, it follows that*

$$\chi * T(f\mu_G) = \int_G \chi(g)l_{g^{-1}}^* T(f\mu_G) \, dg = \int_G hf\mu_G \ .$$

Proof Let $m : G \times G \longrightarrow G$ denote the multiplication $(g, y) \mapsto gy$. Then we obtain

$$\chi * T(f\mu_G) = \int_G \chi(g)T_y(f(gy)\mu_G(y)) \, dg$$

$$= \chi \otimes T[(f \circ m)\mu_G(y) \otimes \mu_G(g)]$$

$$= T_y[\chi(f(gy)\mu_G(g)\mu_G(g)]$$

$$= T_y\left(\int_G \chi(g)f(gy) \, dg \, \mu_G(y)\right)$$

$$= T_y\left(\int_G \chi(gy^{-1})f(g)\Delta_G(y) \, dg \, \mu_G(y)\right)$$

$$= f \otimes T[\chi(gy^{-1})\Delta_G(y)\mu_G(g) \otimes \mu_G(y)]$$

$$= \int_G T_y[\chi(gy^{-1})\Delta_G(y)\mu_G(y)]f(g) \, dg$$

$$= \int_G h \, f \, dg \ .$$

For related calculations and facts about $\chi \otimes T$ see ([15],17.10.3.). $\qquad\square$

b. Smoothing p-currents on G

In order to smooth p-currents, where $p > 1$, it is convenient to introduce a left-invariant Riemannian metric $<,>$ on G. We also let $<,>$ denote the induced metric on the exterior algebra of differential forms. Let $((\mu_i))$ denote an orthonormal basis of $\mathfrak{g}^*$. It follows that (μ_I), $I = (i_1, \ldots, i_p)$, $1 \leq i_1 < \ldots < i_p \leq n$, form an orthonormal basis of $\bigwedge_p \mathfrak{g}^*$.

The operator $*$ is defined by the equation

$$< \omega, \alpha > \mu_1 \wedge \cdots \wedge \mu_n = \omega \wedge *\alpha .$$

Let $\alpha = \sum'{}a_I \mu_I$. For any ordered p-tuple I let I' denote the uniquely determined ordered (n-p)-tuple such that $(I, I') = \sigma(1, \ldots, n)$, where σ denotes a permutation. Let $\epsilon_I = sgn\sigma$. Then

$$*\alpha = \sum'_I \epsilon_I a_I \mu_{I'} .$$

For a p-current T on G we define the distributions T_I, $\#I = p$ by

$$T_I(\alpha \mu_G) = T(\alpha * \mu_I) .$$

Thus $T_I(\alpha \mu_I \wedge *\mu_I) = T(\alpha * \mu_I)$, and we define the coefficient functions

$$h_I(g) := T_{I,y}[\chi(gy^{-1})\Delta_G(y)\mu_G(y)].$$

Proposition 1.1.3 *Let* $T_\chi = \sum'_I h_I \mu_I$. *Then* T_χ *represents the smoothing* $\chi * T$, *i.e., for* $\varphi = \sum'_J b_J * \mu_J$ *an* $(n - p)$-*form with compact support in* G

$$\int_G \chi(g) l^*_{g^{-1}} T(\varphi) \, dg = \int_G T_\chi \wedge \varphi .$$

Proof

$$\int_G T_\chi \wedge \varphi = \sum'_{I,J} \int_G h_I b_j \mu_i \wedge *\mu_J$$

$$= \sum'_I \int_G h_I b_I \mu_I \wedge *\mu_I$$

$$= \sum'_I \int_G \chi(g) l^*_{g^{-1}} T_I(b_I \mu_G) \, dg$$

$$= \sum'_I \int_G \chi(g) l^*_{g^{-1}} T(b_I * \mu_I) \, dg$$

$$= \int_G \chi(g) l^*_{g^{-1}} T(\sum'_I b_I * \mu_I) \, dg$$

$$= \int_G \chi(G) l^*_{g^{-1}} T(\varphi) \, dg$$

Hence, the convolution $\chi * T$ applied to φ is given by integration against T_χ. □

1.2 Smoothing of currents on a homogeneous space

In this section X is a differentiable manifold which is the homogeneous space of the connected real Lie group G, i.e., $X = G/H$, where H is a closed subgroup. Let $((V_1,\ldots,V_n))$ be a basis of $\mathfrak{g}$ with $((V_1,\ldots,V_k))$ a basis of $\mathfrak{h}$. Take $((\mu_1,\ldots,\mu_n))$ to be the dual basis of left invariant 1-forms. Fix the forms

$$\mu_0 = \mu_1 \wedge \cdots \wedge \mu_k$$
$$\mu^\perp = \mu_{k+1} \wedge \cdots \wedge \mu_n$$
$$\mu_G = \mu_1 \wedge \cdots \wedge \mu_n$$
$$\mu_H = \mu_0|H .$$

Assume that X is orientable, i.e., there exists a smooth function

$$u : G \longrightarrow \mathbb{R}^{>0}$$

such that

$$r_h^* u = \Delta_G(h) \cdot \Delta_H^{-1}(h)u ,$$

for all $h \in H$. The orientability of X is of course no additional condition if X is a complex manifold, or if H is a discrete subgroup of G.

If u is such a function, then

$$\mu_X = u\mu^\perp$$

pushes down to X and thus defines a volume form there. Since $\Delta_H^{-1}(h) = u^{-1}r_h^* u\Delta_G^{-1}(h)$, it follows that $r_h : H \longrightarrow H$ preserves the orientation on H induced by μ_H. We have $r_h^* \mu_0 = \Delta_H(h^{-1})\mu_0 + \alpha$, where $\alpha|gH = 0$ for all $g \in G$. Hence μ_0 defines a bundle orientation of the H- principle bundle $G \longrightarrow G/H = X$, which is preserved by the right action of H.

Let $\Omega_q(G)$ denote the space of compactly supported q-forms on G and define

$$\Omega_q^r(G) = \{\omega \in \Omega_q(G) : \omega = \underset{\substack{\#I=q \\ \#I\cap\{1,\ldots,k\}\leq r}}{{\sum}'} a_I\mu_I\} .$$

Then $\omega \in \Omega_q^r$ if and only if $\omega(g;\xi_1,\ldots,\xi_q) = 0$, whenever more than r of the ξ_i's are annihilated by $d\pi$. For $q \geq k$ let

$$\Omega_q^H(G) := \{\omega \in \Omega_q(G) : \omega = \underset{\substack{\#I=l \\ I\cap\{1,\ldots,k\}=\emptyset}}{{\sum}'} a_I\mu_0 \wedge \mu_I\} ,l = q - k .$$

We have the decomposition

$$\Omega_q(G) = \Omega_q^H(G) + \Omega_q^{k-1}(G) ,\omega = \omega^H + \omega^\perp.$$

a. The fiber integral

If $\varphi \in \Omega_q(G)$, then integration over the fibers of $\pi : G \longrightarrow G/H$ defines a form $\int_H \varphi \in \Omega_{q-k}(X)$ on the base. This is given by Fubini's Theorem:

$$\int_G \pi^*\omega \wedge \varphi = \int_X \omega \wedge \int_H \varphi$$

for every $\omega \in \Omega_{n-q}(X)$.

If T is a p-current, then its *pull-back* is defined by

$$\pi^*T(\varphi) = T(\int_H \varphi) \quad , \varphi \in \Omega_{n-p}(G) .$$

Basic Properties.

   (i) *If T is a p-current of order r, then π^*T is also a p-current of order r.*

   (ii) *$r_h^*\pi^*T = \pi^*T$ for all $h \in H$.*

   (iii) *$l_g^*\pi^*T = \pi^*l_g^*T$.*

   (iv) *$\pi^*T(\varphi) = 0$ for all $\varphi \in \Omega_{n-p}^{k-1}(G)$.*

   (v) *$d\pi^*T = \pi^*dT$.*

Proof   Direct verification.   □

b. Pushing down

Let T be a current on X and define its smoothing on X by

$$\chi * T := \int_G \chi(g) l_{g-1}^* T \, dg.$$

Our main goal here is to prove the following:

Proposition 1.2.1 *The smooth differential form $(\pi^*T)_\chi$ on G is the pullback of a smooth differential form T_χ on X and in the sense of currents we have*

$$T_\chi = \chi * T .$$

This is the consequence of following two lemmas.

Lemma 1.2.2 *For $p \leq dim(X)$ let ω be a smooth p-form on G. Then ω is a pullback of a p-form on X if and only if $r_h^*\omega = \omega$ for all $h \in H$ and*

$$\int_G \omega \wedge \varphi = 0$$

for all $\varphi \in \Omega_q^{k-1}$, $q = n - p$.

Proof Let $\omega = \pi^*\omega'$. Of course $r_h^*\omega = \omega, \forall h \in H$, follows immediately and, if $\varphi \in \Omega_q^{k-1}$, then $\int_H \varphi = 0$ and $\in \pi^*\omega' \wedge \varphi = 0$ is a consequence of Fubini's Theorem.

Conversely, since ω is supposed to be right H-invariant, it is enough to show that it is in $\Omega_q^0(G)$. For this we decompose ω as

$$\omega = \omega_0 + \omega_1 \ ,$$

where for ordered p-tuples I

$$\omega_0 = \sideset{}{'}\sum_{I \cap \{1,\dots,k\}=\emptyset} a_I \mu_I \quad ; \quad \omega_1 = \sideset{}{'}\sum_{I \cap \{1,\dots,k\}\neq\emptyset} a_I \mu_I \ .$$

Then for $q = n - p \geq k$ and $\varphi = \varphi^H + \varphi^\perp \in \Omega_q(G)$ we have

$$\omega_1 \wedge \varphi_H = \omega_0 \wedge \varphi^\perp = 0 \ .$$

Suppose $\omega_1 \neq 0$, i.e., $\omega \notin \Omega_p^0(G)$. Then there is a multi-index I_0 such that $\#I_0 = p$, $I_0 \cap \{1,\dots,k\} \neq \emptyset$ and $a_{I_0} \neq 0$ in the description of ω_1 above. Let $U \subset G$ be a relatively compact open subset of G such that $a_{I_0}|U \neq 0$. Choose a multi-index J such that $\mu_{I_0} \wedge \mu_J = \epsilon_{I,J}\mu_G$, where $\epsilon_{I,J}$ is the sign of the permutation which maps (I, J) to $(1,\dots,n)$. Let b_J be a smooth positive function with support contained in U. Then we have $b_J \in \Omega_q^{k-1}(G)$ and $\omega_1 \wedge b_J\mu_J = \epsilon_{I,J}a_{I_0}b_J\mu_G$. Since $a_{I_0} \neq 0$ and $b_J \geq 0$, the function $a_{I_0}b_J$ does not change its sign on G, i.e., we have

$$\int_G \omega \wedge b_J \, \mu_J = \pm \int_G a_{I_0}b_J \, dg \neq 0.$$

$\square$

Lemma 1.2.3 *Let ω be a smooth p-form on G and T_ω the current induced by ω. If T_ω satisfies Basic Properties (ii) and (iv) in the previous section, then ω is the pullback of a smooth p-form ω' on X. If $\omega = \pi^*\omega'$ and $\varphi' = \int_H \varphi$ is a test form, then $T_\omega(\varphi) = T_{\omega'}(\varphi')$.*

Proof It is clear that *(ii)* implies $r_h^*\omega = \omega$. So by Lemma 1.2.2, we have $\omega = \pi^*\omega'$ for a smooth form ω' on X. Moreover

$$T_\omega(\varphi) = \int_G \omega \wedge \varphi = \int_X \int_H \omega \wedge \varphi = \int_X \omega' \wedge \int_H \varphi = \int_X \omega' \wedge \varphi' = T_{\omega'}(\varphi') \ .$$

$\square$

Proof of Proposition 1.2.1 By Proposition 1.1.3 we have $(\pi^*T)_\chi = \chi * \pi^*T$ in the sense of currents. Thus for $h \in H$

$$r_h^*(\pi^*T)_\chi = r_h^*(\chi * \pi^*T) = \chi * r_h^*\pi^*T = \chi * \pi^*T \ .$$

Moreover for $\varphi \in \Omega_{n-p}^{k-1}(G)$ it is clear that $\chi * \pi^* T(\varphi) = 0$, since $l_g^* \varphi \in \Omega_{n-p}^{k-1}$, i.e., $\pi^* T(\varphi) = 0$. Hence $(\pi^* T)_\chi$ satisfies the conditions of Lemma 1.2.3. The formula $T_\chi = \chi * T$ is an immediate consequence of the formula $(\pi^* T)_\chi = \chi * \pi^* T$ and the formula in Lemma 1.2.3 $\qquad\qquad\qquad\qquad\qquad\qquad\qquad\qquad\qquad\qquad\qquad\qquad\square$

Remark 1.1 *(Richthofer) Assume that $\int_G \chi = 1$. Then the differential form T_χ is cohomologous to T.*

Proof Every T is cohomologous to some smooth form ω.i.e., there exists a $(p-1)$-current R so that

$$T(\varphi) - \int_X \omega \wedge \varphi = R(d\varphi)$$

for every test form φ. Then for any compactly supported *closed* test form φ we have

$$\int_X T_\chi \wedge \varphi = \int_G \chi(g) T(l_g^* \varphi) \, dg$$

$$= \int_G \chi(g) \int_X \omega \wedge l_g^* \varphi \, dg$$

$$= \int_G \chi(g) \int_X l_g^*(\omega \wedge \varphi) \, dg \, ,$$

because, by the connectedness of G, the form $l_g^* \omega$ is cohomologous to ω. Moreover, we have $\int_X l_g^*(\omega \wedge \varphi) = \int_X \omega \wedge \varphi$ and $\int_G \chi(g) \, dg = 1$. Thus it follows that

$$\int_X T_\chi \wedge \varphi = \int_X \omega \wedge \varphi \, ,$$

and by Poincare duality T_χ is cohomologous to ω. $\qquad\qquad\qquad\qquad\qquad\qquad\square$

c. The explicit description of T_χ

Using the formula for $(\pi^* T)_\chi$ in 1.1.3, we now give Richthofer's explicit description of $T_\chi \wedge \omega$.

Proposition 1.2.4 *Let T be a p-current on $X = G/H$ and ω an $(m-p)$-form on X. Then the form*

$$\eta(gH) = T_{x'} \left[\left(\int_H \chi(gy^{-1}) \Delta_G(y) u^{-1}(g) \mu_0(y) \wedge l_{gy^{-1}}^* \pi^* \omega(y) \right)(x') \right] \mu_X(gH)$$

is well-defined, and for any compactly supported function α on X we have

$$\int_X \alpha \cdot T_\chi \wedge \omega = \int_X \alpha \cdot \eta \, .$$

Proof First we choose a function $\theta : G \longrightarrow \mathbb{R}^{>0}$ with fiber compact carrier such that

$$\int_H \theta\mu_0 \equiv 1 \,.$$

Then we have $\int_H \theta\mu_0 \wedge \pi^*(\alpha\omega) = \alpha\omega$. Let $S = \pi^*T$. We obtain

$$\int_X \alpha T_\chi \wedge \omega = \int_X T_\chi \wedge \alpha\omega$$

$$= \int_G \chi(g) l^*_{g^{-1}} T(\alpha\omega)\, dg$$

$$= \int_G \chi(g) T(l^*_g(\alpha\omega))\, dg$$

$$= \int_G \chi(g) S\, l^*_g(\theta\mu_0 \wedge \pi^*\alpha\, \pi^*\omega)\, dg$$

$$= \int_G S_y(\chi(g)\pi^*\alpha(gy)\theta(gy)\mu_0(y) \wedge l^*_g\pi^*\omega(y))\, dg$$

$$= \int_G S_y(\chi(gy^{-1})\pi^*\alpha(g)\theta(g)\Delta_G(y)\mu_0(y) \wedge l^*_{gy^{-1}}\pi^*\omega(y))\, dg$$

$$= \int_G \pi^*\alpha(g)\theta(g) S_y(\chi(gy^{-1})\Delta_G(y)\mu_0(y) \wedge l^*_{gy^{-1}}\pi^*\omega(y))\, dg$$

$$\overset{*}{=} \int_G \theta(g)\mu_0(g) \wedge$$

$$\wedge S_y(\chi(gy^{-1})\Delta_G(y)u^{-1}(g)\mu_0(y) \wedge l^*_{gy^{-1}}\pi^*\omega(y))\pi^*\alpha\mu_X(g)$$

$$= \int_X \left(\int_H \theta(g)\mu_0(g) \wedge \right.$$

$$\left. \wedge S_y(\chi(gy^{-1})\Delta_G(y)u^{-1}(g)\mu_0(y) \wedge l^*_{gy^{-1}}\pi^*\omega(y)) \right) \pi^*\alpha\mu_X(g)$$

$$= \int_X S_y(\chi(gy^{-1})\Delta_G(y)u^{-1}(g)\mu_0(y) \wedge l^*_{gy^{-1}}\pi^*\omega(y))\alpha\mu_X$$

$$= \int_X \alpha\eta \,.$$

Equation $*$ is valid since $dg = u^{-1}(g)\mu_0(g) \wedge \pi^*\mu_X(g)$. The last two equations follow from the fact that the function

$$f(g) = S_y(\chi(gy^{-1})\Delta_G(y)u^{-1}(g)\mu_0(y) \wedge l^*_{gy^{-1}}\pi^*\omega(y))$$

is invariant by the right H-action. For this, note that $r^*_h u = \Delta_G(h)\Delta_H(h^{-1})u$. Thus, we can calculate

$$f(gh) = S_y(\chi(ghy^{-1})\Delta_G(y)u^{-1}(gh)\mu_0(y) \wedge l^*_{ghy^{-1}}\pi^*\omega(y))$$

$$= T_{x'}\left[\left(\int_H \chi(ghy^{-1})\Delta_G(y)(r^*_h u)^{-1}\mu_0(y) \wedge l^*_{ghy^{-1}}\pi^*\omega(y)\right)(x')\right]$$

$$= T_{x'}\left[\left(\int_H \chi(ghy^{-1})\Delta_G(y)\Delta_G(h^{-1})\Delta_H(h)u^{-1}(g)\mu_0(y)\wedge\right.\right.$$

$$\left.\left.\wedge\, l^*_{ghy^{-1}}\pi^*\omega(y)\right)(x')\right]$$

$$= T_{x'}\left[\left(\int_H r^*_{h^{-1}}(\chi(gy^{-1})\Delta_G(y)u^{-1}(g)\mu_0(y) \wedge l^*_{gy^{-1}}\pi^*\omega(y))\right)(x')\right]$$

$$= f(g) \ .$$

Note here that, restricted to the fibers of $G \longrightarrow G/H$, we have $r^*_h\mu_0 = \Delta_H(h^{-1})\mu_0$, and $l^*_{gy^{-1}}\pi^*\omega(y;\xi) = \omega(\pi(g); d\pi \circ dl_g \circ dl_{y^{-1}}\xi)$. Thus

$$r^*_{h^{-1}}(l^*_{gy^{-1}}\pi^*\omega)(y;\xi) = \pi^*\omega(yh^{-1}; dr_{h^{-1}}\xi)$$

$$= \omega(\pi(g); d\pi \circ dr_{h^{-1}} \circ dl_g \circ dl_h \circ dl_{y^{-1}}\xi)$$

$$= \omega(\pi(gh); d\pi \circ dl_{ghy^{-1}}\xi)$$

$$= l^*_{ghy^{-1}}\pi^*\omega(y;\xi) \ .$$

Hence the proof is complete $\qquad\qquad\square$

By the choice of a suitable basis of **h** we can explicitly compute the fiber integral in the expression of η in Proposition 1.2.4.

Let $\xi_1,\ldots,\xi_k$ be a basis of **h** such that $\mu_0(\xi_1 \wedge \cdots \wedge \xi_k) \equiv 1$.

Proposition 1.2.5 *Let* $s : X \longrightarrow G$ *be an arbitrary section and* ω *an* (m–p)-form *on X. Then*

$$T_\chi \wedge \omega(gH) = T_x \int_H \chi(ghs(x)^{-1})\Delta_G(s(x)h)u^{-1}(g)l^*_{gh^{-1}s(x)^{-1}}\omega(x)dh\,\mu_X(gH).$$

Proof Since $\pi^*\omega$ vanishes on $\xi_1,\ldots,\xi_k$ and $\mu_0(\xi_1 \wedge \cdots \wedge \xi_k) \equiv 1$, it is clear that the retrenchment of $\chi(gy^{-1})\Delta_G(y)u^{-1}(g)\mu_0(y) \wedge l^*_{gy^{-1}}\pi^*\omega(y)$ to the fiber over x is given by $\chi(gy^{-1})\Delta_G(y)u^{-1}(g)l_{gy^{-1}}\omega(x)$. Hence, provided that $\pi(y_0) = x$, we obtain

$$\int_H \chi(gy^{-1})\Delta_G(y)u^{-1}(g)\mu_0(y) \wedge l^*_{gy^{-1}}\pi^*\omega(y)(x)$$

$$= \int_H \chi(gh^{-1}y_0^{-1})\Delta_G(y_0h)u^{-1}(g)l^*_{gh^{-1}y_0^{-1}}\omega(x)\,dh$$

and it is easy to check that the last integral does not depend on the choice of the particular y_0. $\qquad\qquad\square$

1.3 Smoothing positive currents on a complex homogeneous space

Here we present Richthofer's main theorem on smoothing positive p, p-currents on a complex manifold X which is homogenous under a Lie group G of holomorphic transformations. Roughly, it states that the differential form representing a smoothing has as much positivity as can be expected. Before giving a precise statement (*Theorem 1.3.1*), it is necessary to formulate the various notions of positivity (see also, e.g.,[24]).

As usual we refer to elements of $\bigwedge^{p,q} TX^*$ as p,q-covectors and smooth sections as p,q-forms. For $x \in X$ let h be a hermitian metric so that in local coordinates $h(x) = \sum dz_j \otimes d\bar{z}_j$. Define

$$\omega_h := \frac{i}{2} \sum_j dz_j \wedge d\bar{z}_j$$

and

$$\tau_n := (\frac{i}{2})^n dz_1 \wedge d\bar{z}_1 \wedge \cdots \wedge dz_n \wedge d\bar{z}_n = \omega_h^n \ .$$

A p,p-covector μ at X represents a complex p-plane in TX_x^* if and only if there exists an orthogonal projection $\pi_x = (\pi_1, \dots, \pi_p)$ onto a p-plane in $TX_x^{1,0}$ such that

$$\mu = (\frac{i}{2})^p \pi_1 \wedge \bar{\pi}_1 \wedge \cdots \wedge \pi_p \wedge \bar{\pi}_p \ .$$

Let $P_{\mathbb{C}}(p, n)_x$ denote the set of all covectors in $\bigwedge^{p,p} TX_x^*$ which represent p-planes in TX_x.

An n,n-covector α is called positive if $\alpha = \lambda \tau_n(x)$, where $\lambda \geq 0$, and *strictly positive* if $\lambda > 0$.

A real p,p-covector α in x is called positive if there exist covectors (α_I) in $P_{\mathbb{C}}(p, n)_x$ and positive real numbers λ_I such that

$$\alpha = \sum \lambda_I \alpha_I \ .$$

By duality, *a real k,k-covector in x is called weakly positive* if for any positive p,p-covector α (p+k=n) the n,n-covector $\alpha \wedge \beta$ is positive. The set of positive (resp., weakly positive) covectors in x will be denoted by $SP_x^{p,p}$ (resp., $WP_x^{k,k}$).

Let $*$ denote the operator $\bigwedge^{p,q} TX_x^* \longrightarrow \bigwedge^{n-p,n-q} TX_x^*$, given by

$$(\alpha, \beta)\tau_n(x) = \alpha \wedge *\beta \ ,$$

where $(\ ,\ )$ denotes the scalar product induced by h. With this notation, using the obvious identifications, the convex cones $SP_x^{p,p}$ and $WP_x^{k,k}$ are the positive duals to each other.

The cone $SP_x^{p,p}$ is closed in $\bigwedge^{2p} TX_x$ since it is generated by the closed set $P_{\mathbb{C}}(p, n)_x$. It also has non- empty interior in the real p,p-covectors. The same is true for $WP_x^{p,p}$. *A real p,p covector β is called positive definite* if it is contained in the interior of $WP_x^{p,p}$, i.e., if $\alpha \wedge \beta$ is strictly positive for any positive (n-p,n-p) covector α. All of the definitions given thus far are independent of the choice of the Hermitian metric h. Thus, they pointwise carry over to forms. Note that a $1, 1$-form is positive definite (resp., semi-definite) in the above geometric sense if and only if the associated symmetric form is positive definite (resp., semi-definite) in the sense of linear algebra.

Positivity for p,p-currents is defined in a similar way. *A p,p- current T on X is real* if $T(\bar{\eta}) = \overline{T(\eta)}$ for any test form η. *A real p,p-current T is called positive* if $T(\eta) \geq 0$ for every positve test form η. *A real p,p-current T is called positive definite* at $x \in X$ if $T(\eta) > 0$ for every positive testform η, which does not vanish in x.

From now on we assume that, in addition to having compact support, the function χ is *non-negative* on G. It follows that if T is positive (resp., positive definite) at x, then the same is true for the smoothed current $\chi * T$.

Prop. 1.2.1 shows that the current $\chi * T$ is given by integration against a smooth differential form T_χ. The main goal of this section is to show that T_χ inherits positivity from T.

Theorem 1.3.1 *(Richthofer [51]) Let T be a closed positive p,p-current which is positive definite at $x_0 \in X$ and let U denote the interior of the support of χ. Then T_χ is a closed weakly positive p,p-form which is positive definite on $U \cdot x_0$.*

Proof By Prop. 1.2.5, for an (n–p,n–p)-form ω,

$$T_\chi \wedge \omega(gH) = T_x \int_H \chi(ghs(x)^{-1}) \Delta_G(s(x)h) u^{-1}(g) l^*_{gh^{-1}s(x)^{-1}} \omega(x) dh \, \mu_X(gH).$$

This representation for $T_\chi \wedge \omega$ shows that T_χ is a p,p-form, because T vanishes on all forms which are not of bitype (n–p,n–p). The fact that T_χ is closed follows from the fact that $(\pi^* T)_\chi$ is closed and the properties of the fiber integral. The fact that T_χ is weakly positive follows from the definition of $\chi * T$.

For the definiteness property, let H be the isotropy group at x_0 and $s(x_0) = e$. For $x \in X$ fixed, let

$$f(h) = \chi(ghs(x)^{-1}) \Delta_G(s(x)) \Delta_G(h) u^{-1}(g) l^*_{gh^{-1}s(x)^{-1}} \omega(x) \ .$$

Then $\int_H f(h) \, dh \in SP^{p,p}_x \setminus \{0\}$ if and only if $\chi | g\pi^{-1}(x)$ is not identically zero, for in this case there is an $h \in H$ such that $\chi(g(s(x)h)^{-1}) \neq 0$.

Now suppose $\omega(gx_0) \neq 0$. If $\chi(g) \neq 0$ then $\chi | g\pi^{-1}(x_0)$ does not vanish identically. Thus for $x = x_0$ we have $\int_H f(h) \, dh \in SP^{p,p}_x \setminus \{0\}$. Therefore, the interior of the support of the form

$$x \mapsto \int_H \chi(ghs(x)^{-1}) \Delta_G(s(x)h) u^{-1}(g) l^*_{gh^{-1}s(x)^{-1}} \omega(x) \, dh$$

contains x_0. Thus we obtain

$$T_\chi \wedge \omega(gx_0) =$$
$$= T_x \left(\int_H \chi(ghs(x)^{-1}) \Delta_G(s(x)h) u^{-1}(g) l^*_{gh^{-1}s(x)^{-1}} \omega(x) \, dh \right)$$
$$> 0 \ ,$$

and the proof is complete. $\qquad\square$

The following was our main motivation for considering the smoothing problem in the first place.

Corollary 1.3.2 *Suppose $X = G/H$ admits a closed positive 1,1-current which is positive at one point $x_0 \in X$. Then, given a relatively compact open set V in X, there exists on X a closed semi-positive define $1,1$-form ω on X which is positive definite on V.*

Proof Just choose χ so that $U \cdot x_0 \supset V$. $\qquad\qquad\qquad\qquad\qquad\qquad\qquad\qquad$ $\square$

Since the *existence of a $1,1$-form for every V* (as in the above *Corollary*) seemed to be the right notion for the applications, we coined an expression for such manifolds: *weakly Kähler*. However, the following observation shows that a new notion is not needed.

Proposition 1.3.3 *(F. Berteloot) A weakly Kählerian manifold X is Kählerian.*

Proof Let $\{V_k\}$ be a locally finite covering of X by contractible relatively compact open sets. Since X is weakly Kählerian, we have ω_k for every V_k as in *Cor. 1.3.2*. For every $k, l \in \mathbb{N}$ there exists a smooth plurisubharmonic function φ_{kl} on V_l so that $\omega_k|V_l = \frac{i}{2}\partial\bar{\partial}\varphi_{kl}$. Note that φ_{kk} is strictly plurisubharmonic on V_k and that $\varphi_{kl} - \varphi_{kj}$ is pluriharmonic on the intersection V_{lj}.

We may assume that φ_{kl} is defined in a neighborhood of $\overline{V}_l$ and let $\|\varphi_{kl}\|_l^s$ be its H^s-norm there. Define $M_k := max_{l \leq k}\|\varphi_{kl}\|_l^s$ and $\lambda_k := (2^k M_k)^{-1}$. Then $\psi_l := \sum \lambda_k \varphi_{kl}$ converges in H^s for every l. If s is big enough, then ψ_l is at least C^2. Furthermore $\psi_l - \psi_j$ is pluriharmonic and $\omega := \frac{i}{2}\partial\bar{\partial}\psi_j$ is globally defined and positive-definite. In order to obtain a smooth form, apply *Theorem 1.3.1*. $\qquad\qquad\qquad\qquad\qquad\qquad\qquad$ $\square$

2 Hypersurfaces in homogeneous complex manifolds

We consider here complex manifolds which are homogeneous under actions of Lie groups of holomorphic transformations. If X is such a manifold, G is such a group, $x_0 \in X$, and $H := Iso_G(x_0) = \{g \in G : g(x_0) = x_0\}$, then X is naturally G-equivariantly identifiable with the coset space G/H. If G is complex and is acting holomorphically, then H is complex and the natural complex structure on the coset space agrees with that of X.

The first results on hypersurfaces in homogeneous spaces were proved in the case where X is compact and G is abelian, i.e., compact tori. Here the theory of periodic hypersurfaces in $\mathbb{C}^n$, which culminated in the precise description of such a hypersurface as the zero set of a Θ-function (Theorem of Appell-Humbert, see e.g.,[44]), led to some of the first important results in the theory of several complex variables.

One of the main goals in the general case has been to determine natural, at least dense, sets of globally defined objects which determine the hypersurfaces $\mathcal{H}(X)$, e.g., Θ-functions, algebraic representation functions which are sections of particular bundles, holomorphic functions,. . . . In this chapter we give a sampling of results in this direction. Due to lack of time we have postponed writing a systematic treatment.

2.1 Separation by hypersurfaces and Kählerian homogeneous spaces

When considering a homogeneous manifold from the point of view of analytic hypersurfaces, one is led to identifying the points which lie in every hypersurface. In the case of compact homogeneous manifolds, Grauert and Remmert ([22]) introduced the following *hypersurface reduction*: For $x \in X$, let $A(x)$ be the intersection of all hypersurfaces $F \in \mathcal{H}(X)$ with $x \in F$. They showed that $x \sim y$ if and only if $x \in A(y)$ defines a holomorphic equivalence relation and the quotient $X \longrightarrow X/\sim$ is given by an equivariant fibration $G/H \longrightarrow G/J$, where J is also a closed complex subgroup of G.

If X is non-compact, the construction of a hypersurface reduction is a more complicated matter. The best results in this regard are proved in ([45]):

> *Let X be a complex manifold which is homogeneous with respect to a real Lie group of holomorphic transformations, $X = G/H$. Define x and y to be separated if and only if there exists a hypersurface which contains x and does not contains y and vice versa, and $x \sim y$ whenever x and y are not separated. This defines an equivalence relation with quotient given by a holomorphic homogeneous fibration $\pi : X = G/H \longrightarrow G/J = X/\sim := Y$ onto a complex manifold. The group J is a closed subgroup of G and is complex if G is complex. As desired, $\pi^*(\mathcal{H}(Y) = \mathcal{H}(X)$ and Y is hypersurface separable in the above sense. Furthermore, the map $\pi : X \longrightarrow Y$ is a holomorphic fiber bundle with $H^0 \triangleleft J$ and J/H^0 a complex Lie group acting holomorphically on the fiber.*

Remark *X is hypersurface separable if and only if $A(x) = \{x\}$ for all $x \in X$. We say that X is locally hypersurface separable if x is isolated in $A(x)$ for all $x \in X$. Since this notion is stable under the process of going to a covering space, it is more useful.*

There is also a reduction given by the meromorphic functions ([30]), and there are examples of hypersurface separable homogeneous manifolds which are not meromorphically separable ([45]).

Up to this point in time, the various reductions have served as useful tools in the subject. However, there are a number of basic questions which need to be answered. For example:

    1. *If X has a hypersurface, does it have a non-constant meromorphic function?*

In fact, most results on hypersurfaces in homogeneous manifolds seem to be proved indirectly. In this regard, the following theorem has been of central importance (see e.g.[9], [21], [46]).

Theorem 2.1.1 *Let X be a locally hypersurface separable homogeneous manifold. Then X is Kählerian.*

Proof If X is locally hypersurface separable, then a *general position* argument shows that there are $n := dim(X)$ hypersurfaces $H_1, \ldots, H_n$ which have a common point x_0 and local coordinates $(z_1, \ldots, z_n)$ so that $H_k = \{z_k = 0\}, k = 1, \ldots, n$, near x_0 (see [9]). Let H be the union of these hypersurfaces and let T be the current on X which is defined by integration over H. It follows that T is a closed positive current which is positive-definite at x_0. As a consequence of *Thm. 1.3.1 and Prop. 1.3.3*, X is Kählerian. $\qquad\square$

2.2 Compact homogeneous spaces

Let X be a compact complex space. In this case the group $Aut_{\mathcal{O}}(X)$ is a complex Lie group acting holomorphically in the compact-open topology. Thus, when dealing with compact homogeneous manifolds, we assume that G is a complex Lie group acting holomorphically.

Using complex analytic methods, Grauert and Remmert ([22]) proved the following

Theorem 2.2.1 *A compact hypersurface separable homogeneous complex manifold is projective algebraic.*

Of course projective algebraic manifolds are Kählerian and therefore the following theorem of Borel and Remmert ([11]) is applicable.

Theorem 2.2.2 *A compact homogeneous Kähler manifold is a product* $X = T \times Q$ *of a complex torus and a rational homogeneous manifold.*

Corollary 2.2.3 *A compact hypersurface separable homogeneous complex manifold is a product of an abelian variety and a rational homogeneous manifold.*

Remark *It should be underlined that the result of Grauert and Remmert shows via general methods that a hypersurface separable torus is projective algebraic, i.e., an abelian variety. In order to prove this Theorem of Lefschetz by classical methods one must go through a good bit of Θ-theory (see e.g. [44]). One way or the other, one certainly needs the finite-dimensionality of $H^1(X, \mathcal{O})$ and, for finer results, the fact that topologically trivial line bundles come from representations of the fundamental group.*

Our goal here is to give a proof of the Theorem of Grauert and Remmert by using smoothing of currents, i.e., *Theorem 1.3.1*: If X is hypersurface separable, then it is Kählerian. By *Theorem 2.2.2* it is a product $T \times Q$ and, by the classical Θ-theory, T is an abelian variety, e.g., X is projective algebraic.

In the following sections we outline the basics on compact Kählerian homogeneous manifolds and, using elementary concepts from symplectic geometry, prove *Theorem 2.2.2*. Thus, we have given a proof which corresponds to our hope of understanding hypersurfaces by understanding the globally defined analytic objects which define them—in this case Θ-functions.

2.3 Compact Kählerian homogeneous spaces

Using the hypersurface reduction and the smoothing procedures discussed in the previous sections, we are led to the study of homogeneous Kähler manifolds. In this section we give a proof of the classification theorem of Borel and Remmert in the compact case.

a. The basic examples

1.) Tori

If V is an n-dimensional complex vector space and Γ is a discrete additive subgroup of rank $2n$, then $X = V/\Gamma$ is a *compact torus*. We refer to Γ as a *lattice* in V. Of course the complex structure on X depends on the position of Γ in V, e.g., if $n > 1$ and Γ is generic, then X only has constant meromorphic functions.

Let H be a positive-definite hermitian product on the vector space V. Of course H is Γ-invariant and gives rise to a hermitian metric on X. The associated $(1,1)$-form ω_h satifies $d\omega_h = 0$ and in particular X is Kählerian.

Remark *In this case the metric h is invariant. However we emphasize that when we refer to a homogeneous Kähler manifold, we only assume that the underlying manifold is Kählerian.*

2.) Homogeneous rational manifolds

Suppose that X is a compact manifold embedded in a complex projective space $\mathbb{P}_N$. Let $A \cong PSl_{N+1}(\mathbb{C})$ be the group of holomorphic automorphisms of $\mathbb{P}_N$ and assume that

$$G \subset Stab_A(X) := \{g \in A : g(X) = X\}$$

acts transitively on X, i.e., there is a G-equivariant embedding of X in some projective space.

Let R be the radical of G, i.e., the maximal connected normal solvable subgroup. Since R is solvable, it stabilizes a flag $\{*\} \subset \mathbb{P}_1 \subset \ldots \mathbb{P}_n$ of linear subspaces. Of course some intersection $\mathbb{P}_k \cap X$ contains an isolated point p, and, since R is connected, $Rp = p$. Using the fact that R is normal in G, it follows that $Rg(p) = g(p)$ for all $g \in G$. But G acts transitively on X. Hence we conclude that R acts trivially on X.

Proposition 2.3.1 *Let X be a compact complex manifold which is homogeneous with respect to an almost effective G-action. Suppose that there is a G-equivariant embedding of X in some projective space. Then G is semi-simple.*

Proof The above argument shows that the radical R acts trivially, and, since the action is almost effective, it follows that $R = \{e\}$. $\qquad\qquad\square$

Let X and G be as in the *Proposition* and let B be a maximal connected solvable subgroup of G, i.e., B is a *Borel subgroup*. The same argument as above shows that B fixes some point in X. Thus, if $x \in X$ and $P := G_x$ is the isotropy group at x, then P contains a Borel subgroup of G, i.e., P is a *parabolic* subgroup. We summarize this as follows.

Proposition 2.3.2 *The compact homogeneous complex manifolds X which can be equivariantly embedded in a complex projective space are of the form $X = G/P$, where G is semi-simple and P is parabolic.*

In fact the converse statement is also true: *If $X = G/P$ as above, then X is compact and can be equivariantly embedded in a complex projective space.* Such results follow immediately from the foundations of the theory of algebraic groups (see e.g.[10],[31]). Furthermore, via the theory of root systems, one has a precise description of P. For example, if $P = L \cdot U^+$ is a Levi-decomposition, where U^+ is the unipotent radical of P and L is a maximal reductive subgroup, then there is another parabolic subgroup $P^- = L \cdot U^-$ which is conjugate to P such that $U^- \cap P = \{e\}$ and $U^- \cdot P$ is Zariski open in G. Hence, if $P = G_x$, then $U^- x \cong U^-$ is Zariski open in G/P. Now U^- is, as an algebraic variety, a copy of affine space and the identification $U^- x \cong U^-$ is algebraic. Thus, it follows that $X = G/P$ is rational. Therefore we refer to such varieties as *homogeneous rational manifolds*.

Note that, since $\mathbb{C}^n \cong U^- x$ is Zariski dense, X is simply-connected. Hence every maximal compact subgroup K of G acts transitively on X. Note further that an embedding $X \hookrightarrow \mathbb{P}_N$ yields a Kähler structure on X. Thus, by averaging, we have an invariant Kähler metric for any such K.

b. The classification theorem

Our goal here is to prove the Theorem of Borel and Remmert (*Thm. 2.2.2*) on the classification of compact homogeneous Kähler manifolds. Our proof is based on elementary symplectic techniques which we apply to the basic fibration of X.

1.) The Tits fibration

The beginning point of many classification programs is the reduction from the abstract to the concrete which is given by the Tits fibration. Although there is a version of this for real groups (see [27]), since our application here is for compact manifolds, we restrict to the case of complex groups.

Let $X = G/H$ be the homogeneous space of a complex Lie group, i.e., G is complex and H is a closed complex subgroup. Then the *normalizer* $N := N_G(H)^0 = \{g \in G : gH^0 g^{-1} = H^0\}$ is likewise complex. We refer to the fibration $G/H \longrightarrow G/N$ as the *Tits fibration*. The fiber $F := N/H = N/H^0/H/H^0 = M/\Gamma$ is homogeneous under the action of a complex Lie group M having discrete isotropy Γ.

The base of the Tits fibration is equivariantly realized in a projective space. For this we regard $\mathfrak{h}$ as a point in the Grassmann manifold of complex k-planes, $k := dim(\mathfrak{h})$, in $\mathfrak{g}$. Of course G acts on this Grassmann manifold via its adjoint representation on $\mathfrak{g}$, and $Iso_G(\mathfrak{h}) = N$, i.e., G/N is the G-orbit of the point $\mathfrak{h}$. Embedding via the Plücker map gives us a concrete equivariant realization of the base G/N in projective space. Thus, if for example G/H is compact, then the base of the Tits fibration is a homogeneous rational manifold.

2.) Symplectic notation

Let (M, ω) be a *symplectic manifold*, i.e., ω is a non-degenerate 2-form on the differentiable manifold M. Let $\mathcal{E}(M)$ denote the space of C^∞ functions on M. For $f \in \mathcal{E}(M)$ the associated *hamilton field* V_f is determined by contraction, $df = i_{V_f}\omega$, and the induced Lie algebra structure on $\mathcal{E}(M)$ is defined by $\{f, g\} := \omega(V_f, V_g)$. Let $Ham_{loc}(M)$ be the set of vector fields V on M so that the Lie derivative $L_V\omega$ vanishes. It follows that $Ham_{loc}(M)$ is a Lie subalgebra of the space of all vector fields and $h : \mathcal{E}(M) \longrightarrow Ham_{loc}(M)$, $f \mapsto V_f$, is a Lie morphism. Finally, let $Ham(M) := Im(h)$ and note that

$Ham_{loc}(M)/Ham(M) \cong H^1_{deR}(M)$.

If K is a Lie group of symplectic diffeomorphisms, i.e., $k^*\omega = \omega, \forall k \in K$, then we have a natural Lie morphism $\kappa : \mathfrak{k} \longrightarrow Ham_{loc}(M)$ given by the K-action. The action is said to be *Poisson* if there is a *lifting* Lie morphism $\lambda : \mathfrak{k} \longrightarrow \mathcal{E}(M)$ so that $\kappa = h \cdot \lambda$. If for example $\mathfrak{k}$ is semi-simple, there is a unique lifting (see e.g. [23]).

Suppose we have a Poisson action via a lifting $\lambda : \mathfrak{k} \longrightarrow \mathcal{E}(M)$. The *moment map* $\mu : M \longrightarrow \mathfrak{k}^*$ with respect to λ is defined by $\mu(x)(\xi) = \lambda(\xi)(x)$. We refer to $f_\xi := \lambda(\xi)$ as the *ξ-th coordinate function* of μ. The following fact is an important consequence of the definitions:

$$df_\xi(x) = 0 \text{ if and only if } \xi = 0 :$$

3.) Proof of the classification theorem

Let X be a Kähler manifold which is homogeneous with respect to a complex Lie group G. Let K be the semi-simple part of a maximal compact subgroup of G. By averaging over K, we may assume that the Kähler form ω is K-invariant. We regard (X, ω) as a symplectic manifold equipped with a Poisson K-action. Since K is semi-simple, there exists a unique lifting.

Proof of Theorem 2.2.2 Let $X = G/H \longrightarrow G/N := Q$ be the Tits fibration with fiber $F = M/\Gamma$. If $G = R \cdot S$ is a Levi-Malcev decomposition, then, since R acts trivially on the homogeneous rational manifold Q, we write $Q = S/P$. Our first goal is to show that the parabolic subgroup P acts trivially on F.

For this let $P = L \cdot U$ be a Levi-decomposition of P. Since L is reductive, it is the complexification $K_L^{\mathbb{C}}$ of a maximal compact subgroup K_L. By conjugating appropriately, we may assume that $K_L \subset K$; in particular ω is K_L-invariant. For ξ in the Lie algebra of K_L we have the ξ-th coordinate f_ξ of the moment map. Now the fiber F is symplectic with respect to the restricted form. Thus, we may regard f_ξ as a coordinate function for the moment map of F with respect to the K_L-action.

Since F is compact, there exists a point $x \in F$ where $df_\xi = 0$, i.e., $\xi(x) = 0$. Now Γ is discrete and therefore the holomorphic tangent bundle of F is trivial. Thus, a holomorphic vector field which vanishes somewhere on F must vanish identically. Hence K_L acts trivially on F and consequently so does its complexification L.

Let $I_P = \{g \in P : g(x) = x \forall x \in F\}$ be the *ineffectivity* of the P-action of F. Of course I_P is a normal subgroup in P. But elementary aspects of the structure theory for parabolic groups (see e.g.[31]) show that the only normal subgroup which contains L is P itself. Thus $I_P = P$ and P acts trivially on F. Since P is an S-isotropy in the base Q, it follows that the S-orbits in the bundle space X are sections and the Tits fibration realizes X as a product $X = F \times Q$.

It remains to show that the fiber F is a torus. This the content of the following remark of H. C. Wang ([56]).

Proposition 2.3.3 *Let X be a compact Kähler manifold with trivial holomorphic tangent bundle. Then X is a torus.*

Proof If T_X is trivial, then the Lie algebra $\mathfrak{g}$ of holomorphic vector fields on X is n-dimensional, $n := dim(X)$. Now X is compact. Thus the fields in $\mathfrak{g}$ can be globally integrated. Since $\mathfrak{g}$ spans the tangent space $T_{X,x}$ at each $x \in X$, it follows that $Aut_{\mathcal{O}}(X)^0 =$

G is an n-dimensional complex Lie group which acts transitively on X, i.e., $X = G/\Gamma$, where Γ is a discrete cocompact subgroup of G.

Suppose now that X is Kähler; in particular the holomorphic 1-forms on X are closed. If $((V_1, \ldots, V_n))_{\mathbb{C}}$ is a basis for $\mathfrak{g}$ and $((\mu_1, \ldots, \mu_n))_{\mathbb{C}}$ is the dual basis of 1-forms, the fact $d\mu_i = 0\ \forall i$ is equivalent to $\mu_i([V_i, V_j]) = 0\ \forall i, j, k$. This of course implies that $[V_i, V_j] = 0$ $\forall j, k$, i.e., G is abelian. $\qquad\qquad\qquad\square$

2.4   A guiding conjecture

A number of years ago D. Akhiezer posed an extremely interesting question which, no matter how it is answered, is very important for the study of the complex analytic objects related to hypersurfaces. We formulate this as

Akhiezer's conjecture Let $X = G/H$ be a homogeneous space of a complex Lie group with hypersurface reduction $G/H \longrightarrow G/J$. Suppose

  i. H is not contained in a proper parabolic subgroup

  ii. $\mathcal{O}(X) \cong \mathbb{C}$.

Then J contains the commutator subgroup G'.

Roughly speaking, this means that if the hypersurfaces are not related to rational functions, i.e., those coming from G/P for some parabolic subgroup P, nor to holomorphic functions, then they come from an abelian group, where we can hope to understand the relationship to Θ-functions.

Under the assumptions of the conjecture, G/J is an abelian group with only constant holomorphic functions. We refer to such groups as *Cousin groups*.[3] We consider these groups, in particular the question of existence of Θ-functions, in the next section.

If X is compact, then the conjecture is an immediate consequence of the Theorem of Grauert and Remmert (*Thm. 2.2.1*). Akhiezer himself proved it in the case where G is nilpotent ([2]). In Section 5 of this chapter we outline the proof of K. Oeljeklaus and W. Richthofer in the solvable case ([46]).

Akhiezer also proved this for the case where G is semi-simple, but under restrictions on the nature of the isotropy ([3]). The author and G. Margulis proved this for the general semi-simple case ([26]). Finally, if the Levi-Malcev decomposition $G = R \cdot S$ is a product, i.e., $S \triangleleft G$, then the results for the solvable and semi-simple cases can be pieced together to prove the conjecture ([47]). The case where the Levi-Malcev decomposition is non-trivial remains a mystery and is certainly one of the most interesting problems in the area of complex analysis on homogeneous spaces.

---

[3] P. Cousin considered the 2-dimensional case in [14].

Using smoothing of currents it is possible to give an elementary proof of Akhiezer's conjecture in the case where G is semi-simple. For this, recall that a semi-simple group G carries a unique affine algebraic structure as an algebraic group. Thus, if H is any subgroup, then we can speak of the complex Zariski hull $I := \overline{H}$. Now, if I is any algebraic subgroup of a semi-simple group G, then I is either reductive or is contained in a proper parabolic subgroup (see e.g. [31]). If I is reductive, then G/I has many non-constant holomorphic functions– in fact it is Stein (see e.g. [Mat]). Hence, under the assumptions of the conjecture, it follows that H is Zariski dense. Therefore, it is enough to prove the following result which was proved by other methods in [26].

Proposition 2.4.1 *Let G be a semi-simple complex Lie group, H a Zariski dense complex subgroup, and $X := G/H$. Then $\mathcal{H}(X) = \emptyset$.*

Proof Since the adjoint representation is algebraic, the normalizer N of H^0 in G is an algebraic subgroup and therefore the density of H implies $H^0 \triangleleft G$. Of course we may assume that the G-action is almost effective. Hence we may assume that H is discrete. Furthermore, using the hypersurface reduction, we may assume that X is hypersurface separable.

Since H is Zariski dense, there exists a semi-simple element $\gamma \in H$ which generates a discrete group $\Gamma \cong \mathbb{Z}$. Finally, we consider $X' := G/\Gamma$. Now $X' \longrightarrow X$ is a covering map. Hence X' is locally hypersurface separable and is therefore Kählerian (*Thm. 1.3.1*). Let ω be the Kähler form and note that, since $b_2(X') = 0$, we have $\omega = d\mu$, where μ is a 1-form on X'.

Let A be the algebraic closure of Γ in G. By replacing γ by an appropriate multiple, we may assume that A is connected. Since γ is semi-simple, it follows that $A = (\mathbb{C}^*)^n$ and the A-orbit of the neutral point in X' is a *Cousin group* $C := A/\Gamma = \mathbb{C}^n/\Lambda$, where Λ is an additive subgroup of rank $n + 1$. Let $K \cong \mathbb{R}^{n+1}/\Lambda$ be the maximal compact subgroup in C. Of course K is a torus. Note that the universal cover $\mathbb{R}^{n+1}$ of K is naturally embedded in $\mathbb{C}^n$ and therefore contains a *complex* 1-parameter group which we call $\mathbb{C}$.

After all of these definitions, the proof is a simple matter. By averaging over K, we may assume that μ is K-invariant. On a $\mathbb{C}$-orbit we therefore have $\mu = a\,dz + b\,d\bar{z}$, where a and b are constants. Hence, ω vanishes along the $\mathbb{C}$-orbits, contrary to ω being positive-definite. $\square$

Remark *In the language of invariant theory, the above result states that if G is semi-simple and H is Zariski dense, then there are no invariant hypersurfaces in G: $\mathcal{H}(G)^H = \emptyset$. The significantly more general result $\mathcal{H}(G)^H = \mathcal{H}(G)^{\overline{H}}$ for arbitrary H will be discussed in Section 6.*

In closing this section we should note that, under further assumptions on the geometry of X, Akhiezer's conjecture can often be proved. For example, the classification of hypersurface separable homogeneous manifolds having more than one end as a topological space ([21]) yields the conjecture in that case.

2.5 Hypersurfaces in abelian groups

Although our primary goal here is to discuss the abelian case, we begin by introducing the notation for nilpotent groups where the situation is somewhat similar, but where several interesting problems are still open.

Let G be a simply-connected nilpotent complex Lie group and let $\Gamma \subset G$ be a discrete subgroup. Then there exists a unique connected real Lie subgroup $G(\Gamma)$ of G with $G(\Gamma)/\Gamma$ compact (see [39]). If $\mathfrak{g}(\Gamma)$ is its Lie algebra, we define $\mathfrak{m}(\Gamma) := \mathfrak{g}(\Gamma) \cap i\mathfrak{g}(\Gamma)$ and note that the corresponding subgroup $M(\Gamma)$ is a normal subgroup of $G(\Gamma)$. It would seem that the $M(\Gamma)$-action on the CR-manifold $G(\Gamma)/\Gamma$ should determine much of the function theory on the complex homogeneous space G/Γ. Here we comment on the abelian case.

From now on in this section $G = (\mathbb{C}^n, +)$; in particular G is abelian. In this case $X = G/\Gamma$ is itself a complex abelian Lie group with maximal compact subgroup $K = G(\Gamma)/\Gamma$. The group X is called a *Cousin group* if $\mathcal{O}(X) \cong \mathbb{C}$. The following observation ([42]) reduces most questions of global function theory to the case Cousin groups.

Proposition 2.5.1 *There is a unique splitting* $X = A \times B \times C$ *of* X *into a product of complex abelian Lie groups, where* $A = \mathbb{C}^n$, $B = (\mathbb{C}^*)^n$, *and* C *is a Cousin group.*

Remark *An abelian complex Lie group X is a Cousin group if and only if $\mathfrak{g}(\Gamma) + i\mathfrak{g}(\Gamma) = \mathfrak{g}$ and the subgroup $M := M(\Gamma)$ is dense in K.*

It would be desirable to study hypersurfaces in Cousin groups with methods of Θ-function theory analogous to those in the case of compact complex tori. However, as Cousin himself realized ([14]),this is in general not possible: *There are hypersurfaces in Cousin groups which are not zero-sets of Θ-functions.* From the Fourier series point of view, this is due to *small denominator* convergence difficulties which lead to the infinite-dimensionality of $H^1(X, \mathcal{O})$. Nevertheless we do have the following result ([26]).

Proposition 2.5.2 *Suppose that a topologically trivial line bundle L on a Cousin group has a holomorphic section $\sigma \not\equiv 0$. Then L is holomorphically trivial.*

Using this result, it is possible to reduce some problems to Θ-theory. To exemplify this, we now explain some unpublished ideas of K. Oeljeklaus and W. Richthofer ([48]).

Let L be a holomorphic line bundle on X and $J : \Gamma \times \mathbb{C}^n \longrightarrow \mathbb{C}^*$ its factor of automorphy. Writing $J(\gamma, z) = e^{2\pi i a(\gamma, z)}$, define in the usual way $E : \Gamma \times \Gamma \longrightarrow \mathbb{Z}$ by

$$E(\gamma_1, \gamma_2) := a(\gamma_2, z + \gamma_1) + a(\gamma_1, z) - a(\gamma_1, z + \gamma_2) - a(\gamma_2, z).$$

Also let E denote the extension to a real bilinear skew-symmetric form on $G(\Gamma)$. Of course E represents the Chern class $c(L)$ and we have a uniquely determined hermitian form H on $\mathbb{C}^n$ with $Im(H)|G(\Gamma) \times G(\Gamma) = E$.

Proposition 2.5.3 *([5]) Suppose that H is positive-definite on M. Then there is a lattice $\Lambda \subset G = \mathbb{C}^n$ so that $\Gamma \subset \Lambda$ and G/Λ is an abelian variety. In particular $X = G/\Gamma$ possesses a non-degerate Θ-function. Furthermore, X can be realized as a principal algebraic $(\mathbb{C}^n \times (\mathbb{C}^*)^m)$-bundle over an abelian variety.*

Remark *The Cousin groups in the above Proposition are often called quasi-abelian varieties. It should be underlined that the realizations as bundles is in general far from being unique.*

Proposition 2.5.4 [4] *A Cousin group is quasi-abelian if and only if it is locally hypersurface separable.*

We prove this as a consequence of an observation on $Stab_G(F) := \{g \in G : g(F) = F\}$, where $F \in \mathcal{H}(X)$. For this it is convenient to introduce some notation.

As above, let L be the line bundle of F and H the hermitian form which is derived from a factor of automorphy for L. For $v \in M$, let $V := < v >_{\mathbb{C}}$ and let T be the topological closure of the V-orbit of the neutral point $e \in K$. Of course there exists a unique $\mathbb{R}$-vector space W of $\mathbb{C}^n$ with $T = W/W \cap \Gamma$. Finally, let $W^{\mathbb{C}} := W + iW$ be the smallest complex subspace which contains W.

Proposition 2.5.5 *If $H(v, v) = 0$, then $W^{\mathbb{C}} \subset Stab_G(F)$.*

Proof Define $\chi : K \times K \longrightarrow S^1$ by $(k_1, k_2) \mapsto e^{2\pi i E(k_1, k_2)}$. Since $E|\Gamma \times \Gamma$ is $\mathbb{Z}$-valued, it follows that χ is well-defined. Of course $\chi|(V/V \cap \Gamma) \times (V/V \cap \Gamma) \equiv 1$. Since $(V/V \cap \Gamma) \times (V/V \cap \Gamma)$ is dense in $T \times T$ and χ is continuous, $\chi|T \times T \equiv 1$. Thus $E|W \times W \equiv 0$.

Now consider $X' := W^{\mathbb{C}}/W \cap \Gamma$. Since $T = W/W \cap \Gamma$ is compact and contains a dense V-orbit, it follows that X' is a Cousin group. We regard $W^{\mathbb{C}}/W^{\mathbb{C}} \cap \Gamma$ as a $W^{\mathbb{C}}$-orbit in G/Γ and, for $g \in G$, define $\tau_g : W^{\mathbb{C}}/W^{\mathbb{C}} \cap \Gamma \longrightarrow G/\Gamma$ as its translation by g. Define $p : W^{\mathbb{C}}/W \cap \Gamma \longrightarrow W^{\mathbb{C}}/W^{\mathbb{C}} \cap \Gamma$ to be the natural projection and $\iota_g := \tau_g \cdot p$.

Since E represents $c(L)$ and $E|W \times W \equiv 0$, $\iota_g^*(L)$ is a topologically trivial line bundle on X'. Now $\tau_g(F)$ defines a section in $\iota_g^*(L)$. Thus, as a consequence of *Prop. 2.5.2*, for every $W^{\mathbb{C}}$-orbit B, either $F \supset B$ or $F \cap B = \emptyset$. In other words, $W^{\mathbb{C}} \subset Stab_G(F)$. $\square$

Corollary 2.5.6 *Let X be a Cousin group and suppose that there exists $F \in \mathcal{H}(X)$ with $Stab_G(F)$ discrete. Then X is quasi-abelian.*

Proof By *Prop. 2.5.5* the hermitian form H associated to $L(F)$ is definite. By going to the dual bundle if necessary, we therefore have a bundle with H positive-definite on M. The result follows from *Prop. 2.5.3*. $\square$

The proof of *Prop. 2.5.4* is now immediate, because X being *locally hypersurface separable* is equivalent to the existence of $F \in \mathcal{H}(X)$ containing irreducible components $F_1, \ldots, F_n$, $n := dim(X)$, such that $\bigcap_1^n F_j$ contains an isolated point. This of course implies that $Stab_G(F)$ is discrete. $\square$

---

[4]This result has been independently proved by Abe ([4]) and Catanese and coauthor ([12])

2.6 Hypersurfaces in solv-manifolds

In this section G is a solvable complex Lie group, H is a closed complex subgroup, and $X := G/H$. We refer to X as being a *solv-manifold*. If G is nilpotent, then there are two basic methods for understanding the complex analytic structure of X which can not be applied in the general solvable case:

  i. One considers quotients $G/H \longrightarrow G/JH$, where J is a central subgroup. This is basically an inductive approach.

  ii. One reduces to the case of discrete isotropy, $X = G/\Gamma$, and uses the existence of a unique connected real subgroup $G(\Gamma)$ which contains Γ as a cocompact subgroup.

Even when Γ is contained in such a $G(\Gamma)$, in the case of a solv-manifold $X = G/\Gamma$, it is not easy to understand the hypersurfaces $\mathcal{H}(X)$.

Example Let $G_1 \cong \mathbb{C}$ and $G_2 \cong \mathbb{C}^2$ be abelian. Define a semi-direct product $G := G_1 \cdot G_2$, with $G_2 \triangleleft G$, by a 1-parameter subgroup $\rho : G_1 \longrightarrow Aut(G_2) \cong Gl_2(\mathbb{C})$. We choose ρ with $\rho(1) \in Sl_2(\mathbb{Z})$. Thus we have the *integral points* $G_\mathbb{Z}$ and *real points* $G_\mathbb{R}$ as the corresponding semi-direct products. We write $\rho(1) = e^A$ and assume that the eigenvalues of A are not purely imaginary. In this example this is only a *genericity* assumption. However, in general, this is a fundamental condition.

Let $\Gamma := G_\mathbb{Z}$ and consider the commutator fibration $G/\Gamma \longrightarrow G/G'\Gamma$. Here $G' = G_2$ and this realizes X as the total space of a fiber bundle $\pi : X \longrightarrow Y$ with fiber $F \cong \mathbb{C}^* \times \mathbb{C}^*$, base $Y \cong \mathbb{C}^*$, and structure group $\mathbb{Z}$. It can be shown that $\pi^* \mathcal{O}(Y) = \mathcal{O}(X)$ ([38]) and therefore we have a homogeneous example where the answer to the *Question of Serre* is negative: *The fiber and base are Stein, but all holomorphic functions on the total space come from the base.* (See [53],[16] for the first examples of such phenomena.) In fact, one can embed in this example a domain which fibers over an annulus with fiber a bounded domain and again the functions come from the base ([13]).

It can also be shown that the hypersurfaces come from the base: $\pi^*(\mathcal{H}(Y)) = \mathcal{H}(X)$.
Proof There are no intermediate fibrations to the commutator fibration $\pi : X \longrightarrow Y$. Thus it is enough to show that X is not locally hypersurface separable. If it were, then it would be Kählerian. Let ω denote the pull-back of the Kähler form to G. Since $G_\mathbb{R}/\Gamma$ is compact, we may assume that ω has a strictly plurisubharmonic potential φ which is $G_\mathbb{R}$-invariant. But such functions are also G'-invariant ([38]), contrary to the strict plurisubharmonicity $\square$

Using a process which reduces the higher-dimensional case to examples of this type, Loeb proved the following fundamental result ([38]).

Theorem 2.6.1 *Let G be a simply-connected complex solvable Lie group with real form $G_\mathbb{R}$. Let Γ be a discrete cocompact subgroup of $G_\mathbb{R}$. Then the following are equivalent:*

  (i.) G/Γ is Kähler.

  (ii.) G/Γ is Stein.

(iii.) $Spec(\mathrm{ad}_{\mathfrak{g}_{\mathbb{R}}})$ *is not purely imaginary.*

Remark *The proofs in the low-dimensional examples involve delicate applications of the maximum principle. This can also be regarded from the point of view of harmonic analysis, in particular the theory of entire vectors and generalized Paley-Wiener phenomena (see [19],[50]).*

Motivated by *Theorem 2.6.1*, the author and E. Oeljeklaus proved a technical result which, along with Loeb's theory and basic facts on algebraic hulls, leads to results for general solv-manifolds ([29])

Lemma 2.6.2 *Let G be a connected complex solvable Lie group, Γ a discrete ssubgroup, and $X = G/\Gamma$. Suppose that X is Kähler and that Γ contains a regular element of G. Then Γ contains a nilpotent subgroup $\widehat{\Gamma}$ of finite index.*

An element $g \in G$ is *regular* if the algebraic closure of the cyclic group generated by $Ad(g)$ contains a maximal torus in the algebraic closure of $Ad(G)$.

Using this lemma and involved Lie algebra computations, K. Oeljeklaus and W. Richthofer developed a structure theory for Kählerian solv-manifolds ([46]). The following is the main result.

Theorem 2.6.3 *Let X be a Kählerian solv-manifold. Then, if $X = G/H \longrightarrow G/J := Y$ is the holomorphic reduction, it follows that*

    i. Y is Stein

    ii. The fiber $F = J/H$ is a Cousin group.

The *holomorphic reduction* is defined by the analytic equivalence relation $x \sim y$ *if and only if $f(x) = f(y)$ for all $f \in \mathcal{O}(X)$.*

Corollary 2.6.4 *(Akhiezer's Conjecture for solv-manifolds) Let X be a hypersurface separable solv-manifold with $\mathcal{O}(X) \cong \mathbb{C}$. Then X is a Cousin group.*

Proof Since X is hypersurface separable, it is Kählerian (*Thm. 2.1.1*), and $\mathcal{O}(X) \cong \mathbb{C}$ implies that $J = G$ in the above theorem. $\qquad\qquad\square$

2.7 Hypersurfaces in homogeneous manifolds of semi-simple groups

If a group H acts holomorphically on a complex manifold X, then $\mathcal{H}(X)^H$ denotes the set of invariant hypersurfaces. If H is a subgroup of an algebraic group G, then $\overline{H}$ denotes its algebraic closure. In this language, the basic result for semi-simple groups ([9]) can be stated as follows:

Theorem 2.7.1 *Let G be a semi-simple complex Lie group and $H \subset G$ a subgroup. Then*

$$\mathcal{H}(G)^H = \mathcal{H}(G)^{\overline{H}}.$$

Remark *This is equivalent to the following statement in the context of homogeneous spaces:*

> *Let X be a complex homogeneous space of a complex semi-simple Lie group G. Then the hypersurface reduction is given by $X = G/H \longrightarrow G/\overline{H}$.*

As in so many cases, this statement on hypersurfaces is a consequence of Kähler theory.

Theorem 2.7.2 *Let X be the homogeneous space of a complex semi-simple Lie group, $X = G/H$. Then X is Kählerian if and only if H is algebraically closed.*

In order to derive the result on hypersurfaces as a consequence, it is enough to prove this for H discrete ([9]). The general case is proved by similar techniques ([8]).

The proof of this involves an analysis of a potential φ for the pull-back of the Kähler form to the semi-simple group G. Of course this function is plurisubharmonic and can be assumed to be periodic with respect to a maximal compact subgroup in G. If H were not Zariski closed, then would be an element γ in H which generates a copy of $\mathbb{Z} =: \Gamma$ in G whose Zariski A is not contained in H. Furthermore, $A\Gamma$ is one of three possible types.

We regard the pull-back of φ to A and utilize its periodicity properties, in particular those coming from the fact that the Kähler form is Γ periodic. The case of $A \cong Sl_2$ is handled by [7] and [26]. The proof is completed by a reduction via *Kiselman's minimum principle* to the Sl_2-case.

In the context of group actions, the minimum principle ([34]) can be stated as follows: *Let $G = \mathbb{C}^*$ act holomorphically on a complex manifold X and let φ be a plurisubharmonic function on on X which is K-invariant, $K = S^1$. Then the function $min(\varphi)$ defined by taking the minimum over the G-orbits, $min(\varphi)(x) := min_{g \in G}\varphi(g(x))$, is likewise plurisubharmonic.* It should be noted that, in the context of invariant theory, there is a version of this for reductive groups. It would seem that such a technique would help in understanding the transfer of information from the *Kempf-Ness Variety* to the categorical quotient.

3   Hypersurfaces in compact almost homogeneous manifolds

In this chapter X denotes a connected compact complex manifold and G is a connected complex Lie group acting holomorphically on X. The manifold is said to be *almost homogeneous* with respect to this action if, for some point $x_0 \in X$, the orbit $Gx_0 =: \Omega$ is open. It follows that $E := X \backslash \Omega$ is a proper analytic subset of X.

3.1 Algebraic Reduction

In order to discuss the hypersurfaces in X, it is appropriate to first consider the notion of an *algebraic reduction*.

Recall that the field $\mathfrak{M}(X)$ of meromorphic functions on X is an algebraic function field of transcendence degree $a := a(X)$ at most $dim(X)$. We refer to $a(X)$ as the *algebraic dimension* of X. If $a(X) = dim(X)$, then X is called a *Moishezon* manifold.

Let $f_1, \ldots, f_a$ be a transcendence basis and take $g \in \mathfrak{M}(X)$ so that

$$\mathfrak{M}(X) = \mathbb{C}(f_1, \ldots, f_a)[g].$$

The minimal polynomial of this algebraic extension defines a hypersurface Y in $\mathbb{P}_{a+1}$ with $\mathfrak{M}(X) \cong \mathfrak{M}(Y)$. The natural map given by the functions $f_1, \ldots, f_{a+1}, g$ (as quotients of sections of some line bundle) is a meromorphic map $\pi : X \dashrightarrow Y$ with

$$(1) \qquad\qquad \pi^*\mathfrak{M}(Y) \cong \mathfrak{M}(Y) \cong \mathfrak{M}(X).$$

We refer to this map as *an* algebraic reduction. It is defined by (1), surjectivity, and the Moishezon property of the base, and is unique only up to *bimeromorphic* maps.

It is a basic result that *at most finitely many hypersurfaces in* $\mathcal{H}(X)$ *are transversal to* π, i.e., with finitely many exceptions, if H is an irreducible hypersurface in X, then $\pi(H)$ is a hypersurface in Y (see [54]).

For almost homogeneous spaces, using the cycle space along with an analysis of the albanese fibration, E. Oeljeklaus (unpublished) has proved the existence of an equivariant reduction.

Proposition 3.1.1 *Let X be almost homogeneous with respect to G. Then there exists a G-almost homogeneous projective algebraic manifold Y and a G-equivariant meromorphic map $\pi : X \dashrightarrow Y$ which is an algebraic reduction.*

It should be noted that if $dim(X) = 2$, i.e., X is a *surface*, and $a(X) = 0$ *or* 1, then the algebraic reduction is a holomorphic map and is unique up to biholomorphic morphisms. For $a = 0$ this is obvious, and, if in the case $a = 1$ a reduction had a point of indeterminacy, then X would contain a curve with positive self-intersection number. This in turn yields an ample bundle on X, contrary to $a = 1$.

If X is a surface with $a = 2$, then it is projective algebraic. This is not true in higher dimensions. For example, there are almost homogeneous 3-folds, e.g., with respect to $G = (\mathbb{C}^*)^3$, which are Moishezon but are not projective algebraic.

Let us take a quick look at the classification (due to Potters ([49]), see also [27]) of almost homogeneous surfaces from the point of view of hypersurfaces. Here we only consider the case where X is minimal in the sense that it contains no curves with self-intersection number -1.

3.2 Almost homogeneous surfaces

Let X be a compact surface and suppose that V_1, V_2 are holomorphic vector fields on X which are generically independent. Then $\sigma := V_1 \wedge V_2$ is a non-trivial anti-canonical section. If $\{\sigma = 0\} = \emptyset$, then $\mathfrak{g} = ((V_1, V_2))_{\mathbb{C}}$ *is* the Lie algebra of global holomorphic fields on X and the associated group G acts transitively on X, i.e., $X = G/\Gamma$, where Γ is discrete.

Now let $((\omega_1, \omega_2))$ be a basis of the invariant holomorphic 1-forms and note that by Stokes' Theorem

$$\int |d\omega_j|^2 = \int d\omega_j \wedge d\bar{\omega}_j = 0, j = 1, 2.$$

Hence $d\omega_j = 0$, $j = 1, 2$, G is abelian, and X is a torus. Thus, if X is almost homogeneous and is not a torus, then $\Gamma(X, K^{-1}) \neq 0$. Therefore, X is either a ruled surface or in Kodaira's class VII.

If X is ruled and $b_1(X) = 0$, then it is a *Hirzebruch surface* and all such surfaces are almost homogeneous.

In general a compact almost homogeneous manifold X is an equivariant bundle over $Alb(X)$ (see e.g., [27]). In the 2-dimensional case it can be shown that such a bundle is almost homogeneous if and only if X is the compactification of a topologically trivial line bundle over its 1-dimensional albanese. Of course ruled surfaces are projective algebraic. Thus, there are plenty of hypersurfaces.

If X is in *Class VII*, then one must use the algebraic reduction along with special details of the Kodaira classification to show that X is a Hopf surface, i.e., the universal cover of X is $C^2 \setminus \{0\}$. The fundamental group $\pi_1(X)$ contains a central subgroup $\Gamma \cong \mathbb{Z}$ which is of finite index and is generated by a contraction $\gamma \in Aut(C^2) \setminus \{0\})^5$.

In the case $a(X) = 1$ the contraction γ is linear, $(z, w) \mapsto (\lambda z, \lambda w), |\lambda| \leq 1$, and the algebraic reduction realizes X as a principal torus bundle over $\mathbb{P}_1$. The only irreducible hypersurfaces in X are the fibers of this bundle.

Since in general there are only finitely many hypersurfaces transversal to the algebraic reduction, if $a(X) = 0$, then $\mathcal{H}(X)$ is finite. In the almost homogeneous case there are at most two curves in X. The case of two curves occurs only when X is a special equivariant compactification of a Cousin group $G = \Omega$. Here the curves correspond to the *ends* of G.

An interesting example of the case where X contains only one curve occurs when γ is *non-linear*:

$$\gamma(z, w) := (\alpha^m x + \beta y^m, \alpha y),$$

where $\alpha, \beta \in \mathbb{C}$ with $0 < |\alpha|, |\beta| < 1$ and $m \in \mathbb{N}$. Let Γ be the group generated by γ and $X := \mathbb{C}^2 \setminus \{0\}/\Gamma$. The group $G := Aut_{\mathcal{O}}(X)^0$ is the centralizer of γ in $Aut_{\mathcal{O}}(X)$: $G = \mathbb{C}^* \times \mathbb{C}$, *where* $(\alpha, \beta) \in G$ *acts as* γ *above.*

Note that the axis $\{y = 0\} = \widetilde{E}$ is Γ-stable and G acts transitively on its complement. As a result $E = \widetilde{E}/\Gamma$ and $\Omega \cong \mathbb{C}^* \times \mathbb{C}^*$. Of course X contains no other curves, because if a curve has non-trivial intersection with the open orbit, then $card\mathcal{H}(X) = \infty$ and $a(X) \geq 1$.

---

[5] The contractions were classified in 1911 by M.S.Lattes [35]

3.3   Rational surfaces

For the purpose of applications in the following sections and for general information for the reader, we mention here some concrete properties of almost homogeneous rational surfaces.

Let X be a rational almost homogeneous surface. If it is minimal, then X is either $\mathbb{P}_1 \times \mathbb{P}_1$, $\mathbb{P}_2$, or a Hirzebruch surface Σ_n, $n \geq 2$. Of course $\mathbb{P}_1 \times \mathbb{P}_1$ (*resp.*, $\mathbb{P}_2$) is homogeneous with respect to $Sl_2 \times Sl_2(resp., Sl_3)$. The surface Σ_n is obtained as a compactification of the hyperplane section bundle H^n over $\mathbb{P}_1$ by adding the *section at infinity*. The standard action of Sl_2 on $\mathbb{P}_1$ lifts to H_n and therefore we have an action of Sl_2 on Σ_n.[6] This has three orbits:

i. The open orbit, which fibers via $\pi : \Sigma_n \longrightarrow \mathbb{P}_1$ as the associated principal $\mathbb{C}^*$-bundle,

ii. $E_\infty \cong \mathbb{P}_1$, which is a section with $E_\infty \cdot E_\infty = -n$,

iii. E_0, which is a section with $E_0 \cdot E_0 = +n$.

The case of the *quadric* $\Sigma_0 = \mathbb{P}_1 \times \mathbb{P}_1$ is special. For our purposes it is important to consider it to be equipped with the diagonal Sl_2-action. Let σ be the involution defined by $(p, q) \mapsto (q, p)$. The quotiont $\Sigma_0/((\sigma))$ is just $\mathbb{P}_2$ equipped with the Sl_2-action coming from the adjoint representation.

Let $B_+(resp., B_-)$ be the group of upper-(resp., lower-) triangular matrices in Sl_2 and let $T := B_+ \cap B_- \cong \mathbb{C}^*$ be the group of diagonal matrices. When studying Sl_2-actions, it is often useful to *linearize the T-action in neighborhoods of its fixed points*.

a. Linearization at fixed points

Let G be a Lie group of holomorphic transformations of a complex space X. If G has a fixed point $x_0 \in X$, then it has a natural linear representation on the Zariski tangent space $T_{x_0}X$. This is not necessarily faithful, e.g., in the case of the group of translations of $\mathbb{P}_1$, $z \mapsto z + a$, it is trivial.

If G is reductive, i.e., $G = K^{\mathbb{C}}$ is the complexification of a maximal compact subgroup, the tangent representation is in fact faithful. Furthermore, there is a holmorphic coordinate chart U at x_0 in which K *is* acting linearly, and, for any $g \in G$, there exists $\widetilde{U} = \widetilde{U}(x_0) \subset U$ so that $g(\widetilde{U}) \subset U$ and the action of g is defined by the induced linear representation of $G([33])$.

We apply these considerations to a $\mathbb{C}^*$-action on a smooth complex manifold X. In *linear* coordinates this is given by

$$\alpha(z_1, \dots, z_n) = (\alpha^{m_1} \cdot z_1, \dots, \alpha^{m_n} \cdot z_n),$$

where x_0 corresponds to $0 \in \mathbb{C}^n$. If $m_j < 0$ (*resp.*, $m_j > 0$) for all j, we refer to x_0 as an *attracting (resp., repulsing)* fixed point. Otherwise we refer to it as being *mixed*.

b. The T-action on the quadric

---

[6]For details on the full automorphism group see e.g. [27].

The quadric $\Sigma_0 = \mathbb{P}_1 \times \mathbb{P}_1$ fibers equivariantly over both factors, each of which has two T-fixed points: "0" and "∞". Of course one is attracting and the other is repulsing, and this characterizes the nature of the T-action near its four fixed points in Σ_0: Two fixed points are contained in the diagonal and of these one is repulsing and the other is attracting. The other two fixed points are $(+,-)$-mixed, i.e., $\alpha(z,w) = (\alpha^n \cdot z, \alpha^m \cdot w)$ with $n > 0$ and $m < 0$. In fact $n = 2$ and $m = -2$

In the quotient $\mathbb{P}_2 = \Sigma_0/((\sigma))$, there are only three fixed points, since the mixed ones are identified by σ .

c. The T-action on $\Sigma_n, n \geq 1$

The $\mathbb{P}_1$-bundle $\Sigma_n \longrightarrow \mathbb{P}_1$ is of course T-equivariant with T acting in the usual way on the base. Thus, the only T-stable fibers are those over "0" and "∞" and T has two fixed points in each of these fibers. The two in E_0 are mixed, one in E_∞ is repulsing, and the other in E_∞ is attracting.

It is useful to introduce the following notation: The attracting fixed point q, which is in fact also a B_--fixed point, is called the *distingiushed* T-fixed point in Σ_n. The section E_∞ *(resp., E_0)* is referred to as the *outer (resp., inner) Sl_2-orbit*. The unique attracting T-fixed point in the quadric Σ_0 and in its quotient $\Sigma_0/((\sigma))$ is also called distinguished.

3.4 Higher-dimensional almost homogeneous spaces

From the point of view of analytic hypersurfaces the study of compact almost homogeneous manifolds X, $dim(X) \geq 3$, is far from complete. There are, however, some general results. For example, *the smaller E is, the larger the algebraic dimension*:

Proposition 3.4.1 *([27])* $a(X) \geq codim(E) - 1$.

In case E contains an *isolated point* or contains a *curve as an isolated component*, there are complete classification results ([1],[27],[28],[36]) which show that, except for simple Hopf-type constructions, X is projective algebraic.

It would seem that if all *building blocks* of the manifold X are of an algebraic nature, then X should also be of an algebraic nature.

Question Let X be a compact complex manifold which is almost homogeneous with respect to a complex Lie group G. Suppose that G is linear algebraic and that all isotropy groups $G_x, x \in X$, are algebraic subgroups. Is X Moishezon?

In the following sections, as a step in the direction of a classification theory in the 3-dimensional case, we will show that the answer to this *Question* is positive in the case where $dim(X) = 3$.

Under geometric conditions on the orbits or under group-theoretic restrictions, it is possible to obtain more information on the algebraic nature of X. For example, we mention the following observation of C. Gellhaus ([18]).

Proposition 3.4.2 *Suppose that the open orbit Ω is acyclic as a topological space. Then X is equivariantly bimeromorphically equivalent to $\mathbb{P}_n$ with Ω equivalent to $\mathbb{A}^n$.*

Even if the group is of a simple algebraic nature, e.g., $G = (\mathbb{C}^*)^2$, as we have seen with the examples of Hopf surfaces defined by non-linear contractions, X may be highly non-algebraic and may have very few hypersurfaces. In this regard, in the case of abelian groups, interesting phenomena are discussed by F. Lescure ([37]). It should be mentioned that one can construct pathology for any algebraic group, e.g., G is an algebraic group acting algebraically on the open orbit and $\mathcal{H}(X)$ is finite. However, it seems that the group $(\mathbb{C}^*)^n$ always plays a fundamental role.

3.5 Almost homogeneous 3-folds

In any problem involving homogeneous or almost homogeneous spaces it is prudent to begin by looking at all possible equivariant fibrations (see [27] for a unified approach). Certain manifolds can only be fibered in trivial ways, e.g., the Hopf surfaces with $a(X) = 0$, and must therefore be handled by other methods. In the case where X is a 3-dimensional compact almost homogeneous manifold, the most interesting case where no fibrations are available is that where $G = Sl_2$. Here $\Omega_X \cong Sl_2/\Gamma$, where Γ is discrete.

Example Let $G := Sl_2$ act diagonally on $Y := \mathbb{P}_1 \times \mathbb{P}_1$ and let C be a 1-dimensional torus. Note that G acts on every principal C-bundle X over Y. Provided we choose the bundle appropriately, G has an open orbit $\Omega = G/\Gamma$ in X, where $\Gamma \cong \mathbb{Z}$.

Question Are there essentially different examples in the case where Γ is infinite?

Our main goal is to show that if Γ is finite, then X is Moishezon. This is a consequence of the following result.

Theorem 3.5.1 *(Characterization of Moishezon Sl_2-manifolds) Let X be a compact 3-dimensional complex almost homogeneous manifold. Assume that for all $x \in X$ the isotropy $Iso_{Sl_2}\{x\}$ is an algebraic subgroup of Sl_2. Then X is Moishezon and the Sl_2-action is algebraic.*

We prove this result in the following section. Before doing so, we derive here the main consquence.

Corollary 3.5.2 *Let X be a compact complex manifold which is almost homogeneous with respect to $Sl_2(\mathbb{C})$. Suppose that the isotropy groups in the open orbit are finite. Then X is Moishezon.*

Proof Note that, since the isotropy group Γ in the open orbit Ω is finite, it follows that Ω has one end as a topological space. Therefore $E = X \backslash \Omega$ is connected.

Let V_1, V_2, V_3 be three independent vector fields coming from the Sl_2-action. It follows that E is defined by the vanishing of $\sigma := V_1 \wedge V_2 \wedge V_3$. In particular the irreducible components of E are all 1-codimensional. From the classification of almost homogeneous surfaces we see that either E is itself a Hopf surface, which is homogeneous under the Sl_2-action, or *every* Sl_2-isotropy group is algebraic.

If E is a homogeneous Hopf surface, then $b_2(E) = 0$ and consequently the normal bundle N_E is topologically trivial. Hence, since we can topologically embed a tubular neighborhood U of the 0-section of N_E as an open neighborhood of E in X, it follows that $b_1(\Omega \cap U) = 2$. Since Γ is finite, this is impossible. $\qquad\square$

Remarks 1) *This result is the main point of the Bochum Dissertation of M. Martin ([41]).*
2) *There is a combinatorial classification of Moishezon manifolds which are almost homogeneous with respect to algebraic Sl_2-actions ([43]). Such manifolds are not always projective algebraic.*

3.6   Characterization of Moishezon Sl_2-manifolds

Let X be a 3-dimensional compact complex manifold which is almost homogeneous with respect to a holomorphic Sl_2-action. Recall that a subgroup $B \subset Sl_2$ is a Borel subgroup if and only if it is conjugate to B_+. Here we give a proof of *Theorem 3.5.1* which characterizes the Moishezon case. We begin with some preparation and reformulation.

Proposition 3.6.1 *The manifold X is Moishezon if and only if, for every Borel group B, every B-orbit is Zariski open in its closure.*

Proof If X is Moishezon, then there is an equivariant algebraic reduction $\pi : X \to Y$ onto a 3-dimensional projective algebraic variety where Sl_2 acts as an algebraic group (*Prop. 3.1.1*). Since every B-orbit in Y is Zariski open in its closure and π is birational, it follows that every B-orbit in X is Zariski open in its closure as well.

Conversely, let $\pi : X \to Y$ be an algebraic reduction. If the generic π-fiber F were positive-dimensional, then we could find a Borel group B and a B-orbit which is transversal to F at some point of $F \cap \Omega$. This is also true for all nearby B-orbits in Ω. Since the B-orbits are Zariski open in their closures, we produce in this way infinitely many hypersurfaces in X which are transversal to F. Thus the generic fiber of an algebraic reduction must be 0-dimensional and consequently X is Moishezon. $\square$

Since conjugation does not change any phenomena and, since the above *Proposition* can be strengthened to allow a few exceptions, in order to prove *Theorem 3.5.1* it is enough to prove the following result of M. Martin ([41]).

*Sl_2-**Lemma*** *With possibly countably many exceptions, every B_+-orbit is Zariski open in its closure.*

Remark *Of course after this Lemma is proved, it follows that X is Moishezon. Thus, there are no exceptions.*

The proof of the *Sl_2-Lemma* follows from a detailed analysis of the behavior of the B_+-orbits near T-fixed points in E. We begin with preparations for the proof by recalling that, after desingularizing, an irreducible component F of E is one of the following (see Sections 2 and 3):

   i. $\mathbb{P}_1 \times \mathbb{P}_1$ with diagonal Sl_2-action; $\mathbb{P}_2$ with the Sl_2-action induced by the adjoint represention.

ii. A Hirzebruch surface $\Sigma_n, n = 1, 2, \ldots$.

iii. A product $P = \mathbb{P}_1 \times C$, where C is a smooth curve and Sl_2 acts on P by only acting on the first factor.

Remark 1) *By looking at the possible 3-dimensional representations of Sl_2, one shows via a linearization argument that Sl_2 has no fixed point in E. Thus, for example, $\mathbb{P}_2$ with the action of Sl_2 which fixes a point can not occur.*
2) *In addition to the usual notion, we refer to any B_--fixed point in a component of type P in $iii.$ as being distinguished. Thus every component of E contains distinguished fixed points.*

Note that in $i.$ and $ii.$ B_+ has an open orbit in F. Thus, if a closure $\overline{B_+(x)}$ has non-empty intersection with this open orbit, then $\overline{B_+(x)} \supset F$. Furthermore, if $B_+(x)$ is not Zariski open in its closure, then $\overline{B_+(x)} \backslash B_+(x)$ must be bigger than a 1-dimensional analytic set. Hence, either $\overline{B_+(x)}$ contains a component F of E as above, or there is a component P which is of type $iii.$ and $\overline{B_+(x)}$ contains infinitely many of the $\mathbb{P}_1$'s. In either case we have a *distinguished* B_--fixed point q in $\overline{B_+(x)} \cap E$.

Let (x, y, z) be coordinates at q such that the T-action is linearized in the sense of Section 1.a.. We choose the x-axis to be the direction of the closed Sl_2-orbit at q in E. Consequently,

$$\alpha(x, y, z) = (\alpha^2 \cdot x, \alpha^k \cdot y, \alpha^\delta \cdot z).$$

If $q \in P$, then we take the y-axis to be the direction of the T-fixed point set in P, i.e., in this case the action is given by $(x, y, z) \mapsto (\alpha^2 \cdot x, y, \alpha^\delta \cdot z)$. In the other cases we may take $k > 0$, since q is distinguished. Note, however, that the y-axis may not lie in F.

Proposition 3.6.2 *Let $x \in \Omega$ and let $q \in E$ be a distinguished T-fixed point in $\overline{B_+(x)}$. Suppose that $\delta \geq 0$. Then there exists a neighborhood $U = U(q)$ so that $B_+(x) \cap U$ is a connected subset of $U \backslash E$.*

Proof   Since all exponents are non-negative, we may choose U so that if $(x, y.z) \in U$, then $\alpha(x, y, z) \in U$ for $|\alpha| \leq 1$.
  If $\delta > 0$, then

$$\lim_{\alpha \longrightarrow 0} \alpha(x, y, z) = (0, y, 0)$$

$(resp., = (0, 0, 0) = q)$ if $k = 0$ $(resp., k \neq 0)$. If $\delta = 0$, then $k > 0$ and

$$\lim_{\alpha \longrightarrow 0} \alpha(x, y, z) = (0, 0, z).$$

Take $p_1, p_2 \in U$ and $\alpha, \beta \in T$ with $|\alpha|, |\beta|$ arbitrarily near 0. Let u_t be the transformation in $Aut_\mathcal{O}(X)$ which corresponds to the unipotent matrix $\left(\begin{smallmatrix} 1 & t \\ 0 & 1 \end{smallmatrix}\right)$. The idea is to choose α, β and a curve u_t appropriately so that, for $|t| \leq |t_0|$, $u_t(\alpha(p_1)) \subset U$ and $u_{t_0}(\alpha(p_1)) = \beta(p_2)$.
  If p_j corresponds to the matrix $\left(\begin{smallmatrix} \alpha_j & \mu_j \\ 0 & \alpha_j^{-1} \end{smallmatrix}\right)$, $j = 1, 2$, then $t_0 := (\alpha\alpha_1)(\beta\mu_2 - \alpha\mu_1)$ is the desired value. It is therefore enough to determine $U' \subset U$ so that

a) $\alpha(U) \subset U'$ for $|\alpha|$ small enough, and

b) $u_t(U') \subset U$ for $|t| \leq |t_0|$.

If q is attracting, i.e., $k, \delta > 0$, then it is obvious that such a set exists. If the T-fixed point set is 1-dimensional, e.g., the y-axis, then U' must be taken to be an appropriate neighborhood of the T-fixed point set in U. $\qquad\square$

Proposition 3.6.3 *Under the assumptions of Prop. 3.6.2 there exits an open neighborhood $U = U(q)$ so that $\overline{B_+(x)} \cap U$ is a closed analytic subset of U.*

Proof Let $p \in U$ as above and $C := \overline{\{\alpha(p) : |\alpha| \leq 1\}}$ be the *closure of the orbit of p*, $|\alpha| \leq 1$. Take W to be a neighborhood of the identity in B_+ so that the restriction of the action map

$$\varphi : W \times C \longrightarrow X$$

has image in U. Let $q_0 := \lim_{\alpha \to 0} \alpha(p)$. Now, the holomorphic map φ has constant fiber-dimension equal to 1. Hence there is a neighborhood $V = V(e, q_0) \subset W \times U$ and $U' \subset U$ so that $\varphi(V)$ is a closed analytic subset of $U' = U'(q_0)$.

Now $A := \varphi(V) \backslash E$ is a closed analytic subset of $\Omega \cap U$ and is therefore a component of $B_+(x) \cap U$. By *Prop. 3.6.2* we may choose U' smaller so that $B_+(x) \cap U'$ is connected, i.e., we may assume that $A = B_+(x) \cap U'$ and consequently $\overline{B_+(x)} \cap U' = \varphi(V)$ is closed. But if $\overline{B_+(x)} \cap U$ were not closed for U sufficiently small, then there would exist such a point q_0 where it is not locally closed, contrary to the above argument. $\qquad\square$

We now turn to the case $\delta < 0$ and again carry out an argument via linearization at a T-fixed point.

Consider a sequence $\{p_n\} \subset U \cap B_+(x)$ and assume that $p_n \longrightarrow q \in U \cap E$. We may choose linear coordinates with $q = (0, 0, 0)$ and $p_n = (x_n, y_n, z_n)$. Take $\{\alpha_n\}$ with $\alpha_n \longrightarrow 0$ so that

$$\alpha_n^\delta \cdot z_n = r$$

is fixed for all $n \in \mathbb{N}$. Since $\delta < 0$, this is possible. Furthermore, note that in local coordinates $\alpha_n(p_n) \longrightarrow (0, 0, r)$. The only condition is $|r| < \epsilon$ so that

$$Z := \{(0, 0, r) : |r| < \epsilon\} \subset U.$$

We refer to Z as *an open piece of the z-axis* and summarize this as follows.

Lemma 3.6.4 *If $\delta < 0$ and $q \in \overline{B_+(x)}$ is a distinguished T-fixed point, then $\overline{B_+(x)}$ contains an open piece of the z-axis at q.*

The following result allows us to restrict our attention to the Hirzebruch surfaces in the complement E of the open orbit.

Proposition 3.6.5 *Suppose that uncountably many different orbits $B_+(x_\alpha), \alpha \in A$, are not Zariski open in their closures. Then there is a component F of E, which after desingularization, is a Hirzebruch surface, with $F \subset B_+(x_\alpha), \alpha \in A'$, where A' is an uncountable subset of A.*

Proof Suppose that uncountably many of the closures $\overline{B_+(x_\alpha)}$ intersect a component P of type iii in infinitely many $\mathbb{P}_1$'s. Then there exists a $\mathbb{P}_1$–call it Q– and an uncountable set $A' \subset A$ with $\overline{B_+(x_\alpha)} \supset Q$ and such that $\overline{B_+(x_\alpha)}$ is not locally closed at every point of Q for all $\alpha \in A'$.

Let $q \in Q$ be a distinguished T-fixed point. By *Prop. 3.6.3* $\delta < 0$, and therefore by *Lemma 3.6.4* $\overline{B_+(x_\alpha)}$ contains an open piece Z of the z-axis at q for all $\alpha \in A'$. In particular, since the B_+-orbits in Ω are disjoint, it follows that Z is contained in another component F of E. Since q is *not* a B_+-fixed point, $B_+(q')$ is open in F for every $q' \in Z\backslash\{q\}$. Hence $F \subset \overline{B_+(x_\alpha)}$ for all $\alpha \in A'$.

Finally, note that q is a mixed fixed point for the T-action on F and that the Sl_2-orbit of q is closed. Consequently, a desingularization of F is a Hirzebruch surface. This completes the proof in the case where uncountably many closures are not locally closed somewhere along a component of type iii..

If uncountably many closures contain a component whose desingularization is of type i., i.e., the quadric Σ_0 or the quotient $\Sigma_0/((\sigma))$, then *all* of them have a distinguished T-fixed point q in common. By the same argument as in the previous case, they all contain Z which in turn is contained in a component F whose desingularization is a Hirzebruch surface. $\square$

We are now in a position to complete the

Proof of the Sl_2-Lemma Suppose that uncountably many different B_+-orbits $B_+(x_\alpha)$ contain a component F in their closures. By *Prop. 3.6.5* we may assume the desingularization of F is a Hirzebruch surface. Let $q \in F =: F_1$ be its distinguished point. For exactly the same reasons as in the proof of *Prop. 3.6.5*, every one of these closures contains a piece Z of the z-axis at q which itself is contained in an adjacent component F_2. It then follows that F_2 is contained in every one of the closures. Of course the desingularization of F_2 is a Hirzebruch surface. So we go to its distinguished point and repeat the argument, obtaining a sequence of surfaces $F_1, F_2, \ldots$.

Observe that, at a distinguished point q_n in the *outer Sl_2-orbit* in F_n,

$$\alpha(x, y, z) = (\alpha^2 \cdot x, \alpha^k \cdot y, \alpha^\delta \cdot z),$$

where $k > 0$ and $\delta < 0$. Since q_n is attracting in F_n, the xy-plane agrees with F_n in the linearization and, except for the z-axis, no other T-orbit in X contains q_n in its closure. Hence,

    * the only components of E at q_n are F_n and F_{n+1}.

Now the chain of Hirzebruch surfaces $F_1, F_2, \ldots$ must be infinite, because otherwise some open piece Z of the z-axis would be contained in the open orbit Ω. This is contrary to the orbits $B_+(x_\alpha)$, $\alpha \in A$, being disjoint. Thus, for some N, it follows that $F_1 = F_N$, i.e., we have a *ring* of components. Therefore, from (*) and the fact that E is connected, it follows $E = F_1 \cup \ldots \cup F_N$. Arguing as we did above with the signs of $k\ \delta$, and noting that they are just reversed at the repulsing fixed points, we see that, except for the orbits in E, no T-orbits have points of E in their closures. This however is impossible, because the T-orbits in Ω are closed and non-compact in Ω! Therefore the assumption that there are uncountably many B_+-orbits which are not Zariski open in their closures has led us to a contradiction and the proof is complete. $\square$

4   Subvarieties of higher codimension

We refer to a complex manifold as being *group-theoretically parallelizable* if it is of the form $X = G/\Gamma$, where G is a complex Lie group and Γ is a discrete subgroup. In this chapter we discuss compact subvarieties in such manifolds. In particular we describe the main results of [32]. As motivation we begin with several examples where group-theoretically parallelizable manifolds play an important role.

Example 1. Let X be a complex manifold which is homogeneous under the action of a *real* Lie group G of holomorphic transformations. Then we have the *hypersurface reduction* $h :$ $X \longrightarrow Y$ which is an equivariant holomorphic fiber bundle $G/H \longrightarrow G/J$. Let $F = J/H$ be the fiber. It can be shown that H^0 is normal in J and that $F = J/H^0/H/H^0 = L/\Gamma$, where L is a *complex Lie group acting holomorphically on F with discrete isotropy* ([45]).

Example 2. Let $X = G/H$, where G is a $\mathbb{C}$-algebraic group and H is a Zariski dense complex subgroup. Since the adjoint representation of G is algebraic, if J is any connected subgroup in G, then the normalizer of J in G is algebaic. Hence $H^0 \lhd G$, and, under the mild assumption that G is acting almost effectively, it follows that $H = \Gamma$ is discrete.

Example 3. Let X be an n-dimensional compact complex manifold and assume that the complex tangent bundle TX is holomorphically trivial, i.e., X is parallelizable in the sense of complex geometry. Since X is compact, the holomorphic vector fields can be globally integrated, and therefore the triviality of TX implies that $G := Aut_{\mathcal{O}}(X)^0$ acts transitively with discrete isotropy.

4.1   Existence and non-existence of subvarieties

The following remark reflects the *non-existence* side of the subject.

Proposition 4.1.1 *Let Z be a compact Kähler manifold which is a subvariety of a group-theoretically parallelizable complex manifold $X = G/\Gamma$. Then, for every smooth point $z \in Z_{reg}$, the tangent space $T_z Z$ is an abelian Lie subalgebra of the Lie algebra $\mathfrak{g}$ of G.*

Proof   Let $i : Z \hookrightarrow X$ be the natural injection. For $v, w \in T_z Z$ there are uniquely determined invariant holomorphic vector fields $\alpha, \beta \in \mathfrak{g} = \Gamma(X, TX)$ such that $\alpha(i(z)) = i_*(v)$ and $\beta(i(z)) = i_*(w))$. Let $\omega \in \mathfrak{g}^* = \Gamma(X, TX^*)$ be an invariant holomorphic 1-form. Since Z is Kählerian, $di^*\omega(z) = 0$. Hence,

$$0 = d\omega(z)(i_*(v), i_*(w)) = d\omega(\alpha, \beta) = 2^{-1}\omega([\alpha, \beta]).$$

Since this is true for all $\omega \in \mathfrak{g}^*$, it follows that $[\alpha, \beta] = 0$ for all $v, w \in T_z Z$. $\square$

Remark *Since the property all holomorphic forms are closed is a bimeromorphic invariant, the Proposition is more generally valid for varieties Z in Fujiki's $Class(C)$. ([17])*

Question Using the abelian Lie algebras $T_z Z$, is it possible to have some control over a *hull of Z in X*?

In general it is very difficult to prove *existence* results for subvarieties Z in X, e.g., even in the case of hypersurfaces in tori! However, in the non-abelian case, if X is compact, then certain complex subgroups have closed orbits. For example, we have the following observation (see [32]).

Proposition 4.1.2 *Let Γ be a discrete cocompact subgroup of a complex Lie group G. Let $B \subset G$ be a complex abelian subgroup with $B\Gamma$ closed. Then there exists a maximal abelian subgroup $A \supset B$ with $A\Gamma$ closed, i.e., the A-orbit of the neutral point in $X = G/\Gamma$ is a closed submanifold in X.*

Remark *Except in the case where G is itself abelian, an application of the Proposition for $B = \{e\}$ yields proper subvarieties.*

If X is a complex space, then $\mathfrak{M}(X)$ denotes the set of meromorphic functions on X. It is often the case that $\mathfrak{M}(X) \cong \mathbb{C}$, even when X has many subvarieties. This is not the case for tori.

Corollary 4.1.3 *Let X be a compact complex-parallelizable manifold. Then X has no subvarieties if and only if it is a simple torus* [7] *with $\mathfrak{M}(X) \cong \mathbb{C}$.*

Proof By the above *Proposition 4.1.2* it is enough to prove that a simple torus with $\mathfrak{M}(X) \cong \mathbb{C}$ has no positive-dimensional proper subvarieties. Assuming that Z is such a subvariety, we produce a hypersurface. This is enough, because a hypersurface separable compact torus is projective algebraic (see *Thm. 2.2.1*)

We may assume that Z is of minimal positive dimension and is irreducible. Of course Z is non-singular. Furthermore, assuming $dim_{\mathbb{C}}(Z) \geq 2$, if $i : Z \longrightarrow X$ is the natural injection, for every holomorphic 1-form ω, the restriction $i^*(\omega)$ either vanishes identically or nowhere. Thus $T_p Z = T_q Z$ for all $p, q \in Z$. It follows that Z is a linear subtorus, contrary to X being simple. Hence Z is a curve.

If $A, B \subset X$ are subvarieties, then, by the *Proper Mapping Theorem*, the sum $A + B$ is likewise a subvariety. Let $n := dim_{\mathbb{C}}(X)$ and let H be the subvariety defined by adding Z to itself $(n - 1)$-times. Then H is the desired hypersurface. $\square$

Remarks 1) *It should be noted that the same argument as in the proof of Cor. 4.1.3 shows that a submanifold of a torus which, as an abstract complex manifold, is a torus, is embedded as a linear subtorus.*
2) *Using the fibration theory developed in ([27]), it is possible to prove the following more general statement: Let X be a homogenous space of a complex Lie group, and assume that it has no subvarieties. Then X is either a simple torus or $X = S/\Gamma$, where S is semi-simple and Γ is discrete and Zariski dense.*

---

[7] A torus is called simple if it has no linear subtori.

In general, if X is a complex manifold, it is of course interesting to know which global objects *live* on X. For example, what are the vector bundles on X? In this regard, J. Winkelmann ([57]) has proved the following result.

Existence of vector bundles *Every compact complex manifold possesses a non-trivial holomorphic vector bundle.*

4.2 The induced Iitake-fibration

Let Z be a compact complex manifold. Recall that the *Kodaira-dimension*, $\kappa(Z)$, is defined to be $-\infty$, if no power of the canonical bundle has a non-zero section, and otherwise is defined to be the transcendence degree of the quotient field of the canonical ring. Since $\kappa(Z)$ is a bimeromorphic invariant, it is possible to discuss this notion on irreducible reduced complex spaces, e.g., such a space is said to be of *general type* if $\kappa(X) = dim(X)$.

Using the canonical ring, it is possible to construct a meromorphic *Iitake-reduction* (see [54]):

Theorem 4.2.1 *Let Z be an irreducible reduced complex space. Then there exists a mero-morphic map $\psi : Z \dashrightarrow Y$ onto a compact complex space Y with $dim_{\mathbb{C}}Y = \kappa(Z)$. There is a subset U of Z such that*

  i. U is the complement of at most countably many nowhere dense analytic subsets

  ii. For all $u \in U$ the fiber $F_u = \psi^{-1}(\psi\{u\})$ is irreducible and smooth with $\kappa(F_u) = 0$.

Our goal here is to give a description of the *Iitake-reduction* for a compact subvariety Z of a group-theoretically parallelizable manifold $X = G/\Gamma$.

Lemma 4.2.2 *Let Z be a compact irreducible subvariety of $X = G/\Gamma$. Then $\kappa(Z) \geq 0$ and $\kappa(Z) = 0$ if and only if there is a closed complex subgroup L in G such that Z is an L-orbit.*

Proof Let $p \in Z_{reg}$. If $k := dim(Z)$, then consider the evaluation map $\epsilon : \bigwedge^k \mathfrak{g}^* \longrightarrow \bigwedge^k T_p Z^*$. This is obviously not the zero-map. Therefore $\kappa(Z) \geq 0$.

If $T_p(Z) \neq T_q(Z)$ as subspaces of $\mathfrak{g}$, then there exists $\omega \in \bigwedge^k \mathfrak{g}^*$ with $\epsilon_p(\omega) \neq 0$ and $\epsilon_q(\omega) = 0$ and vice versa. Hence, if $\kappa(Z) = 0$, then, for all $p, q \in Z_{reg}$, $T_p(Z) = T_q(Z)$. Consequently, if the orbit of a 1-parameter subgroup H of G is tangent to Z at $p \in Z_{reg}$, then $H(p) \subset Z$. It follows that $L := Stab_G(Z)$ has an open orbit in Z. But all L-orbits have the same dimension. Thus Z is an L-orbit. $\square$

In the case of a torus, i.e., G abelian, it is intuitively clear how to realize an Iitake-reduction of Z by an equivariant fibration of the ambient space X. It follows from *Lemma 1* that every fiber $F_u, u \in U$, is the orbit of a linear torus subgroup. Since there are only countably many such tori, there is a single such torus T so that the restriction of the fibration $X \longrightarrow X/T$ to Z generically agrees with the given Iitake-reduction (see [54]) for details).

This proof can in fact be carried over to the non-abelian case. However, it is not completely straight-forward (see [32]):

Proposition 4.2.3 *Let G be a complex Lie group and Γ a discrete subgroup. Then there exist at most countably many connected Lie subgroups H in G with $H/H \cap \Gamma$ compact.*

Remark *This statement is not true for real Lie groups.*

We now proceed to the formulation of the result on the *induced Iitake-fibration*. Let $X = G/\Gamma$ be group-theoretically parallelizabe and let Z be an irreducible subvariety of X. We may assume that G is simply-connected so that the map $\pi : G \longrightarrow X, g \mapsto g\Gamma$, gives the universal covering. Furthermore, we may also assume that Z contains the neutral point in X and define Z_0 to be the connected component of $\pi^{-1}(Z)$ which contains the identity.

Regard G to be acting on itself on the right. Let $H =: Stab_G(Z_0)^0$ and $\Gamma_0 := Stab_\Gamma(Z_0)$. We may regard Z to be a subvariety of G/Γ_0 and, since $\Gamma_0 \subset Stab_G(Z_0)$, we have the fibration $G/\Gamma_0 \longrightarrow G/\Gamma_0 H$ which induces the quotient $q : Z \longrightarrow W := Z/H$. This quotient is the desired Iitake-reduction.

Theorem 4.2.4 *([32]) The fibration $q : Z \longrightarrow Z/H := W$ is an Iitake-reduction onto a complex space of general type. Furthermore, modulo the discrete ineffectivity of its action on G/Γ_0, H can be identified with $Aut_\mathcal{O}(Z)^0$.*

Proof Let $\psi : Z \to Y$ and $U \subset Z$ be as in *Theorem 4.2.1*. Let $\pi_0 : G \longrightarrow G/\Gamma_0$ be the natural map. For $z \in Z$ let F_z be the ψ-fiber whenever it is defined.

For $\pi_0(g) \in U$ it follows that $\kappa(F_{\pi_0(g)}) = 0$ and therefore $g^{-1}F_{\pi_0(g)}$ is an orbit of a connected group I_g. Since U is the complement of at most countably many nowhere dense analytic sets, it follows from *Theorem 4.2.4* that $I := I_g$ does not depend on the fiber.

We may assume that the neutral point $e\Gamma_0 \in U$. Hence Γ_0 normalizes I. Furthermore, I stabilizes U. Thus the right-action of I stabilizes Z and $q_1 : Z \longrightarrow Z/I$ is a fibration onto a complex space V.

We claim that V is of general type. To show this it is enough to show that its normalization is of general type. Since the I-action on Z lifts to the normalization, it is therefore enough to handle the case where Z (and therefore W) is normal.

Now let $((Y_1, \ldots , Y_k))$ be a basis of the holomorphic vector fields on Z which are induced from the H-action. For $\omega \in \Gamma(Z, K^m)$ let $C(\omega)$ be the contraction of ω with $Y_1 \wedge \ldots \wedge Y_k$.

For $w \in W_{reg}$ and $v \in T_w W$, let Y_v be an H-invariant field tangent to Z along the fiber over w. Since the difference of two such *lifts* is a linear combination of the Y_i's, the contraction of v with $C(\omega)$ can be defined by contracting $C(\omega)$ with Y_v. In this way we define a map $C : \Gamma(Z, K^m) \longrightarrow \Gamma(V, K^m)$ which is the inverse of the pull-back map defined by q_1. In particular C is an isomorphism, and, since $\kappa(Z) = dim V$, it follows that V is of general type.

The reduction $Z \longrightarrow Z/I = V$ has connected fibers. Thus it is H-equivariant. Recall that the automorphism group of a compact complex space of general type is finite. Consequently H acts trivially on V, i.e., $H \subset I$. By definition $I \subset H$. Hence $V = W$ and W is of general type. The same argument shows that $Aut_\mathcal{O}(Z)^0$ acts trivially on the base and, modulo possible discrete ineffectivity, is equal to H. $\square$

Corollary 4.2.5 *Let $X = G/\Gamma$ be compact and let L be a connected subgroup of G. Let Z be the complex analytic closure of an L-orbit in X. Then Z is the orbit of a closed connected complex subgroup H of G.*

Proof Since the base Z/H of the induced Iitake-reduction is of general type and therefore has only finitely many automorphisms, the L-equivariance of $Z \longrightarrow Z/H$ and the density of the L-orbits imply that H acts transitively on Z. □

Note that in the above procedure we begin with G/Γ, go to G/Γ_0, and end up with $G/\Gamma_0 H$. If we don't assume that Γ is cocompact, the first step does not change the situation, i.e., the manifold remains group-theoretically parallelizable. However, since H may not be normal in G, the quotient Z/H may not be contained in a group-theoretically parallelizable manifold.

Question Let N be the normalizer of H in G. Under what conditions is the image of Z in G/N just a point?

The following gives a partial answer which is sufficient for many applications.

Proposition 4.2.6 *If G is solvable or if G is reductive and $X = G/\Gamma$ is compact, then the image of Z in G/N is a point.*

Proof In fact we show that, under these assumptions, the manifold G/N is holomorphically separable. For this, recall that via the adjoint representation G/N is a G-orbit in some projective space. If G is solvable, then it stabilizes a flag and consequently every orbit is contained in an affine space; in particular G/N is holomorphically separable.

If G is reductive and G/Γ is compact, then, since *all* elements of Γ are semi-simple, it can be shown that the Zariski closure of H is reductive (see [32] Prop. 5.1). This in turn implies that N is reductive. Thus G/N is Stein ([40]). □

Zusatz *If G is reductive and X is compact, then G/N is Stein.*

In the next sections we give several applications of the induced Iitake-fibration.

4.3 Lower estimates on codimension

Let $X = G/\Gamma$, where Γ is discrete, and let Z be a compact irreducible subvariety of X. Recall our remark at the beginning of this chapter (Prop. 1.1): *If Z is in Class(C), i.e., bimeromorphic to a Kähler manifold (see [55]), then, for every $z \in Z$, $T_z Z$ is an abelian subalgebra of* $\mathfrak{g}$. In any concrete situation, this, along with the induced Iitake-fibration, yields a lower estimate on the codimension of Z in X.

For example, if G is solvable, then $H \triangleleft G$ and the image $Z_1 = Z/H$ is contained in a solv-manifold G_1/Γ_1 of the same type. But Z_1 is of general type; in particular it is in Class(C). Thus we immediately obtain an estimate for the codimension of Z_1 in X_1 in terms of the maximal abelian subalgebras in G_1. In order to obtain more precise information, it would be useful to have answers to questions of the following type.

Question Let G be the 3-dimensional nilpotent group of upper-triangular matrices and let Γ be the subgroup of matrices whose entries are in the gaussian integers. Then $X = G/\Gamma$ is compact. Let G' be the commutator subgroup, i.e., the 1-parameter subgroup in the upper right-hand corner. Then $G'\Gamma$ is closed and we have the fibration $X = G/\Gamma \longrightarrow G/G'\Gamma = Alb(X)$. Let C be a curve in $Alb(X)$ and let Z be its preimage in X. Are there curves in Z which are mapped surjectively onto C?

In the semi-simple case it is possible to obtain concrete estimates.

Theorem 4.3.1 *([32]) Let G be a simple complex Lie group, Γ a discrete cocompact subgroup, and $X := G/\Gamma$. Let Z be a subvariety of X. Then $codim(Z) \geq \sqrt{dim(X)}$.*

Sketch of Proof By *Prop.* 4.2.6 the variety Z is contained in an N-orbit, where N is the normalizer of the group H of the induced Iitake-fibration. Now N is reductive, and $codim(N) \geq \sqrt{dim(G)}$ for every proper reductive subgroup N (see [32] Lemma 9.1). $\square$

Remarks 1) *Other estimates of this type can be found in [32].*
2) *The compact manifolds Z which can be embedded in a group-theoretically parallelizable manifold are as abstract manifolds not likely to be of a new nature. For example, if Z is in $Class(C)$, then, at every smooth point of Z, the albanese map $\alpha : Z \longrightarrow Alb(Z)$ has maximal rank.*

Proof The pull-back to a desingularization $\widehat{Z}$ of every invariant holomorphic 1-form is closed. Furthermore, given $z \in Z_{reg}$ and $v \in T_z(Z)$, there exists such form ω with $\omega(z) \neq 0$. $\square$
3) *The lack of information on existence is reflected by the following question.*
Question Let $G = Sl_2(\mathbb{C})$ and take Γ to be cocompact. Are there subvarieties of X which are not orbits?

4.4 The Bloch-conjecture in the non-abelian case

The following *Bloch Conjecture* was proved by Green and Griffiths ([20]): *Let $f : \mathbb{C} \longrightarrow X$ be a holomorphic map of the complex numbers into a compact complex torus. Then the complex analytic closure Z of $f(\mathbb{C})$ is a subtorus.* Using this result, along with the methods developed in the above sections, it is possible to prove the non-abelian analogue.

Theorem 4.4.1 *Let X be a compact complex parallelizable manifold and $f : \mathbb{C} \longrightarrow X$ a holomorphic map. Then the complex analytic closure Z of the image $f(\mathbb{C})$ is an orbit of a complex Lie subgroup of $G := Aut_\mathcal{O}(X)^0$.*

This is an immediate consequence of the following two propositions.

Proposition 4.4.2 *Under the assumptions of Theorem 4.4.1, let $Z \longrightarrow Z/H$ be the induced Iitake reduction and let N be the normalizer of H in G. Assume that G/N is holomorphically separable. Then Z is an H-orbit.*

Proof Since $f(\mathbb{C})$ is dense in the albanese image $\alpha(Z)$, the Theorem of Green and Griffiths implies that $\alpha : Z \longrightarrow Alb(Z)$ is surjective. By *Remark 2)* in the above section and the fact that, if Z is a torus, then it is embedded as an orbit (*Lemma 4.2.2*), it follows that Z is not of general type.

Since G/N is holomorphically separable, the base $W = Z/H$ is contained in an N-orbit which is also group-theoretically parallelizable. Of course the image of $f(\mathbb{C})$ in W is dense. But now W is of general type and is, in particular, in $Class(C)$. Hence, using *Remark 2)* and the Theorem of Green and Griffiths as above, it follows that W is a point,i.e., Z is an H-orbit. $\qquad\qquad\Box$

We say that G is *minimal* for Z if there is no proper subgroup $\widetilde{G}$ of G with Z contained in a $\widetilde{G}$-orbit.

Proposition 4.4.3 *Under the assumptions of Theorem 4.4.1, if G is minimal, then G/N is holomorphically separable.*

Proof Let $G = R \cdot S$ be a Levi-Malcev decomposition. Since Γ is cocompact in G, it follows that $R\Gamma$ is closed (see [52]). Thus the projection Λ_0 of Γ_0 in S is closed. Therefore, as a consequence of the *Zusatz to Prop. 4.2.6* and *Prop. 4.4.2* above, the image of Z in S/Λ_0 is an orbit. Hence, the minimality of G implies that the restriction of $G/\Gamma_0 \longrightarrow S/\Lambda_0$ to Z is surjective.

Now the homogeneous space G/N is a G-orbit via a linear representation on a projective space. Let $\overline{G}$ be the *algebraic closure* with respect to this action in the sense that $G \subset \overline{G} = \overline{R} \cdot S$.

Since $\overline{R}$ is an algebraic group, we have the fibration

$$\overline{G}/\overline{N} \longrightarrow \overline{G}/\overline{R}\cdot\overline{N} = S/I,$$

where I is an *algebraic* subgroup of S. But Λ_0 is cocompact in S and is therefore Zariski dense. Thus, since $\Lambda_0 \subset I$, it follows that $I = S$ and G/N is contained in an $\overline{R}$-orbit. Of course $\overline{R}$ stabilizes a flag and consequently such orbits are holomorphically separable. (*They are even affine.*) In particular, G/N is holomorphically separable. $\qquad\qquad\Box$

Bibliography

[1]   Ahiezer,D.N.: Dense orbits with two ends. Math. USSR, Izvestija II (1977), 293–307

[2]   Ahiezer,D.N.: Invariant analytic hypersurfaces in complex nilpotent Lie groups. Ann. Glob. Analysis and Geometry 2 (1984), 129–140

[3]   Ahiezer,D.N.: Invariant meromorphic functions on complex semi-simple Lie groups. Invent. Math. 65 (1982), 325–329

[4]   Abe,Y.: Holomorphic sections of line bundles over (H,C)-groups, Manu. Math.,60 (1988), 379-385

[5] Andreotti,A.; Gherardelli,F.: Estensioni commutative di Varieta' Abeliane. Quaderno manoscritto del Centro di Analisi Globale del CNR, Firenze (1972), 1–48

[6] Berteloot,F.: Habilitation, Universite de Lille (1993)

[7] Berteloot,F.: Fonctions plurisubharmonique sur $Sl_2(\mathbb{C})$-invariant par un sous-groupe monogène. J. d'analyse Math. 48 (1987), 267–276

[8] Berteloot,F.: Existence of a Kähler metric on semi-simple homogeneous manifolds, C.R. de L'Acad. des Sc.,Tome 305(1987), 809-812

[9] Berteloot,F.; Oeljeklaus,K.: Invariant Plurisubharmonic Functions and Hypersurfaces on Semisimple Complex Lie Groups. Math. Ann. 281 (1988), 513–530

[10] Borel,A.: Linear algebraic groups. New York, Benjamin (1969)

[11] Borel,A.; Remmert,R.: Über kompakte homogene Kählersche Mannigfaltigkeiten. Math. Ann. **145** (1962), 429–439

[12] Capocase,F.; Catanese,F.: Periodic meromorphic functions, Acta Math.,166(1991), 27-68

[13] Coeure,F; Loeb,J.-J.: A counterexample to the Serre problem with a bounded domain in $\mathbb{C}^2$ as fiber, Ann. of Math.(2)122(1985)

[14] Cousin,P.: Sur les fonctions triplement periodique de deux variables, Acta Math. 10(1910), 105-232

[15] Dieudonné,J.: Grundzüge der modernen Analysis Bd.5. Vieweg-Verlag, Braunschweig Wiesbaden 1979

[16] Demailly,J.-P.: Un example de fibré holomorphe non de Stein à fibre $\mathbb{C}$ ayant pour base le disque ou le plan. Invent. Math. 48 (1978), 293–302

[17] Fujiki,A.: On automorphism groups of compact Kähler manifolds. Invent. math. **44** (1978), 225–258

[18] Gellhaus,C.: Äquivariante Kompaktifizierungen des $\mathbb{C}^n$, MZ 206 (1991), 211-217

[19] Goodman,R.: Analytic and entire vectors for representations of Lie groups, Trans. AMS 143(1969), 55-76

[20] Green,M.; Griffiths,P: Two Applications of Algebraic Geometry to Entire Holomorphic Mappings. The Chern Symposium 1979, Springer Verlag (1988)

[21] Gilligan,B.; Oeljeklaus,K.; Richthofer,W.: Homogeneous complex manifolds with more than one end. Can. J. Math. Vol. XLI, No.1 (1989), 163–177

[22] Grauert,H.; Remmert,R.: Über kompakte homogene komplexe Mannigfaltigkeiten. Arch. Math. 13 (1962), 498–507

[23] Guillemin,V.; Sternberg,S.: Symplectic techniques in physics. Cambridge Univ. Press 1984

[24] Harvey,R.: Holomorphic Chains and their Boundaries, Proc. of Symp. in Pure Math. XXX Part 1 (1977), 303-382

[25] Huckleberry,A.T.: Analytic hypersurfaces in homogeneous spaces. Publication Institut Elie Cartan, Journees Complexes Nancy 82 (1983), 134–153

[26] Huckleberry,A.T.; Margulis,G.A.: Invariant analytic hypersurfaces. Invent. Math. 71 (1983), 235–240

[27] Huckleberry,A.T.; Oeljeklaus,E.: Classification Theorems for Almost Homogeneous Spaces. Publication de l'Institut Elie Cartan, Nancy, Janvier 1984, 9. 178 pp.

[28] Huckleberry,A.T.; Oeljeklaus,E.: A Characterization of Complex Homogeneous Cones Math. Z. 170 (1980), 181–194

[29] Huckleberry,A.T.; Oeljeklaus,E.: On holomorphically separable complex solvmanifolds. Annales de l'Institut Fourier (Grenoble), Tome XXXVI - Fascicule 3 (1986), 57–65

[30] Huckleberry,A.T.; Snow,D.: A Classification of Strictly Pseudoconcave Homogeneous Manifolds. Ann. Scuola Norm. Sup. Pisa 8 (1981), 231–255

[31] Humphreys,J.E.: Linear Algebraic Groups. Springer-Verlag, New York 1975

[32] Huckleberry.A.; Winkelmann,J.: Compact subvarieties in paralleliziable complex manifolds, Math. Ann. 295, 469-483(1993)

[33] Kaup,W.: Reelle Transformationsgruppen und invariante Metriken auf komplexen Räumen. Invent. Math. 3 (1967), 43–70

[34] Kiselman,C.O.: The partial Legendre transformation for plurisubharmonic functions. Invent. Math. 49 (1978), 137–148

[35] Lattes,M.S.: Sur les formes reduites des transformations ponctuelles a deux variables, Comptes Rendue 152(1911), 1566-1569

[36] Lescure,F.: Sur les compactifications équivariantes des groupes commutatifs. Annales de l'Institut Fourier 38.4 (1988), 93–120

[37] Lescure,F.: Compactifications équivariantes par des courbes. Bull. Soc. Math. France 115, Mém. 26 (1987)

[38] Loeb,J.: Actions d'une forme de Lie réelle d'un groupe de Lie complexe sur les fonctions plurisousharmoniques. Annales de l'Institut Fourier, 35-4 (1985), 59–97

[39] Matsushima,Y.: On discrete subgroups and homogeneous spaces of nilpotent Lie groups. Nagoya Math. J. 2, 95–110, (1951)

[40] Matsushima,Y.: Espaces homogènes de Stein des groupes de Lie complexes I. Nagoya Math. J. 16 (1960), 205–218

[41] Martin,M.: Über den Körper meromorpher Funktionen auf einer kompakten $Sl_2(\mathbb{C})$-fast-homogenen komplex-dreidimensionalen Mannigfaltigkeit, Dissertation, Ruhr Universität Bochum (1992)

[42] Morimoto,A.: On the classification of non-compact abelian Lie groups, Trans, AMS123-(1966),200-228

[43] Moser,L.: Dissertation, Universite de Lausanne (1989)

[44] Mumford,D.: Abelian Varieties, Studies in Math. 5, London, Oxford Univ. Press (1974)

[45] Oeljeklaus,K.: Hyperflächen und Geradenbündel auf homogenen komplexen Mannigfaltigkeiten. Schriftenr. Math. Inst. Univ. Münster, Serie 2, Heft 36 (1985)

[46] Oeljeklaus,K.; Richthofer,W.: On the Structure of Complex Solvmanifolds. J. Diff. Geom. 27 (1988), 399–421

[47] Oeljeklaus,K.; Richthofer,W.: Recent Results on Homogeneous Complex Manifolds   in: Complex Analysis III. Pric. Univ. Maryland 1985-86, 78–119, Lecture Notes in Math. 1277

[48] Oeljeklaus,K.; Richthofer,W.: Complex Solv-manifolds, Bochum Preprint (1985)

[49] Potters,J.: On almost homogeneous compact complex analytic surfaces. Invent. math. 8 (1969), 244–266

[50] Penny,R.: Entire vectors and holomorphic extensions of representations, Trans. AMS 198(1974), 107-121

[51] Richthofer,W.: Currents in Homogeneous Manifolds. Bochum preprint (1987)

[52] Raghunatan,M.S.: Discrete subgroups of Lie groups. Erg. Math. Grenzgeb. 68, Springer (1972)

[53] Skoda,H.: Fibrés holomorphes à base at à fibre de Stein. Invent. Math. 43 (1977), 97–107

[54] Ueno,K.: Classification theory of algebraic varietes and compact complex spaces. Lecture Notes in Mathematics 439, Springer-Verlag Berlin 1975

[55] Varouchas,J.: Kähler spaces and proper open morphisms. Math. Ann. **283** (1989), 13–52

[56] Wang,H.C.: Complex parallelisable manifolds. Proc. Am. Math. Soc 5 (1954), 771–776

[57] Winkelmann,J.: Every compact complex manifold admits a non-trivial holomorphic vector bundle, Revue Romaine de Math. Pure et Appl, v.38 n.7(1993)

A multidimensional Jordan residue lemma with an application to Mellin-Barnes integrals

Mikael Passare, August Tsikh and Oleg Zhdanov

The classical Jordan lemma states that if a function ψ is continuous on the real axis with a holomorphic continuation to the upper half plane $\Pi_+ = \{z = x + iy; y > 0\}$ except for a finite number of points $\{a\} \subset \Pi_+$, and if $\psi(z)$ tends to zero as $|z| \longrightarrow \infty$ in the closed half plane $\bar{\Pi}_+$, then

$$(0.1) \qquad \int_{-\infty}^{\infty} \psi(x)\, e^{i\alpha x}\, dx = 2\pi i \sum_{\{a\}} \operatorname{res}_a \omega,$$

where α is an arbitrary positive number and ω denotes the differential form $\psi(z)\, e^{i\alpha z}\, dz$. The proof of (0.1) is based on the fact that, by the Cauchy formula, the right hand side is equal to a sum of integrals $\int_{-R}^{R} \omega + \int_{C_R^+} \omega$, with C_R^+ being the upper semicircle of radius R centered at the origin. The condition that ψ should decrease as R goes to infinity guarantees that

$$(0.2) \qquad \int_{C_R^+} \omega \longrightarrow 0, \qquad \text{as} \quad R \longrightarrow \infty,$$

from which (0.1) follows. It is clear that the same formula also holds for an infinite set $\{a\}$ of singular points as long as (0.2) is valid for some sequence $\{R_k\}$ tending to infinity. In this form the idea of the Jordan lemma is applicable not only to integrals of Fourier type such as (0.1), but also to the kind of integrals that arise as inverse Mellin transforms, a particular example of which is provided by the so called Mellin-Barnes integrals

$$(0.3) \qquad \varphi(t) = \frac{1}{2\pi i} \int_{\gamma - i\infty}^{\gamma + i\infty} \frac{\prod_j \Gamma(a_j z + b_j)}{\prod_k \Gamma(c_k z + d_k)}\, t^{-z}\, dz,$$

see [2], where Γ denotes the Euler gamma function, the constants a_j, b_j, c_k, and d_k are real, and $\gamma \in \mathbb{R}$ is chosen so that the vertical line of integration in (0.3) does not intersect the poles of the integrand. From Stirling's formula

$$(0.4) \qquad \Gamma(s + i\tau) \sim \sqrt{2\pi}\, |\tau|^{s-1/2}\, e^{-\pi|\tau|/2}, \qquad \text{as} \quad |\tau| \longrightarrow \infty,$$

it follows that, if $\sum_j |a_j| > \sum_k |c_k|$, the integral (0.3) converges for every real $t > 0$, and in fact also for t in some sector containing the positive real axis. Let us recall that integrals of the form (0.3) play an important role in the theory of special functions, since it is by means of such integrals that the general solution to the hypergeometric equation is obtained, see [2, p.11]. Using the above mentioned version of the Jordan lemma with the upper half plane Π_+ replaced by the right or left half plane $\Pi_{\gtrless} = \{\operatorname{Re} z \gtrless \gamma\}$, one can prove that for any $t > 0$ the integral (0.3) is equal to the sum of residues in $\Pi_<$ if the quantity

$$\Delta = \sum a_j$$

is positive, and equal to minus the sum of residues in $\Pi_>$ if $\Delta < 0$. If however $\Delta = 0$, then the integral (0.3) may be computed as the sum of residues in $\Pi_<$ only for t in some interval like $0 < t < K_1$, and as minus the sum of residues in $\Pi_>$ only for t in an interval $K_2 < t < \infty$.

In this paper we offer a certain multidimensional generalization of the Jordan lemma in the form discussed above. The corresponding analogue of property (0.2) will be assumed to hold not just for one single differential form ω, but for several, in fact $2^n - 1$, such forms. This is due to the fact that the real subspaces $\mathbb{R}^n = \mathbb{R}^n + i0 \subset \mathbb{C}^n$ and $\gamma + i\,\mathbb{R}^n$, with $\gamma \in \mathbb{R}^n$, appear as the distinguished boundaries of polyhedra with $2^n - 1$ faces coming together. In the case $n = 1$ there is just one face of the polyhedron Π_+ that meets the axis $\mathbb{R}$, namely the polyhedron Π_+ itself. In the first section we consider an abstract version of the Jordan lemma for integrals along distinguished boundaries of unbounded polyhedra and settle the question to what extent such integrals may be computed as sums of Grothendieck residues, see [1] and [6] for a background on this notion. In section 2 the results of the first section are applied in calculating double Mellin-Barnes integrals.

We wish to mention that our paper has been conceived much under the influence of Oleg Marichev, who gently insisted on the quest for explicit formulas for double Mellin-Barnes integrals, and also furthered this end through numerous discussions. The authors express their gratitude to him and to Alexander Yuzhakov. We also point out that to a large extent the ideas in this paper are connected with the results in [3] and [4], where the problem of inverting Mellin transforms of so-called residue functions was addressed.

1   An abstract multidimensional Jordan lemma

Before giving the formulation of the generalized Jordan residue lemma that we propose to prove, we should explain with what residues we intend to work. Let there be given in $\mathbb{C}^n$ a meromorphic differential form

$$(1.5) \qquad \omega = \frac{h(z)\,dz}{f_1(z)\cdots f_n(z)}, \quad dz = dz_1 \wedge \cdots \wedge dz_n,$$

with poles along the divisors

$$(1.6) \qquad D_j = \{f_j = 0\}, \quad j = 1, \ldots, n.$$

In each point a of the intersection $\cap_{j=1}^{n} D_j$, which is assumed to be a discrete set, there is defined, see [1] or [6], a local Grothendieck residue

$$(1.7) \qquad \operatorname{res}_a \omega = \frac{1}{(2\pi i)^n} \int_{\Gamma_a} \omega,$$

where $\Gamma_a = \{z \in U_a; |f_j(z)| = \varepsilon, j = 1, \ldots, n\}$ is a cycle in some small neighborhood U_a of the point a, oriented according to the inequality $d(\arg f_1) \wedge \cdots \wedge d(\arg f_n) \geq 0$. Notice that the residue (1.7) is computable in terms of a finite number of the Taylor coefficients at a of the functions h and $f_1, \ldots, f_n$.

In this section we consider the question of when the integral

$$(1.8) \qquad \frac{1}{(2\pi i)^n} \int_{\sigma} \omega$$

of the meromorphic form (1.5), along the distinguished boundary σ of some polyhedron Π, is equal to the sum of the residues (1.7) at all points a in the polyhedron Π. In order to obtain an answer to this question for the case of bounded polyhedra $\Pi \subset \mathbb{C}^n$ we first give the exact definitions. By a polyhedron Π we mean the inverse image $g^{-1}(G)$ under a proper holomorphic mapping $g : \mathbb{C}^n \longrightarrow \mathbb{C}^n$ of a domain G of the form $G_1 \times \cdots \times G_n$, where each $G_j \subset \mathbb{C}$ is a domain with piecewise smooth boundary. In view of the properness of the mapping g the polyhedron Π is bounded if the domains G_j are bounded. To every multi-index $K = \{k_1, \ldots, k_p\} \subset \{1, \ldots, n\}$ we associate the corresponding face

$$\sigma_K = \{z; g_k(z) \in \partial G_k, k \in K, g_j(z) \in G_j, j \notin K\}.$$

Definition 1 *The family of divisors (1.6) is said to be compatible with the polyhedron Π if*

$$(1.9) \qquad D_j \cap \sigma_j = \emptyset, \qquad \text{for every} \quad j = 1, \ldots, n.$$

Theorem 1 *If the polyhedron Π is bounded and the family $\{D_j\}$ of polar divisors of the meromorphic form (1.5) is compatible with Π, then the cycle $\sigma = \sigma_{12\ldots n}$ is homologous in the complement of $\{D_j\}$ to a sum $\sum_{a \in \Pi} \Gamma_a$, and for the corresponding integral (1.8) the formula*

$$\frac{1}{(2\pi i)^n} \int_{\sigma} \omega = \sum_{a \in \Pi} \operatorname{res}_a \omega$$

holds.

The proof of Theorem 1 is given in the thesis [5]. To treat the corresponding question for unbounded polyhedra it is necessary to introduce yet another notion. By means of the functions f_j we form the partition of unity

$$\rho_j = \frac{|f_j|^2}{\|f\|^2}, \qquad \text{where} \quad \|f\|^2 = |f_1|^2 + \cdots + |f_n|^2.$$

To each multi-index $J = \{j_1, \ldots, j_p\} \subset \{1, \ldots, n\}$, with $1 \leq p \leq n$, we associate the corresponding bidegree $(n, p-1)$ differential form

$$(1.10) \qquad \xi_J = \sum_{j \in J} (-1)^{(j,J)-1} \rho_j \, \bar{\partial} \rho_J[j] \wedge \omega,$$

where for $j \in J$ we have $\bar{\partial} \rho_J[j] = \bar{\partial} \rho_{j_1} \wedge \ldots [j] \cdots \wedge \bar{\partial} \rho_{j_p}$, and (j, J) denotes the position of j in J. Observe that the form ξ_J is singular only on the set $N \cup (\cup_{k \notin J} D_k)$, where by N is meant the set $\{\|f\| = 0\} = D_1 \cap \cdots \cap D_n$. From this it follows that if the divisors $\{D_j\}$ satisfy the compatibility condition (1.9), then for $p \leq n - 1$ the form ξ_J is regular on the face σ_{J°, where $J^\circ = \{1, \ldots, n\} \setminus J$. For $p = |J| = n$ the face $\sigma_{J^\circ} = \sigma_\emptyset$ is equal to the open polyhedron Π and ξ_J has singularities in the points $N \cap \Pi$.

Definition 2 *We shall say that the differential form ξ_J satisfies the Jordan condition on the face σ_{J° if there exists a sequence $\{R_k\}$ of positive real numbers converging to infinity, such that*

$$(1.11) \qquad \lim_{k \longrightarrow \infty} \int_{S_{R_k} \cap \sigma_{J^\circ}} \xi_J = 0,$$

where S_R is a sphere of radius R centered at some point of the distinguished boundary $\sigma = \sigma_{12\ldots n}$ of the polyhedron Π.

In the case $n = 1$ there is just one form ξ_1, which is equal to ω, and the condition (1.11) coincides with (0.2).

Theorem 2 **(The Jordan lemma)** *Let the integral (1.8) be convergent. If the polar divisors (1.6) of the meromorphic form (1.5) are compatible with the polyhedron Π, and for every multi-index J the corresponding form ξ_J satisfies the Jordan condition on the face σ_{J°, then*

$$(1.12) \qquad \int_\sigma \omega = (2\pi i)^n \sum_{a \in N \cap \Pi} \mathrm{res}_a \, \omega.$$

Proof We introduce the forms $\omega_J = (p - 1)! \, d\xi_J = p! \, \bar{\partial} \rho_J \wedge \omega$, where $p = |J|$ as before. We also agree that $\omega_\emptyset = \omega$. A direct computation, see [1, p.654], shows that for $p = n - 1$ the form ω_J coincides with the form

$$\Omega = (n - 1)! \frac{h \sum_{k=1}^n (-1)^{k-1} \bar{f}_k \, d\bar{f}[k] \wedge dz}{\|f\|^{2n}}.$$

Now we argue as follows. First we claim that

$$\int_\sigma \omega = \int_{\partial \Pi} \Omega,$$

and then using the fact, see [1, p.651], that

$$\frac{1}{(2\pi i)^n} \int_{\partial U_a} \Omega = \mathrm{res}_a \, \omega,$$

where U_a is a small neighborhood of the point a, together with the Jordan condition and the meromorphic form Ω being closed, we obtain the desired equality (1.12). The claim is a consequence of the following statement.

Lemma 1 *It is possible to orient the faces σ_K in such a way that the relation*

$$\sum_{|J|=p-1}\int_{\sigma_{J^\circ}}\omega_J = \sum_{|I|=p}\int_{\sigma_{I^\circ}}\omega_I$$

holds for any $p = 1, 2, \ldots, n-1$

Proof of the lemma. Making use of the property $\rho_1 + \cdots + \rho_n = 1$ we can write

$$\sum_{|J|=p-1}\int_{\sigma_{J^\circ}}\omega_J = \sum_{|J|=p-1}(p-1)!\int_{\sigma_{J^\circ}}\Big(\sum_{k\in J^\circ}\rho_k\,\bar\partial\rho_J + \sum_{i\in J}\rho_i\,\bar\partial\rho_J\Big)\wedge\omega$$

$$= \sum_{|J|=p-1}(p-1)!\int_{\sigma_{J^\circ}}\Big[\sum_{k\in J^\circ}\rho_k\,\bar\partial\rho_J + \sum_{i\in J}(-1)^{(i,J)-1}\rho_i\Big(\sum_{j\in J}\bar\partial\rho_j\Big)\wedge\bar\partial\rho_J[i]\Big]\wedge\omega.$$

Now we observe that $\sum_{j\in J}\bar\partial\rho_j = -\sum_{k\in J^\circ}\bar\partial\rho_k$ and deduce that the last sum above equals

(1.13)
$$\sum_{|J|=p-1}(p-1)!\int_{\sigma_{J^\circ}}\Big[\sum_{k\in J^\circ}\rho_k\,\bar\partial\rho_J + \sum_{i\in J}(-1)^{(i,J)}\rho_i\Big(\sum_{k\in J^\circ}\bar\partial\rho_k\Big)\wedge\bar\partial\rho_J[i]\Big]\wedge\omega.$$

Notice that

$$\partial\sigma_{I^\circ} = \sum_{j\in I}(-1)^{\big(j,(I\setminus j)^\circ\big)-1}\sigma_{(I\setminus j)^\circ},$$

and let us fix a pair (J, k), where J is a multi-index of length $p-1$, and $k \in J^\circ$. In the integral in (1.13) that corresponds to the index J we single out the terms containing the functions ρ_i for $i \in I = J \cup \{k\}$, that is, the integral

$$\int_{\sigma_{J^\circ}}\Big[\rho_k\,\bar\partial\rho_J + \sum_{i\in J}(-1)^{(i,J)}\rho_i\,\bar\partial\rho_k\wedge\bar\partial\rho_J[i]\Big]\wedge\omega.$$

The face σ_{J° occurs in the boundary $\partial\sigma_{I^\circ} = \partial\sigma_{(I\cup k)^\circ}$ with sign $(-1)^{(k,J^\circ)-1}$. Furthermore, we have

$$d\Big\{(-1)^{(k,J^\circ)-1}\Big[\rho_k\,\bar\partial\rho_J + \sum_{i\in J}(-1)^{(i,J)}\rho_i\,\bar\partial\rho_k\wedge\bar\partial\rho_J[i]\Big]\wedge\omega\Big\} = p\,\bar\partial\rho_I\wedge\omega.$$

Hence, we may rewrite (1.13) in the form

$$\sum_{|I|=p}(p-1)!\sum_{i\in I}\int_{\pm\sigma_{(I\setminus i)^\circ}}\big[\rho_{i_1}\,\bar\partial\rho_I[i_1] - \cdots + (-1)^{p-1}\rho_{i_p}\,\bar\partial\rho_I[i_p]\big]\wedge\omega$$

$$\sum_{|I|=p}(p-1)!\sum_{i\in I}\int_{\pm\sigma_{(I\setminus i)^\circ}}\xi_I.$$

Here each term in the outer sum may be written as a limit:

$$(p-1)! \sum_{i \in I} \int_{\pm \sigma_{(I \setminus i)^\circ}} \xi_I = (p-1)! \lim_{R_k \longrightarrow \infty} \sum_{i \in I} \int_{\pm \sigma_{(I \setminus i)^\circ} \cap B_{R_k}} \xi_I,$$

where B_R is a ball of radius R centered at some point in the distinguished boundary σ of the polyhedron Π. Now, the boundary of the chain $\sigma_{I^\circ} \cap B_{R_k}$ is given by

$$\sigma_{I^\circ} \cap S_{R_k} + \sum_{i \in I} \pm (\sigma_{(I \setminus i)^\circ} \cap B_{R_k}),$$

and an application of Stokes' theorem therefore yields

$$(p-1)! \sum_{i \in I} \int_{\pm \sigma_{(I \setminus i)^\circ} \cap B_{R_k}} \xi_I = p! \int_{\pm \sigma_{I^\circ} \cap B_{R_k}} \bar{\partial} \rho_I \wedge \omega - (p-1)! \int_{\pm \sigma_{I^\circ} \cap S_{R_k}} \xi_I.$$

Letting R_k tend to infinity in this equation and taking into account the Jordan condition, we find that our original expression is equal to

$$\sum_{|J|=p-1} \int_{\sigma_{J^\circ}} \omega_J = \sum_{|I|=p} p! \int_{\sigma_{I^\circ}} \bar{\partial} \rho_I \wedge \omega = \sum_{|I|=p} \int_{\sigma_{I^\circ}} \omega_I,$$

and the lemma is proved. $\qquad\qquad\qquad\qquad\qquad\qquad\qquad\qquad\qquad\qquad\qquad\quad \square$

$$\square$$

2   Computation of multiple Mellin-Barnes integrals

By a multiple Mellin-Barnes integral we shall understand an integral of the type

$$(2.14) \qquad\qquad \varphi(t) = \varphi_\gamma(t) = \frac{1}{(2\pi i)^n} \int_{\gamma + i \mathbb{R}^n} \Phi(z)\, t^{-z}\, dz,$$

where $t, z \in \mathbb{C}^n$, $dz = dz_1 \wedge \cdots \wedge dz_n$, $t^{-z} = t_1^{-z_1} \cdots t_n^{-z_n}$, and Φ is a special function of the form $\prod_j \Gamma(\ell_j(z)) / \prod_k \Gamma(q_k(z))$ with linear functions

$$\ell_j = \sum_\nu a_{j\nu} z_\nu + b_j, \qquad q_k = \sum_\nu c_{k\nu} z_\nu + d_k.$$

It will be assumed that all coefficients in the linear functions are real, and that the vector $\gamma \in \mathbb{R}^n$ is so chosen, that the space of integration in (2.14) does not intersect the polar hyperplanes

$$S_m^j = \{\ell_j(z) = -m\}, \qquad m = 0, 1, 2, \ldots.$$

From the asymptotic formula (0.4) it follows that the integral (2.14) is absolutely convergent in the domain

$$\{t \in \mathbb{C}^n; \|\arg t\| < \alpha\pi/2\},$$

where $\|\arg t\|^2 = (\arg t_1)^2 + \cdots + (\arg t_n)^2$ and

$$\alpha = \min_{\|y\|=1} \left(\sum_j |\langle a_j, y \rangle| - \sum_k |\langle c_k, y \rangle| \right).$$

The restrictions of the polar hyperplanes S_m^j to the real subspace $\mathbb{R}^n = \mathbb{R}^n + i0$ divides this space into a countable number of connected polyhedra M_j. If $\alpha > 0$ then the domain of convergence of the integral (2.14) is not empty, and for all γ from a fixed polyhedron M_j the values of the integrals coincide. It is important to verify that the integral (2.14) is the inverse Mellin transform of φ, and the following result shows that this is indeed the case.

Theorem 3 *For the integral (2.14) one has the inversion formula*

$$(2.15) \qquad \Phi(z) = \int_{\mathbb{R}_+^n} \varphi_\gamma(t)\, t^{z-1}\, dt, \qquad z \in M_j + i\,\mathbb{R}^n,$$

where M_j is the polyhedron containing γ.

Proof After making the change of variables $t_j = e^{\eta_j}$, $j = 1, \ldots, n$, we get the expression

$$(2.16) \qquad \int_{\mathbb{R}_+^n} \varphi(t)\, t^{z-1}\, dt = \int_{\mathbb{R}^n} d\eta \frac{1}{(2\pi i)^n} \int_{\gamma + i\mathbb{R}^n} \Phi(\zeta)\, e^{-\langle \eta, \zeta - z \rangle}\, d\zeta.$$

Now we introduce spherical coordinates

$$\eta_j = \rho \cdot \chi_j(\theta), \qquad j = 1, \ldots, n,$$

where $\theta = (\theta_1, \ldots, \theta_{n-1})$. We are going to change the order of integration in (2.16), and perform the integration with respect to ρ. First however, we notice that the inner integral does not depend on the choice of $\gamma \in M_j$, and so for each fixed θ we can take γ equal to

$$\operatorname{Re} z + \varepsilon\, \chi(\theta),$$

where $\varepsilon > 0$ is small enough for the sphere $S_\varepsilon \subset \mathbb{R}^n$, with center at the point $\operatorname{Re} z \in M_j$ and of radius ε, to be contained in M_j. With this taken into account, we can rewrite the right hand side in (2.16) as

$$(2.17) \qquad \frac{1}{(2\pi i)^n} \int_\Pi J(\theta)\, d\theta \int_{\operatorname{Re} z + \varepsilon\chi(\theta) + i\mathbb{R}^n} \Phi(\zeta)\, d\zeta \int_0^\infty e^{-\rho\langle \chi(\theta), \zeta - z \rangle}\, \rho^{n-1}\, d\rho,$$

where $J(\theta) \cdot \rho^{n-1}$ is the Jacobian of the spherical coordinate change, and Π denotes the domain of definition for the parameter θ. The integral with respect to ρ is easily computed, and equals

$$(n-1)!/\langle \chi(\theta), \zeta - z \rangle^n.$$

Next we observe that the iterated integral in ζ and θ is nothing but the integral over the $(2n-1)$-dimensional surface $S_\varepsilon + i\,\mathbb{R}^n$ of some differential form containing the factor $\delta(\lambda(\zeta, z)) \wedge d\zeta$, where

$$\delta(w) = \sum_{k=1}^{n} (-1)^{k-1} w_k \, dw_1 \wedge \ldots [k] \cdots \wedge dw_n$$

is the volume form on the unit sphere in $\mathbb{R}^n$, and

$$(2.18) \qquad \lambda(\zeta, z)|_{S_\varepsilon + i\mathbb{R}^n} = \lambda(\operatorname{Re} z + \varepsilon\, \chi(\theta) + i \operatorname{Im} \zeta, z) = \chi(\theta).$$

Indeed, the quantity $\varepsilon^n J(\theta)\, d\theta$ equals the restriction of $\delta(w)$ to the sphere S_ε. Hence, the integral (2.17) is equal to

$$(2.19) \qquad \frac{(n-1)!}{(2\pi i)^n} \int_{S_\varepsilon + i\mathbb{R}^n} \Phi(\zeta) \, \frac{\delta(\lambda(\zeta, z)) \wedge d\zeta}{\langle \lambda(\zeta, z), \zeta - z\rangle^n},$$

where the vector valued function λ is defined by formula (2.18) for ζ in $S_\varepsilon + i\mathbb{R}^n$ and z in $B_\varepsilon + i\mathbb{R}^n$, with $B_\varepsilon \subset \mathbb{R}^n$ denoting the ball with center $\operatorname{Re} z$ and radius ε. Notice that

$$|\langle \lambda(\zeta, z), \zeta - z\rangle| = |\langle \chi(\theta), \varepsilon\, \chi(\theta) + i \operatorname{Im}(\zeta - z)\rangle| \geq \varepsilon\, |\chi(\theta)| = \varepsilon,$$

so the kernel in (2.19) is precisely of the type that occurs in the Cauchy-Fantappiè integral representation formula. Now we view the unbounded domain $B_\varepsilon + i\mathbb{R}^n$ as the limit of the bounded domains $B_\varepsilon + i B_R$, where $B_R \subset \mathbb{R}^n$ is the ball of radius R centered at 0. The integral over the boundary then splits into a sum of integrals over $S_\varepsilon + i B_R$ and $B_\varepsilon + i S_R$, where S_R is the sphere ∂B_R. As $\|\operatorname{Im} \zeta\| = R \longrightarrow \infty$ the integral over $S_\varepsilon + i B_R$ tends to (2.19), whereas the integral over $B_\varepsilon + i S_R$, with suitable continuous extension of the function λ, tends to zero, because Φ decreases rapidly. Consequently, by the Cauchy-Fantappiè integral formula, the integral (2.19) is equal to $\Phi(z)$ and hence formula (2.15) is proved for $z \in B_\varepsilon + i\mathbb{R}^n$. But since z was an arbitrary point in $M_j + i\mathbb{R}^n$, the theorem follows. $\qquad\square$

In the case $n = 2$ we are now going to indicate how to canonically divide the family of polar lines S_m^j into two groups, thereby forming two polar divisors D_1 and D_2 for the meromorphic form

$$(2.20) \qquad \omega = \Phi(z)\, t^{-z}\, dz = \frac{\prod_j \Gamma(\ell_j(z))}{\prod_k \Gamma(q_k(z))} \, t^{-z}\, dz,$$

occurring in (2.14). Recall that $\ell_j(z) = a_{j1} z_1 + a_{j2} z_2 + b_j$ and define the numbers

$$\Delta_\nu = \sum_\nu a_{j\nu}, \qquad \nu = 1, 2.$$

Assume that $\Delta = (\Delta_1, \Delta_2)$ is not the zero vector, and consider the line

$$L = \{x \in \mathbb{R}^2 \,;\, \langle \Delta, x\rangle = \langle \Delta, \gamma\rangle\},$$

which passes through the point γ and has Δ as its normal vector. This line L is divided by the point γ into two rays L_1 and L_2, and we can define:

$$D_k = \text{the union of all lines } S_m^j \text{ not intersecting } L_k, \quad k = 1, 2.$$

Let Π be the half space $\{\langle \Delta, x\rangle \leq \langle \Delta, \gamma\rangle\} + i\mathbb{R}^2$ with boundary $L + i\mathbb{R}^2$.

Theorem 4 *In the case of two variables the integral (2.14) is equal to the sum of residues in the half space* Π *of the meromorphic form (2.20), computed with respect to the divisors* D_1 *and* D_2:

$$\varphi(t) = \sum_{a\in\Pi} \operatorname{res}_a \ \omega.$$

The proof of Theorem 4 is obtained by applying Theorem 2 to a polyhedron $\tilde{\Pi}$, approximating Π from inside and being bounded by nearby rays $\tilde{L}_1$ and $\tilde{L}_2$ through γ. This works because the divisors D_1 and D_2 are compatible with the polyhedron $\tilde{\Pi}$, and the Jordan condition is satisfied.

Bibliography

[1] P. Griffiths and J. Harris, *Principles of algebraic geometry*, Pure Appl. Math., John Wiley & Sons, New York, 1978.

[2] O. Marichev, *Handbook of integral transforms of higher transcendental functions: Theory and algorithmic tables*, Ellis Horwood, Chichester, 1983.

[3] M. Passare and A. Tsikh, O svjazjah meždu lokal'noj strukturoj golomorfnyh otobraženij, mnogomernymi vyčetami i obobščennymi preobrazovanijami Mellina, *Dokl. Akad. Nauk*, **325:4** (1992), 664-667. (English version will appear in *Soviet Math. Dokl.*)

[4] M. Passare and A. Tsikh, *Residue integrals and their Mellin transforms*, Preprint TRITA/MAT-92-0008, Royal Institute of Technology, Stockholm, 1992.

[5] A. Tsikh, *Metody teorii mnogomernyh vyčetov*, Doctoral dissertation, Novosibirsk, 1990.

[6] A. Tsikh, *Multidimensional residues and their applications*, Transl. Math. Monographs 103, Amer. Math. Soc., Providence, 1992.

The first two authors were supported by a joint Swedish/Russian research project financed by the Royal Swedish Academy of Sciences and the Russian Academy of Sciences

Separately meromorphic mappings into compact Kähler manifolds

Bernard Shiffman

It was recently shown by Ivashkovich [2] that meromorphic mappings from a Hartogs domain into a compact Kähler manifold extend to the polydisk. In this note we use Ivashkovich's extension theorem together with methods of [7], [9] to obtain some other generalizations of results of Hartogs [1] for meromorphic mappings into compact Kähler manifolds (Theorem 1, Theorem 4, Corollary 3).

Our basic philosophy, which comes from Hartogs' work, is that if a complex space X has the property that mappings from a Hartogs domain into X extend to the polydisk ("Hartogs Phenomenon"), then Hartogs' Extension Lemma and, as a consequence, strong Hartogs-type theorems hold for mappings into X. Here, a "mapping" can mean either a holomorphic or meromorphic mapping, and a "Hartogs-type theorem" means a theorem on mappings analogous to the theorem of Hartogs that separately holomorphic functions are holomorphic. For the case of meromorphic functions, this philosophy was applied by Rothstein [6] to show that separately meromorphic functions are meromorphic. (See also [10] for a simplified proof of Rothstein's theorem.) Using polynomial approximation methods of Siciak [11], [12], the author [7] showed that the hypotheses of the theorems of Hartogs and of Rothstein can be weakened to assume only that the function is holomorphic or meromorphic, respectively, on almost all coordinate disks (in the sense of measure theory); see Corollaries 1 and 2. Our main result is the generalization (Theorem 1) of this improved form of Rothstein's theorem to meromorphic mappings into compact Kähler manifolds. In fact we show (Theorem 3) that our result is valid for the class of complex spaces that satisfy the meromorphic extension property of Ivashkovich's theorem. In order to apply our philosophy, we show that a general Hartogs extension lemma holds for this class of spaces (Theorem 4). Similar results were obtained by the author [9] for holomorphic mappings into complex spaces, and the methods of this note closely follow those of [9] and [7].

Other Hartogs-type theorems for holomorphic functions have been obtained by Terada [14], Siciak [11], [12], Zaharjuta [16], Nguyen-Zeriahi [5], the author [8], and others; results for meromorphic functions were also obtained by Kazaryan [3], [4] and Uhlenbeck-Yau [15]. (See [10] for a brief survey and additional references.)

We begin with some terminology and notation: We let $\Delta = \{\zeta \in \mathbb{C} : |\zeta| < 1\}$ denote the unit disk in $\mathbb{C}$. A *coordinate disk in* Δ^n is a disk of the form

$$\Delta_j^a = \{(a_1, \ldots, a_{j-1}, \zeta, a_j, \ldots, a_{n-1}) : \ \zeta \in \Delta\},$$

where $(a_1, \ldots, a_{n-1}) \in \Delta^{n-1}$ and $1 \leq j \leq n$. We let $\mathcal{D}_n$ denote the set of coordinate disks in Δ^n. Regarding $\mathcal{D}_n$ as n disjoint copies of Δ^{n-1}, we give $\mathcal{D}_n$ Lebesgue $(2n-2)$-measure. In addition, each $D \in \mathcal{D}_n$ is given Lebesgue 2-measure.

Definition 1 *A subset $E \subset \Delta^n$ is said to be a full subset of Δ^n if $E \cap D$ is a set of full measure in D for almost all coordinate disks $D \in \mathcal{D}_n$. (A subset of a domain $\Omega \subset \mathbb{C}^n$ is said to be of full measure in Ω if its complement has Lebesgue measure 0. If E is of full measure in Δ^n, then E is a full subset of Δ^n, and the converse is true for measurable E.*

Recall that a meromorphic mapping f from a complex space W into a complex space X, denoted $f : W \rightsquigarrow X$, is given by a holomorphic mapping f_0 from a dense open set $W_0 \subset W$ into X with the property that the closure G of the graph of f_0 is an analytic subvariety of $W \times X$ and the projection $\pi : G \longrightarrow W$ is proper. We let I_f denote the set of points of W at which f is not holomorphic (and we can take $W_0 = W - I_f$); I_f is a subvariety of W. (In this note, we shall consider only the case where W is smooth, so that codim $I_f \geq 2$.) If S is a subvariety of W such that I_f does not contain any irreducible components of S, then $f|S$ is a well-defined meromorphic mapping from S into X. (Note that a meromorphic mapping into $\mathbb{CP}^1$ is a meromorphic function, but a meromorphic mapping from a normal complex space W into $\mathbb{C}$ is actually a holomorphic function on W.)

Our main result is the following strong Hartogs-type theorem for meromorphic mappings into compact Kähler manifolds:

Theorem 1 *Let E be a full subset of Δ^n and let $f : E \rightarrow M$ where M is a compact Kähler manifold. Suppose that for almost all coordinate disks $D \in \mathcal{D}_n$, $f|E \cap D$ extends to a holomorphic mapping from D to M. Then there exists a meromorphic mapping $\tilde{f} : \Delta^n \rightsquigarrow M$ such that $f = \tilde{f}$ almost everywhere on E.*

If we consider the case $M = \mathbb{CP}^1$ we obtain:

Corollary 1 ([7, Cor. 1], [15]) *Let $f : E \rightarrow \mathbb{C}$, where E is a full subset of Δ^n. Suppose that for almost all coordinate disks $D \in \mathcal{D}_n$, $f|E \cap D$ extends to a meromorphic function on D. Then there exists a meromorphic function $\tilde{f}$ on Δ^n such that $f = \tilde{f}$ almost everywhere on E.*

In particular, the following improvement of Hartogs' Theorem on separately holomorphic functions is a special case of Corollary 1.

Corollary 2 *Let $f : \Delta^n \longrightarrow \mathbb{C}$ such that $f|D$ is holomorphic for almost all $D \in \mathcal{D}_n$. Then f is equal almost everywhere to a holomorphic function on Δ^n.*

We shall use the following meromorphic extension theorem of Ivashkovich to prove Theorem 1:

Theorem 2 ([2]) *Let D be a domain in a Stein manifold and let $\tilde{D}$ denote the envelope of holomorphy of D. Then every meromorphic mapping $f : D \rightsquigarrow M$ into a compact Kähler manifold M has a meromorphic extension to $\tilde{D}$.*

In fact the only property of M that we need for the proof of Theorem 1 is that it satisfies the conclusion of Theorem 2. We shall call this property the *meromorphic extension property*. As noted in [9], [2], in order to show that a complex space X satisfies the meromorphic extension property it suffices to show that every meromorphic mapping from a Hartogs domain

$$H_n(r) = \left\{ (z_1, \ldots, z_n) \in \Delta^n : \max_{1 \leq j \leq n-1} |z_j| < r \ \text{ or } \ |z_n| > 1 - r \right\}$$

into X has a meromorphic extension to Δ^n (for $n \geq 2$, $0 < r < 1$). To prove that a meromorphic mapping from $H_n(r)$ to a compact Kähler manifold M extends to the polydisk, Ivashkovich uses bounds on the areas of images of disks in M as well as Siu's "Riemann removable singularities theorem" [13] for meromorphic maps into M.

We see from the theorem of Ivashkovich that Theorem 1 is a special case of the following strong Hartogs-type theorem for mappings into complex spaces with the meromorphic extension property:

Theorem 3 *Let E be a full subset of Δ^n and let X be a complex space with the meromorphic extension property. If $f : E \to X$ such that $f|E \cap D$ has a holomorphic extension to D for almost all $D \in \mathcal{D}_n$, then there exists a meromorphic mapping $\tilde{f} : \Delta^n \rightsquigarrow X$ such that $f = \tilde{f}$ almost everywhere on E.*

We give below a general Hartogs extension lemma (Theorem 4), which we shall use to prove Theorem 3. First we describe some more notation: Suppose that U, V, X are sets, $E \subset U \times V$, and $f : E \longrightarrow X$. We write

$$E_z = \{w \in V : (z, w) \in E\}, \quad E^w = \{z \in U : (z, w) \in E\},$$

and we define the mappings $f_z : E_z \longrightarrow X$, $f^w : E^w \longrightarrow X$ by

$$f_z(w) = f^w(z) = f(z, w),$$

for $z \in U$, $w \in V$. In particular if $f : U \times V \longrightarrow X$, then $f_z : V \longrightarrow X$ for $z \in U$ and $f^w : U \longrightarrow X$ for $w \in V$. If U, V, X are complex spaces and $f : U \times V \rightsquigarrow X$ is meromorphic, then we can set $E = U \times V - I_f$ so that the mapping $\hat{f} = f|E : E \longrightarrow X$ is holomorphic. Furthermore, $\hat{f}_z : E_z \longrightarrow X$ extends to a meromorphic mapping $f_z : V \rightsquigarrow X$ whenever E_z is dense in V, and $\hat{f}^w : E^w \longrightarrow X$ extends to a meromorphic mapping $f^w : U \rightsquigarrow X$ whenever E^w is dense in U.

Theorem 4 (General Hartogs Extension Lemma) *Let X be a complex space that satisfies the meromorphic extension property. (For example, suppose X is a compact Kähler manifold.) Let U, V be open sets in $\mathbb{C}^m$, $\mathbb{C}^n$ respectively, and let V_0 be an open subset of V. Let $f : U \times V_0 \rightsquigarrow X$ be a meromorphic mapping. If f_z has a meromorphic extension to V for almost all $z \in U$, then f has a meromorphic extension to $U \times V$.*

The statement in Theorem 4 that f_z has a meromorphic extension to V is understood to mean that f_z is defined (i.e., $(I_f)_z$ is nowhere dense in U) and there exists a meromorphic mapping $\tilde{f}_z : V \rightsquigarrow X$ such that $\tilde{f}_z|V_0 = f_z$.

The proof of Theorem 4 parallels that of the analogous result for holomorphic mappings in [9, Theorem 3]. We begin the proof with the following weak form of the theorem for mappings into arbitrary complex spaces:

Lemma 1 *Let U be an open set in $\mathbb{C}^m$ and let V_0, V_1, V be connected open sets in $\mathbb{C}$ with $V_0 \subset V_1 \subset\subset V$. Suppose that $f : U \times V_0 \rightsquigarrow X$ is a meromorphic mapping into a complex space X. If f_z has a holomorphic extension to V for almost all $z \in U$, then there exist an open subset U_0 of full Lebesgue measure in U and a holomorphic mapping $\tilde{f} : U_0 \times V_1 \longrightarrow X$ such that $f = \tilde{f}$ on $U_0 \times V_0$.*

Proof Choose a nonempty open set $V_0' \subset\subset V_0$ and let $U' = U - \pi_U(I_f \cap (U \times \overline{V_0'}))$, where $\pi_U : U \times V \longrightarrow U$ is the projection. Since $\dim I_f \leq m - 1$, U' has full measure in U. Let $f' = f|U' \times V_0'$; then $f' : U' \times V_0' \longrightarrow X$ is holomorphic and f_z' has a holomorphic extension to V for almost all $z \in U'$. Hence, by [9, Theorem 1], there is an open subset U_0 of full measure in U' and a holomorphic mapping $\tilde{f} : U_0 \times V_1 \longrightarrow X$ such that $f' = \tilde{f}$ on $U_0 \times V_0'$ and hence by the identity principle $f' = \tilde{f}$ on $U_0 \times V_0$. $\square$

Proof of Theorem 4 We can easily reduce the theorem to the case $V_0 = \Delta^n$, $V = \Delta_R^n$, where $\Delta_R = \{z \in \mathbb{C} : |z| < R\}$ for some $R > 1$. We first suppose that $n = 1$: By Lemma 1, for arbitrary $r < R$ there is an open set $U_0 \subset U$ of full measure such that f has a meromorphic extension to the open set $\Omega = (U \times \Delta) \cup (U_0 \times \Delta_r)$. By [9, Lemma 2], the envelope of holomorphy of Ω contains $U \times \Delta_r$ and hence by the meromorphic extension property of X, f has a meromorphic extension to $U \times \Delta_r$. Since $r < R$ is arbitrary, f extends to $U \times \Delta_R$.

We now prove the general case by induction on n: (We remark here that we only need the case $n = 1$ verified above for the proof of Theorem 3.) Suppose that $n \geq 2$ and Theorem 4 has been verified for $V_0 = \Delta^{n-1}$, $V = \Delta_R^{n-1}$. Let $f : U \times \Delta^n \rightsquigarrow X$ be a meromorphic mapping such that f_z has a meromorphic extension to Δ_R^n for all $z \in U - A$, where $A \subset U$ has measure 0. Let B denote the set of points $(z, \zeta) \in U \times \Delta$ such that $\{z\} \times \Delta^{n-1} \times \{\zeta\} \subset I_f$, and write $E = (A \times \Delta) \cup B$. Then E has measure 0 in $U \times \Delta$ and the meromorphic mapping $f(z, \cdot, \zeta) : \Delta^{n-1} \rightsquigarrow X$ is defined and has a meromorphic extension to Δ_R^{n-1} for all $(z, \zeta) \in U \times \Delta - E$. By the inductive assumption (with U replaced by $U \times \Delta$), it follows that f has a meromorphic extension $f' : U \times \Delta_R^{n-1} \times \Delta \rightsquigarrow X$.

Similarly let B' denote the set of $(z, w) \in U \times \Delta_R^{n-1}$ such that $\{(z, w)\} \times \Delta \subset I_{f'}$, and write $E' = (A \times \Delta_R^{n-1}) \cup B'$. Then E' has measure 0 in $U \times \Delta_R^{n-1}$ and the mapping $f'(z, w, \cdot) : \Delta \longrightarrow X$ is well defined and has a holomorphic extension to Δ_R for all $(z, w) \in U \times \Delta_R^{n-1} - E'$. Therefore by the case $n = 1$, f' has a meromorphic extension $\tilde{f} : U \times \Delta_R^{n-1} \times \Delta_R \rightsquigarrow X$. $\square$

In the remainder of this note, we prove Theorem 3 by using Theorem 4 together with the methods of [7].

Definition 2 *Suppose U is an open set in $\mathbb{C}^n$, E is a subset of U of full measure, and X is a complex space. A mapping $f : E \longrightarrow X$ is said to be meromorphic on U if there exists* *a meromorphic mapping $f' : U \rightsquigarrow X$ such that $f = f'$ on $E - I_{f'}$. (Note that we require pointwise equality on $E - I_{f'}$ and not just equality almost everywhere. If $n = 1$, then f is meromorphic on U if and only if f has a holomorphic extension to U.*

Lemma 2 *Let $f : E \rightarrow X$ be a mapping from a set $E \subset \Delta^m \times \Delta$ into a complex space X. Suppose that*

    (i) there is a set A_1 of measure 0 in Δ^m such that E_z is of full measure in Δ and f_z has a holomorphic extension to Δ for all $z \in \Delta^m - A_1$,

    (ii) there is a set A_2 of measure 0 in Δ such that E^w is of full measure in Δ^m and f^w is meromorphic on Δ^m for all $w \in \Delta - A_2$.

Then there exist nonempty open sets $U \subset \Delta^m$, $V \subset \Delta$ and a holomorphic mapping $f' : U \times V \rightarrow X$ such that $f = f'$ on $E \cap (U \times V - A_1 \times A_2)$.

Proof (This proof closely follows that of [7, Lemma 2].) Choose a countable collection $\{\Omega_k\}$ of Stein open subsets of X together with open subsets $\Omega'_k \subset\subset \Omega_k$ such that $\bigcup \Omega'_k = X$. Let $\{U_j\}$ be a countable basis of connected open sets in Δ^m. Let $f^{w*} : \Delta^m \rightsquigarrow X$ denote the meromorphic extension of f^w to Δ^m for $w \in \Delta - A_2$, and let $f_z^* : \Delta \rightarrow X$ denotes the holomorphic extension of f_z for $z \in \Delta^m - A_1$. Consider the sets

$$Q_{j,k} = \{w \in \Delta - A_2 : f^{w*}|U_j \text{ is holomorphic and } f^{w*}(U_j) \subset \Omega'_k\} \ .$$

Since $\bigcup Q_{j,k} = \Delta - A_2$, we can now fix j, k so that $Q_{j,k}$ has positive Lebesgue outer measure (in Δ). Let F denote the set of $w \in \Delta$ such that $Q_{j,k} \cap N$ has positive Lebesgue outer measure for all neighborhoods N of w. Then F is closed in Δ and has positive measure. Choose a compact subset $K_2 \subset F$ with positive measure.

    Write $U' = U_j$. We shall define $f_1 : U' \times K_2 \rightarrow \overline{\Omega'_k} \subset X$ such that f_1^w is holomorphic for all $w \in K_2$ and

$$f_1(z, w) = f_z^*(w) \quad \text{for } (z, w) \in (U' - A_1) \times K_2 \ .$$

To define f_1, we first choose a countable dense subset $\{z^\mu\}$ of $U' - A_1$. Now let $w_0 \in K_2$ be fixed, and choose a sequence of points $w_\nu \in Q_{j,k} \cap (\bigcap_\mu E_{z^\mu})$ with $w_\nu \rightarrow w_0$. Since Ω_k is biholomorphic to a subvariety of Euclidean space (after shrinking Ω_k if necessary) and $f^{w_\nu*} : U' \rightarrow \Omega'_k \subset\subset \Omega_k$, by passing to a subsequence we can assume that $\{f^{w_\nu*}\}$ converges uniformly on compact subsets of U', and we define

$$f_1(z, w_0) = \lim_{\nu \rightarrow \infty} f^{w_\nu*}(z) \quad \text{for } z \in U' \ .$$

Clearly $f_1^{w_0}$ is holomorphic on U'. Since $(z^\mu, w_\nu) \in E$ and $z^\mu \notin A_1$, we also have

$$f_1(z^\mu, w_0) = \lim_{\nu \rightarrow \infty} f(z^\mu, w_\nu) = \lim_{\nu \rightarrow \infty} f_{z^\mu}^*(w_\nu) = f_{z^\mu}^*(w_0)$$

for $\mu = 1, 2, 3, \ldots$. Thus $f_1^{w_0}$ does not depend on the choice of w_ν satisfying the above conditions. In particular, for each fixed $p \in U' - A_1$, by repeating the above argument with $\{z^\mu\}$ replaced by $\{z^\mu\} \cup \{p\}$, we conclude that

$$f_1(p, w_0) = f_p^*(w_0) \,,$$

and thus f_1 is our desired mapping.

Now choose an open set Ω'' with $\Omega_k' \subset\subset \Omega'' \subset\subset \Omega_k$. Since $f_z^*(K_2) \subset \Omega''$ for $z \in U' - A_1$, we can choose as before a connected open set V' with $K_2 \subset V' \subset \Delta$ and a set $P \subset U' - A_1$ such that P has positive Lebesgue outer measure and $f_z^*(V') \subset \Omega''$ for all $z \in P$. We then choose as before a compact set $K_1 \subset U'$ of positive measure such that $P \cap N$ has positive Lebesgue outer measure for every open set N intersecting K_1. We define $f_2 : K_1 \times V' \to \overline{\Omega''}$ as follows: Let $z \in K_1$ be fixed and choose a sequence $\{z^\nu\} \subset P$ such that $z^\nu \to z$ and $f_{z^\nu}^*$ converges uniformly on compact subsets of V'. We then define

$$f_2(z, w) = \lim_{\nu \to \infty} f_{z^\nu}^*(w)$$

for $w \in V'$. Clearly, $(f_2)_z$ is holomorphic on V'. Since $f_{z^\nu}^*(w) = f_1(z_\nu, w)$ for $w \in K_2$, it follows that $f_2 = f_1$ on $K_1 \times K_2$ and hence $(f_2)_z$ does not depend on the choice of $z^\nu \to z$. Furthermore by the same argument as before we conclude that

$$f_2(z, w) = f^{w*}(z) \quad \text{for } (z, w) \in K_1 \times (V' - A_2) \,.$$

We write $S = (U' \times K_2) \cup (K_1 \times V')$ and let $\tilde{f} : S \to \overline{\Omega''} \subset \Omega_k$ be given by f_1 and f_2. The above discussion shows that $\tilde{f}_z$ is holomorphic on V' for all $z \in K_1$ and $\tilde{f}^w$ is holomorphic on U' for all $w \in K_2$. Furthermore,

$$\tilde{f} = f \quad \text{on} \quad E \cap \{[(U' - A_1) \times K_2] \cup [K_1 \times (V' - A_2)]\} \,.$$

By applying [7, Lemma 1] (or [8, Lemma 4]) to $\tilde{f}$ (regarding Ω_k as a subvariety of Euclidean space as before), we can find nonempty connected open sets $U \subset U'$, $V \subset V'$ and a holomorphic mapping $f' : U \times V \longrightarrow \Omega_k$ such that $U \cap K_1$ and $V \cap K_2$ are of positive measure and $f' = \tilde{f}$ on $S \cap (U \times V)$; thus

$$f' = f \quad \text{on} \quad E \cap \{[(U - A_1) \times (K_2 \cap V)] \cup [(K_1 \cap U) \times (V - A_2)]\} \,.$$

In particular, for $z \in U - A_1$, $f_z' = f_z$ on $E_z \cap K_2 \cap V$ and therefore on $E_z \cap V$; similarly for $w \in V - A_2$, $f'^w = f^w$ on $E^w \cap K_1 \cap U$ and therefore on $E^w \cap U$. Thus $f' = f$ on $E \cap (U \times V - A_1 \times A_2)$. $\qquad\qquad\square$

Lemma 3 *Suppose X is a complex space that satisfies the meromorphic extension property. Let $E \subset \Delta^{m+1}$, $A_1 \subset \Delta^m$, $A_2 \subset \Delta$, $f : E \longrightarrow X$ satisfy the hypotheses of Lemma 2. Then $f|(E - A_1 \times A_2)$ is meromorphic on Δ^{m+1} (as in Definition 2).*

Proof Let $U, V, f' : U \times V \longrightarrow X$ be as in the conclusion of Lemma 2. By Theorem 4 and hypothesis (i) of Lemma 2, f' has a meromorphic extension $g : U \times \Delta \rightsquigarrow X$. Since g is holomorphic on $U \times V$, $g_z : \Delta \longrightarrow X$ is defined and holomorphic for all $z \in U$. Since for all $z \in U - A_1$, $g_z = f_z$ on $E_z \cap V$ and therefore on E_z, it follows that

$$g = f \quad \text{on} \quad E \cap ([(U - A_1) \times \Delta] - I_g \,.$$

Then by hypothesis (ii) of Lemma 2, g^w extends to a meromorphic mapping on Δ^n for all $w \in \Delta - A_2$ such that $(I_g)^w \neq \Delta^m$. Again by Theorem 4, g has a meromorphic extension $\tilde{g} : \Delta^m \times \Delta \leadsto X$. Write $I = I_{\tilde{g}}$. For $w \in \Delta - A_2$ such that $I^w \neq \Delta^m$, we have $\tilde{g}^w = f^w$ on $E^w \cap (U - A_1) - I^w$ and therefore on $E^w - I^w$. Thus, $\tilde{g} = f$ on $E - I - (\Delta^m \times A_2)$. Then by hypothesis (i), $\tilde{g}_z = f_z$ on $E_z - I_z$ for $z \in \Delta^m - A_1$. Hence, $\tilde{g} = f$ on $E - I - (A_1 \times A_2)$. $\square$

A variant of Corollary 1 was given in [7, Theorem 1]; the following lemma gives the analogous variant of Theorem 3:

Lemma 4 *Let $E \subset \Delta^n$ and let $f : E \longrightarrow X$, where X is a complex space that satisfies the meromorphic extension property. Suppose there is a set A of measure 0 in $\mathcal{D}_n$ such that*

 (i) *$E \cap D$ is of full measure in D and $f|E \cap D$ has a holomorphic extension to D for all coordinate disks $D \in \mathcal{D}_n - A$,*

 (ii) *$E \cap D = \emptyset$ for all $D \in A$.*

Then f is meromorphic on Δ^n (as in Definition 2).

Proof Let A_j denote the set of $a \in \Delta^{n-1}$ such that $\Delta_j^a \in A$. We use induction on n. The case $n = 2$ is a consequence of Lemma 3 with $m = 1$. (Note that by hypothesis, $E - A_1 \times A_2 = E$.) The inductive step follows from Lemma 3 by the proof of [7, Theorem 1], with $\mathbb{C}$ replaced by X and the phrase "extends meromorphically to" replaced by "is meromorphic on." $\square$

Theorem 3 is now an immediate consequence of Lemma 4 and the following elementary set-theory fact:

Lemma 5 *Let $E \subset \Delta^n$ and suppose there is a set A of measure 0 in $\mathcal{D}_n$ such that $E \cap D$ is of full measure in D for all $D \in \mathcal{D}_n - A$ (and thus E is a full subset of Δ^n). Then there exists a set A' of measure 0 in $\mathcal{D}_n$ such that $A \subset A'$ and a subset $E_0 \subset E$ such that $E - E_0$ has measure 0 in Δ^n, $E_0 \cap D = \emptyset$ for all $D \in A'$, and $E_0 \cap D$ is of full measure in D for all $D \in \mathcal{D}_n - A'$.*

Proof See the proof of [7, Cor. 1]. $\square$

We complete this note by giving the meromorphic analogue of [9, Theorem 6]:

Corollary 3 *Let X be a complex space that satisfies the meromorphic extension property. (For example, suppose X is a compact Kähler manifold.) Suppose that $f : \Delta^n \longrightarrow X$ is a measurable mapping such that for almost all coordinate disks $D \in \mathcal{D}_n$, $f|D$ is equal almost everywhere to a holomorphic mapping from D into X. Then f is equal almost everywhere to a meromorphic mapping $f' : \Delta^n \leadsto X$.*

The hypothesis in Corollary 3 that f is measurable means that $f^{-1}(\Omega)$ is a Lebesgue-measurable set for all open sets $\Omega \subset X$. Corollary 3 follows from Theorem 3 by the argument of the proof of the holomorphic version in [9, Theorem 6] (which employs the proof for the case of meromorphic functions in [7, Cor. 2]). An easy example in [7, p. 241] shows that Corollary 3 would be false if we do not assume that f is measurable.

Bibliography

[1]   F. Hartogs, *Zur Theorie der analytischen Funktionen mehrerer unabhängiger Veränderlichen, insbesondere über die Darstellung derselben durch Reihen, welche nach Potenzen einer Veränderlichen fortschreiten*, Math. Ann. **62** (1906), 1-88.

[2]   S. M. Ivashkovich, *The Hartogs-type extension theorem for meromorphic maps into compact Kähler manifolds*, Invent. Math. **109** (1992), 47-54.

[3]   M. V. Kazaryan, *On functions of several complex variables that are separately meromorphic*, Mat. Sb. **99** (1976), 141; English translation, Math. USSR-Sb. **28** (1976), 481-489.

[4]   M. V. Kazaryan, *Meromorphic continuation with respect to groups of variables*, Mat. Sb. **125** (1984), 384-397; English translation, Math. USSR-Sb. **53** (1986), 385-398.

[5]   T. V. Nguyen and A. Zeriahi, *Familles de polynômes presque partout bornées*, Bull. Sci. Math. **107** (1983), 81-91.

[6]   W. Rothstein, *Ein neuer Beweis des Hartogsschen Hauptsatzes und seine Ausdehnung auf meromorphe funktionen*, Math. Z. **53** (1950), 84-95.

[7]   B. Shiffman, *Complete characterization of holomorphic chains of codimension one*, Math. Ann. **274** (1986), 233-256.

[8]   B. Shiffman, *Separate analyticity and Hartogs theorems*, Indiana Univ. Math. J. **38** (1989), 943-957.

[9]   B. Shiffman, *Hartogs theorems for separately holomorphic mappings into complex spaces*, C. R. Acad. Sci. Paris Sér. I **310** (1990), 89-94.

[10]   B. Shiffman, *Separately meromorphic functions and separately holomorphic mappings*, Proc. Sympos. Pure Math. **52** (1991), 191-198.

[11]   J. Siciak, *Separately analytic functions and envelopes of holomorphy of some lower dimensional subsets of* $\mathbb{C}^n$, Ann. Polon. Math. **22** (1970), 145-171.

[12]   J. Siciak, *Extremal plurisubharmonic functions in* $\mathbb{C}^n$, Ann. Polon. Math. **39** (1981), 175-211.

[13]   Y.-T. Siu, *Extension of meromorphic maps into Kähler manifolds*, Ann. of Math. **102** (1975), 421-462.

[14]   T. Terada, *Sur une certaine condition sous laquelle une fonction de plusieurs variables complexes est holomorphe*, Publ. Res. Inst. Math. Sci. Ser. A **2** (1967), 383-396.

[15]   K. Uhlenbeck and S. T. Yau, *A note on our previous paper: On the existence of Hermitian-Yang-Mills connections in stable vector bundles*, Comm. Pure Appl. Math. **42** (1989), 703-707.

[16]   V. P. Zaharjuta, *Separately analytic functions, generalizations of Hartogs' theorem, and envelopes of holomorphy*, Mat. Sb. **101** (1976), 143; English translation, Math. USSR-Sb. **30** (1976), 51-67.

Partially supported by National Science Foundation Grant No. DMS-9204037.